Decentralization in Environmental Governance

Christian Zuidema skilfully navigates the theoretical and practice pluralism of environmental governance. In worlds plagued by many forms of environmental pollution, there are no ready-made answers. For environmental policymakers, it is a matter of value preference. Zuidema's persuasive proposal for a consequentialist post-contingency approach offers a proactive way for moving beyond mere damage control, to enable more informed policy decisions.

Jean Hillier, RMIT University, Australia

Decentralization in Environmental Governance is a critical reflection on the dangers and risks of governance renewal; warning against one-sided criticism on traditional command and control approaches to planning. The book formulates the arguments that support when and how governance renewable might be pursued, but this attempt is not just meant for practitioners and scholars interested in governance renewal. It is also useful for those interested in the challenge of navigating a plural landscape of diverse planning approaches, which are each rooted in contrasting theoretical and philosophical positions.

The book develops a strategy for making argued choices between alternative planning approaches, despite their theoretical and philosophical positions. It does so by revitalizing the idea that we can contingently relate alternative planning approaches to the circumstances encountered. It is an idea traced to contingency studies of the mid and late 20th century, reinterpreted here within a planning landscape dominated by notions of uncertainty, complexity and socially constructed knowledge. This approach, called 'post-contingency', is both a theoretical investigation of arguments for navigating the theoretical plurality we face and an empirical study into renewing environmental governance. Next to its theoretical ambitions, *Decentralization in Environmental Governance* is a practical in offering a constructive critique on current processes of governance renewal in European environmental governance.

Christian Zuidema is Assistant Professor in Spatial Planning at the Department of Spatial Planning and Environment at the University of Groningen, Netherlands.

New Directions in Planning Theory
Series Editors
Gert de Roo, University of Groningen, The Netherlands
Jean Hillier, RMIT University, Australia
Joris Van Wezemael, University of Fribourg, Switzerland

Routledge's series, New Directions in Planning Theory, develops and disseminates theories and conceptual understandings of spatial and physical planning which address such challenges as uncertainty, diversity and incommensurability.

Planning theories range across a wide spectrum, from questions of explanation and understanding, to normative or predictive questions of how planners should act and what future places should look like.

These theories include procedural theories of planning. While these have traditionally been dominated by ideas about rationality, in addition to this, the series opens up to other perspectives and also welcomes theoretical contributions on substantive aspects of planning.

Other theories to be included in the series may be concerned with questions of epistemology or ontology; with issues of knowledge, power, politics, subjectivation; with social and/or environmental justice; with issues of morals and ethics.

Planning theories have been, and continue to be, influenced by other intellectual fields, which often imbue planning theories with awareness of and sensitivity to the multiple dimensions of planning practices. The series editors particularly encourage inter- and trans-disciplinary ideas and conceptualisations.

Decentralization in Environmental Governance
A post-contingency approach

Christian Zuidema

LONDON AND NEW YORK

First published 2017
by Routledge
2 Park Square, Milton Park, Abingdon, Oxon OX14 4RN

and by Routledge
711 Third Avenue, New York, NY 10017

Routledge is an imprint of the Taylor & Francis Group, an informa business

© 2017 Christian Zuidema

The right of Christian Zuidema to be identified as the author of this work has been asserted in accordance with sections 77 and 78 of the Copyright, Designs and Patents Act 1988.

All rights reserved. No part of this book may be reprinted or reproduced or utilised in any form or by any electronic, mechanical, or other means, now known or hereafter invented, including photocopying and recording, or in any information storage or retrieval system, without permission in writing from the publishers.

Trademark notice: Product or corporate names may be trademarks or registered trademarks, and are used only for identification and explanation without intent to infringe.

British Library Cataloguing in Publication Data
A catalogue record for this book is available from the British Library

Library of Congress Cataloging-in-Publication Data
A catalog record for this book has been requested

ISBN: 978-1-4724-2253-8 (hbk)
ISBN: 978-1-315-59335-7 (ebk)

Typeset in Times New Roman
by Apex CoVantage, LLC

Contents

List of figures and boxes	vi
Preface	vii

1	Searching for environmental quality	1
2	Governing the environment in a world of change	19
3	Navigating the plural	74
4	Making decentralization work	133
5	A European focus on the local	161
6	Beyond the minimum in the Netherlands	215
7	The relevance of a post-contingency approach	268

Bibliography	288
Index	319

List of figures and boxes

Figures

1.1	The structure of the inquiry	15
2.1	The governance triangle	24
3.1	A contingency approach: complexity as a criterion	111
3.2	The intersubjective in a post-contingency approach	113
3.3	Framework for planning-oriented action, with complexity as a criterion	121
4.1	Decentralization in the framework for planning-oriented action	139
4.2	Possible spread of performance of local governance	142
6.1	Transitions in Dutch environmental policy between 1990 and 2009	263

Boxes

3.1	Interpreting function	115
3.2	An example of structure and function	117
3.3	Matching and non-matching configurations	126
5.1	Seven key elements of sustainable urban management	189
6.1	Integrated Environmental Zoning	223

Preface

I was educated as a planner at a time when the communicative turn was still in full swing. Some of the writings I had to study were still clearly embedded in a positivist tradition and a reliance on the rational planning model so common up until the 1980s. Some other writings were clearly embarking on more relativist and constructivist approaches. In the meantime, I began to understand that in society itself, the process of governance was strongly influenced by the abstract and popular notion of neoliberalism. Although sometimes approached critically, privatization, deregulation and decentralization were actively pursued as strategies to improve the process of governing. As a student it was quite confusing, but I also slowly began to see why all these changes occurred. Because the one key message that seemed to be clear was that the 'old' and 'classic' regulatory-based command-and-control types of planning were to be approached sceptically, if not to be considered fully outdated.

Towards the end of my education, I was introduced to more nuanced perspectives on notions such as rationality and governance renewal. Although hard to fully grasp at that time, I felt attracted to these nuanced perspectives. They seemed to suggest that things were not as clear-cut as presented. Sure, the communicative turn highlighted the important message that knowledge and rationality are socially mediated constructs. Sure, neoliberalism showed us that not everything needs to be controlled by governments. And sure, the regulatory-based command-and-control types of planning had their key limits. But mostly, I began to appreciate that none of these arguments was itself without critiques. Instead, I increasingly began to accept that there was no one best way to organize planning and governance. That is, I began to understand the contingency argument that different circumstances beg for different approaches.

In my work as a planning academic this contingency argument always lingered. But I also felt uneasy with the argument. Contingency studies emerged during the 1960s and 1970s. Although they were updated, nuanced and improved in the decades after that, they largely relied on the same assumptions as planning did at the time: a positivist approach. To put it simply: contextual circumstances could be known through detailed inquiry and there were clear and uniform standards to be found upon which organizational structures and strategies would perform best given these circumstances. The result was a strong reliance on an object-oriented

viii *Preface*

approach; i.e. where planners were merely 'discovering the truth that is out there'. Also, the result was that given this 'discovered truth', the choice for the best organizational structures and strategies was a mere logical consequence; i.e. circumstances would determine the approaches chosen.

This rigid contingency argument not only felt too simplistic for me, but also seemed to overlook the important lessons the communicative turn had introduced within the planning theoretical debate. Nevertheless, its basic premise of contingently relating alternative approaches to planning to alternative circumstances encountered stuck with me. Maybe, I considered, an update was possible that did allow for the nuances of post-positivistic or even relativistic thinking?

In the meantime, my ongoing research had required me to review processes of decentralization within the context of environmental governance. The nuances I had appreciated even as a student became increasingly clear to me. Governance renewal turned out to be often pursued without such nuances. Political ideologies, fashionable ideas within public administrations and academics and wishful thinking were no exceptions. Partly, it seemed, this was also due to a lack of theoretical or academic guidance. The planning theoretical landscape, I found, had become characterized by multiple and contrasting theoretical proposals regarding planning and governance. They were proposals often relying on different and sometimes incommensurable philosophical positions, often linked to a duality between realist and relativist perspectives. In the midst of this 'theoretical plurality', it seemed hard to find clear theoretical arguments and academic evidence to counteract optimistic ideological hopes.

When I studied some of the unexpected and undesirable consequences of decentralization, for example, in the context of the Netherlands and the EU that I will address in this book, I did note that many of these consequences were not that unexpected at all. A careful study of literature on decentralization often warned against many of these consequences. Similarly, it seemed evident that the success of decentralization operations would be conditioned by the circumstances encountered; i.e. not every issue or task seemed well suited for decentralized working. It was within this context that I again was drawn to the contingency argument. Maybe there were arguments that could explain when and why we should rely on alternative planning approaches? And maybe these contingency arguments could even help us navigate the theoretical plurality we face. These were exciting and relevant questions that urged me to investigate further.

This book is the result of my investigation into this contingency argument. With what I call a 'post-contingency' approach, I take the contingency argument out of its reliance on only an object-oriented approach and bring it into an intersubjective domain. In doing so, I also attempt to pragmatically bridge the duality between realist and relativist perspectives and provide guidance upon navigating the plural governance landscape. The book can be read in at least two ways. The first is as a theoretical investigation and attempt at navigating the plural governance landscape. It is here where I to invite the reader to think along with my elaboration of a post-contingency approach and its potentials. Second, though, this book is also about explaining why I consider a post-contingency approach relevant. It is in delivering a constructive critique of governance renewal in environmental governance that I hope to do so. It results in me coming back to the nuances that attracted me as a student of planning: most notably that the 'old' and 'classic' regulatory-based command-and-control types of planning still have an important role to play in our 21st-century governance landscape.

Christian Zuidema
Groningen

1 Searching for environmental quality

Wednesday 23 May 2007 was not a date that will be widely remembered for its groundbreaking news. The news in Europe was dominated by the Champions League final between AC Milan and Liverpool, whilst in the US and the Middle East, a new bombing in Iraq dominated the news. Nevertheless, 23 May 2007 might still go into our history books. After all, it is this day that is considered to be the first day in human history that more people lived in urban areas than in rural areas (Wimberley & Fulkerson 2007) – a fact signifying that the 21st century might well be considered the urban century.

In many countries, the urban era arrived many years earlier. In most Western countries, the majority of the population became urbanized even before or soon after the Second World War. The percentage of the urban population in recent decades has reached levels above 80% in countries such as Belgium, Denmark, Germany, Sweden and the United States (UN 2003). Concentrating people and their activities in small areas, such as urban centres, however, has not been without consequences. These consequences include some serious effects on the local environmental quality. Poor air quality, severe congestion, high levels of ambient noise, serious safety risks, excessive resource consumption and large polluted brown field sites are among the immediate problems most towns and cities face. In 2007, for example, the World Health Organization calculated that up to 20% of deaths in Europe could be attributed to bad environmental quality (WHO 2007). Although surrounded by serious uncertainties, the results of this WHO study do illustrate the urgency of addressing the quality of our local environments, something that has not escaped the notice of many governments.

Since its rapid expansion in the 1970s, environmental policy in most of the Western world has been dominated by central government control and legalistic and regulatory-based policies (e.g. Andersen & Liefferink 1997, Jänicke & Jörgens 2006 and Nelissen 1997). These centralized and legalistic policies have not been without effect, as many environmental conditions have improved in recent decades, including at the local level. Emissions of environmental stressors have been substantially reduced, causing a considerable reduction in emissions into air, water and soil. Despite these successes, the centralistic and legalistic approach is increasingly considered to be *only part* of the solution to the puzzle of environmental policy (e.g. CEC 1990, De Roo 2003 and Van Tatenhove et al. 2000).

2 *Searching for environmental quality*

During the past two decades, also in environmental policy, many new instruments and strategies have been implemented (e.g. Busch & Jörgens 2005, Jänicke & Jörgens 2006, Jordan & Lenschow 2008, Jordan et al. 2005, Lemos & Agrawal 2006, Vig & Kraft 2013).

Among the main innovations in renewing environmental policy is the increase in the role of the local level in governing. In this book, I will specifically focus on this increase of the local level in the realm of environmental policy. Nevertheless, in this book I aim to go beyond a study on the motives and consequences of this increase of the local level. Rather, I consider the changes taking place in environmental policy as exemplary for some prime 'shifts in governance' that can be witnessed in much of the Western world and specifically western Europe over the past few decades. The 'shifts in governance' can be traced to an increasingly sceptical attitude towards traditional approaches to governance such as centralized control and legalistic and regulatory-based policies. In response, these traditional approaches are being replaced, sometimes swiftly and rigorously, by new approaches through widespread governance-renewal operations. In this book, I argue that such governance renewal should itself also be approached sceptically. That is, governance renewal should be supported by clear theoretical arguments and by a keen awareness of the possible consequences of such renewal. The aim of this book is not only to develop such arguments and awareness but also to cast doubt upon the merits of a rapid and generic shift away from centralized and regulatory-based governance approaches. In doing so, I draw from my empirical study into such shifts in governance in the realm of environmental policy in western Europe, especially due to the rather rigorous shift away from strong centralized and regulatory-based governance approaches in this policy field.

1.1 Shifts in governance

During most of the 20th century, the exercise of governance, also in the Western world, was left largely to the discretion of formal governments, most notably the central state (see also Pierre & Guy Peters 2000). Reliance on government control was supported by the long-held assumption that (central) governments are able to exercise a high degree of control over social processes and, while doing so, are also best equipped to represent the 'public good'. Although left relatively undisputed for a long time, the past few decades are characterized by much greater scepticism towards this assumption. This scepticism has fuelled some important 'shifts in governance' (see also Van Kersbergen & Van Waarden 2004).

Changing societies

To a large degree, recent 'shifts in governance' in Europe and beyond are a response to changing societal conditions (e.g. Hajer et al. 2004, Hooghe & Marks 2001, Kooiman 1993, Pierre & Guy Peters 2000, Rhodes 2000, 2007 and Stoker 1998). Social fragmentation and power dispersal have come to challenge the supremacy of the central state in controlling policy development and delivery.

Social fragmentation involves an increasing diversity of interests and opinions held by societal actors, undermining widespread societal support for central governments in exercising control over sections of our societies. Governments, therefore, are encountering many different societal or market parties who are claiming their place in the governance process and, as expressed with power dispersal have the resources to exercise influence. In the meantime, societies have also become more mobile, dynamic and, in general, are considered more complex. Such complexity follows the diversification of our economies, ongoing globalization and innovations in, for example, transportation and information and communication technology. Complexity is furthermore caused by the increased interrelatedness of economies, people and problems. On the one hand, interrelatedness manifests itself in the growing mutual dependence between the aforementioned public and private parties. On the other hand, there is also an awareness of the strong interrelation between problems, their causes and effects. Consequently, there are usually multiple and potentially conflicting objectives to which policies should respond, while multiple parties with their own interests and resources are also involved. To summarize, formal governments are facing more eloquent, fragmented, dynamic and interdependent societies and issues. Whilst having to cope with the vast coordinative efforts that follow, formal governments are also facing a decrease in their power base compared to private parties and a decrease in societal support.

Problems to respond to

While governments seek to respond to changing societal conditions, they find it difficult to envision how they should do so. It is not uncommon for governments to reach back to tried-and-tested strategies, i.e. they establish new and additional policies and regulations to regain control. There are, however, serious doubts concerning this strategy.

In the first place, governments are often criticized for being inflexible, fragmented and inefficient (e.g. De Leeuw 1984, Pierre & Guy Peters 2000). Governments are traditionally organized into separate policy fields and departments, each specializing in a distinct set of policy functions and tasks. Although this specialization might in itself be a logical way to organize governments and policies, specialization is also prone to cause fragmentation. Serious coordinative deficits and policy incoherencies can be the consequences of fragmentation, while fragmentation can also frustrate attempts to cope with the interrelatedness of policy problems (e.g. Breheny & Rockwood 1993, De Leeuw 1984, Miller & De Roo 2004, Pierre & Guy Peters 2000 and Van Tatenhove et al. 2000). In the second place, the great extent to which governments rely on regulatory instruments is built on the assumption that governments can control the future direction of our societies. It is an assumption that is incompatible with a social reality characterized by social fragmentation, diminished public support and the dispersal of power. But by holding on to this assumption, many governments turn to proven methods such as additional regulations to increase their degree of control over societies and their problems. As De Leeuw (1984), for example, shows,

4   *Searching for environmental quality*

this strategy is built on the misconception that additional coordinative efforts will always produce more control. Rather, as De Leeuw highlights, continuously adding more regulations and agencies can produce an overload of government policies that, instead of increasing the capacity to govern, increases governing problems such as fragmentation, bureaucracy and coordinative difficulties. As many governments have continuously faced such problems in recent decades, they are gradually acknowledging that merely 'improving' or 'expanding' existing governmental organizations and policies might not be the way forward. Instead, they are starting to accept a need for more radically different modes of governing (e.g. CEC 2001a, Kickert 1993, Nelissen & Raadschelders 1999, Osborne & Gaebler 1992 and RMO 2002). In other words, what we are witnessing is an increased acceptance of the fact that our traditional governments and their instruments are not well suited to coping with the challenges our society is throwing at us (e.g. Van Kersbergen & Van Waarden 2004, Kickert 1993, Pierre 2009 and Rhodes 2000, 2007).

Innovations occurring

Since the 1980s in particular, many central governments and including those in western Europe expanded the societal capacity to govern by involving lower tiers of government and non-government parties. Through strategies such as decentralization, deregulation and privatization, regional and local governments gain in importance, whilst power and responsibility are also transferred to the private sector and the public (e.g. Derksen 2001, Hooghe & Marks 2001, Jessop 1994, Nelissen et al. 1996, Pierre & Guy Peters 2000 and Stoker 1998). The extended use of financial tools, the increased use of partnerships with non-government parties, and the increased participation and involvement of stakeholders and 'civil society' are examples of changes. But while actively pursuing these changes, it is far from clear what kind of allocation of responsibilities and steering arrangements these changes should lead to.

In reflecting on governance-renewal operations, Hajer et al. (2004) witness extensive administrative experimentation, while De Roo (2002) calls it a process of 'trial and error' (see also Allmendinger 2002a, Nelissen et al. 1996, Nelissen & Raadschelders 1999 and Van Tatenhove et al. 2000). As these scholars argue, we are witnessing the rise of a wide diversity of approaches to governance that are manifesting themselves in an equally diverse set of organizational formats. The result is an increasingly plural landscape of governance practices. Illustrated by the ongoing experimentation, it seems we lack criteria or guidelines to decide which of these practices, approaches and formats are desirable or potentially undesirable. In other words, we are not at all sure which kinds of changes should be pursued. Ideally, we would then look to theories on governance to help us formulate these guidelines. This is not as simple as it seems, however. As it turns out, the plurality that has manifested itself in the practice of governance is also manifested in current theoretical perspectives on governance.

Theoretical shifts

Recent changes in the theoretical debates on governance have undermined the idea that the form of societal control exercised by central governments that was predominant during most of the 20th century is possible or even desirable. Instead, we face what Allmendinger (2002b) calls a 'post-positivist planning landscape' that is subject to a continuous condition of what Healey et al. (1979) refer to as 'theoretical plurality'. This theoretical plurality means that we are witnessing a wide variety of different and sometimes incommensurable theoretical proposals regarding the organization of governance.

The rise of the theoretical plurality we face can be linked to some large-scale societal and theoretical changes in the latter part of the 20th century. Often signified by the concept of postmodernism, this involves an increased divergence of philosophical positions and socio-cultural values and beliefs (e.g. Harvey 1989). Apart from their influence on social fragmentation and dynamics, these diverging beliefs have also resulted in a divergence of theoretical perspectives on governance. In other words, different theoretical positions each draw upon different socio-cultural and philosophical values and beliefs. Consequently, it has become difficult to find common guidelines for connecting theoretical positions or choose between them. Instead, we are even facing a widespread theoretical scepticism towards the possibility that scientific inquiry can produce universal and fundamental knowledge to produce such guidelines.

The theoretical scepticism referred to is based on the rather common acceptance, in theoretical debates of the late 20th and early 21st century, that facts and values are intertwined and that multiple and plural visions on what is 'real' and 'rational' coexist. In the wake of the acceptance of this idea, knowledge and rationality are now increasingly considered as socially constructed, i.e. considered to be human-made (e.g. Berger & Luckman 1967, Gergen 1999 and Healey 1997). Social construction is based on the idea that people use different frames of reference when perceiving, interpreting and judging what they see. Social construction thus explains why people can fundamentally disagree upon what is considered 'real', 'rational' and 'good'. Such disagreements have also manifested themselves in the existence of multiple and plural theories on governance, which are inspired by a wide array of different philosophical and socio-cultural values and beliefs (e.g. Allmendinger 2002a). Given these disagreements, we can also explain why it has become problematic to find guidelines for choosing between these beliefs and preferences. After all, if we disagree upon what is 'real' and 'rational', where is the common ground upon which to build these guidelines? The consequence: theory is no longer providing us with ready-made answers regarding how 'best' to govern but, instead, is confronting us with a plurality that is difficult to navigate.

Where plurality leaves us

In recent decades, governance renewal – for example, in the form of decentralization, deregulation and privatization – has made an impact on the governance

6 *Searching for environmental quality*

landscape, also in western Europe. This impact can also be witnessed when it comes to transferring power and authority to the local level, which is my main interest here. Such changes should ideally be well informed by theoretical arguments and the possible consequences of the changes should be well understood and anticipated. Evidence suggests that, today, many changes are pursued without being well informed, while knowledge of their likely consequences is not evident either (e.g. Connerly et al. 2010, De Vries 2000, Fleurke & Hulst 2006, Flynn 2000, Jordan et al. 2005, Prud'homme 1994 and Walberg et al. 2000). Certainly, many Western governments are taking serious risks in renewing governance without clearly knowing the possible consequences.

If we are to be better informed when renewing governance, we can look to theories on governance for inspiration and guidance. The 'theoretical plurality' prevents us from finding clear answers as to which kind of renewal operations we should pursue. Rather, we face multiple and sometimes incommensurable answers about which governance approaches and related renewal operations are desirable. In addition, we seem to lack the ability to find common ground to serve as a starting point if we are to choose between these answers. Hence, we seem entangled in what Offe (1977) predicted would be a 'restless search' to find appropriate governance approaches for the issues we face amidst a plurality of possible approaches (see also Nelissen 2002). We are thus presented with a clear challenge: can we find some kind of common ground to serve as a starting point in choosing between various governance theories and approaches and, hence, navigate the plural governance landscape we face? It is one of the key challenges I will respond to in this book, in order to establish when and how increasing the role of the local level can be considered more or less appropriate. Before doing so, let me first address the changes in environmental policy relating to these shifts in governance, most notably the rise of more proactive and integrated approaches and the related increased role of the local level.

1.2 A Proactive and integrated approach to the environment

Just as in most of the Western world, environmental policy in Europe has been dominated by central government control and legalistic and regulatory-based policies, at least up until the late 1980s (e.g. Andersen & Liefferink 1997). In fact, environmental policy can even be seen as a key example of a policy field dominated by central government control. Not surprisingly, it ran into the typical problems associated with this mode of governance.

During the 1980s, the specialization of environmental policies was already considered to be causing widespread problems of fragmentation and incoherencies in several European countries (e.g. Andersen & Liefferink 1997, Nelissen 1997). Regulations and policies had been developed for separate areas such as air quality, soil remediation, safety risks, noise nuisance, etc. However useful such specialization might be, it led to fragmented policy agendas that made it difficult to deliver cross-sectoral solutions to the many problems encountered and also caused coordinative problems and even policy incoherencies. In the meantime, the drawbacks of a strong focus on legislative instruments also gradually became evident. Dominated by rigid and fixed standards, such a focus on legislative instruments allowed little flexibility for (environmental) government officials to engage in negotiation and bargaining strategies with key stakeholders. It meant that environmental policies were ill adapted to the dispersal of power and fragmentation in society. In addition, the focus on legislative instruments meant that environmental policies were about fulfilling limit values rather than developing more proactive and strategic policies. Most environmental policies were based on setting restrictions on social developments and economic growth that could potentially harm the environment. But in focusing on *controlling the damage* caused by environmentally intrusive human activities, environmental policy remained a *reactive* policy field, relatively marginalized in comparison to other governance activities focussed on, for example, social development, economic growth and spatial planning.

During the past two decades, many European and non-European governments have tried to push environmental policy beyond its reactive character, whilst also addressing the coordinative problems. A main signifier of these attempts is the concept of 'sustainable development'. Traditionally defined as "development that meets the needs of the present without compromising the ability of *future generations to meet* their own needs" in the words of the 1987 Brundtland commission (WCED 1987; p. 43), sustainable development has become a widely acknowledged planning and policy guideline (e.g. Briassoulis 1999, Feitelson 2004 and Selman 1996). Sustainable development signifies a move away from a marginalized environmental policy field focussed on damage control, to a policy field that is more proactive and integrated into other governance processes. It is a response to both the desire and need to deliver more proactive and integrated approaches to our environmental challenges, including those at the local level.

Sustainable development: beyond damage control

Sustainable development implies that addressing environmental aspects should be about more than just controlling damage from environmentally intrusive human activities. Instead of considering environmental quality as only a limiting condition to economic and social development, sustainable development calls for environmental quality to be seen as a basic principle *in addition to* economic and social priorities. The environmental consequences and benefits of any development should thus be a criterion in decision making alongside social and economic consequences and benefits (see also Briassoulis 1999, O'Riordan & Voisey 1998 and WCED 1987).

On the one hand, sustainable development relates to the idea that environmental policy should be more *proactive*. Sustainable development promotes a future-oriented approach, where the idea is to prevent and anticipate future problems. In a proactive approach, environmental qualities are not only considered as limiting conditions to growth and development, but become part of the envisioned outcomes of both strategic and operational plans. On the other hand, sustainable development also pushes for a more *integrated* or *holistic* attitude. After all,

8 *Searching for environmental quality*

sustainable development calls for economic, social and environmental priorities and ambitions to be considered together. It thus acknowledges the interrelatedness of various problems, their causes and effects. Consequently, when pursuing environmental ambitions, the economic and social effects of doing so should also be taken into account. As many environmentally straining activities and land uses are important contributors to economic development and social welfare, a more holistic attitude supposes that solving environmental issues should take account of the diversity of interests and claims that surround these issues.

Beyond the centre

Sustainable development not only shifts attention to a more proactive and integrated approach to environmental issues, but also posits that this should be done on all levels of authority (e.g. Selman 1996). While being dominated by central state control, the local level in most European countries has for a long time been mostly active in implementing national and international environmental policies and regulations (see also Andersen & Liefferink 1997, Nelissen 1997). Sustainable development challenges the idea that central state control is sufficient, and suggests that action should be taken *at all levels* of government and civil society to bring about the desired changes (e.g. Moore & Scott 2005, Pugh 1996). Advanced through the United Nations Local Agenda 21, many localities have responded to this call, i.e. 'Think Globally, Act Locally'. Similarly, the success of the 1994 Aalborg Charter, whereby 2,500 European towns and cities committed themselves to the pursuit of sustainable development, bears witness to the popularity of sustainable development.

While sustainable development calls for environmental initiatives to be taken at all levels of government, there are also other pressing arguments for local environmental initiatives. First, there is the subsidiarity principle: the idea that matters should be dealt with at the lowest level of authority possible, so as to be closest to the actual problem. Many environmental problems have local causes and effects and, therefore, a clearly local manifestation. This results in the argument that the local level should play an important role in addressing these issues. In the meantime, however, there are also other important practical arguments.

Many local and especially urban environmental issues are strongly embedded in their local contexts. Various local stakeholders, historic development paths, detailed local circumstances and specific local societal and political priorities can affect these issues. They are strongly linked to other local issues (interrelatedness), priorities and associated stakeholders (power dispersal) and are surrounded by diverging preferences (social fragmentation). Proactive and integrated policies aim to translate the interrelatedness between various problems and interests into future-oriented and holistic policies. This requires more dynamic policy approaches that can respond to specific local circumstances. These are not just geographic, demographic or economic circumstances. Also, each locality has different problems, priorities and stakeholder interests. It can be problematic for the central state to anticipate to such local specific circumstances and the existing interrelations (Van Kersbergen & Van Waarden 2004, Kickert 1993 and Rhodes 2000). Instead, the proximity of local authorities to local circumstances, stakeholder interests and the 'civil society' puts those authorities at an advantage over central government in terms of responding to the diverse and local manifestations of (environmental) governance issues (e.g. Hooghe & Marks 2001, Kamphorst 2006).

Supposed benefits related to proximity to local circumstances include the idea that local authorities are in a better position to engage in bargaining or collaborative processes with local stakeholders and civil organizations. In addition, familiarity with local circumstances and interests possibly gives local parties an advantage in terms of translating the interrelatedness of issues and interests *in situ* into integrated strategies, i.e. to 'tailor' them to the situation (compare De Roo 2004, Dryzek 1987, Jordan 1999 and Liefferink et al. 2002). Hence, local authorities are assumed to be important players in the development and delivery of more proactive and integrated approaches to the environment. And in doing so, local authorities are important players in helping governance respond to recent societal changes such as the interrelatedness of issues and interests, power dispersal and social fragmentation.

Encouraging the local

While individual towns and cities have already begun to take more responsibility in environmental policy, national and international governments are also involved in advancing the role of the local level (e.g. Andersen & Liefferink 1997, CEC 2004, Creedy et al. 2007). In doing so, two main strategies can be discerned.

First, it is possible to consider local environmental policies as *complementary* to existing national and international policies and regulations. Regulations that are in place then remain and might even be extended further. The work done locally is intended to add value to these national and international regulations. This is, for example, the approach chosen by the European Union (EU). Even in its 1990 Green Paper on the Urban Environment, the European Commission (CEC) noted that "dealing with the problems of the urban environment requires going beyond sectoral approaches" (1990; p. 1). While the CEC continued to issue new sectoral regulations throughout the 1990s and early 21st century, it also responded to its own call in the 1990 Green Paper. If the severity and complexity of urban environmental issues is to be addressed, the CEC suggests, a more integrated approach is needed locally in order to improve the implementation of existing sectoral regulations. Hence, the CEC is now inviting European cities to develop their own systems of local environmental management (CEC 2006, 2007a). It is an invitation inspired by the idea that local environmental management helps to deal with the specificity and interrelated nature of local issues.

Second, local environmental policies can also *function partly as an alternative* to national and international policies and regulations. This is the approach chosen in the Netherlands, for example. Whilst many national and international policies and regulations remain in place to function as safeguards to prevent excessive

10 *Searching for environmental quality*

damage, other policies and regulations are believed to frustrate local attempts to produce cross-sectoral and balanced approaches, i.e. they force local authorities to continue working with fragmented sectoral approaches. In response, decentralization and deregulation have been pursued so as to give the local level greater freedom to set its own course in environmental policy and, hence, pursue the desired integrated and 'tailor-made' policy approaches. In the meantime, the national government urged local parties not just to produce local environmental policies but also to produce a higher local environmental quality as compared to the situation before decentralization and deregulation (VROM 2001, 2004b).

Although pursued in different ways, both strategies show how the local level is increasingly considered a good place for developing and delivering environmental policy. Also, with both strategies the desired effect is that local parties add value to national and international environmental policies and regulations. To be more concrete, the idea is that local parties will produce their own policies that reveal a more proactive and integrated approach to local environmental issues (e.g. CEC 2004, 2006, VROM 2001). In the meantime, however, there are serious doubts as to whether the increased role of the local level *can* and *will* actually deliver the desired effects.

Doubts

Experiences in practice are showing that it is not easy to pursue proactive and integrated approaches in environmental policy, including at the local level. On the one hand, environmental priorities have a relatively 'weak profile' as compared to more development-oriented priorities. Environmental benefits can, for example, be hard to express in financial terms (such as noise nuisance), are often invisible (as with safety risks), diffuse (as with air pollution), are highly subjective (odour) or focus on a long time horizon (e.g. sustainability). Economic growth, social development or, for example, financial costs are easier to envision and often relate to the short term. In the meantime, many environmental issues are manifesting themselves as social dilemmas. If one town reduces its air pollution, it alone bears the cost, whilst all adjacent towns benefit. Similarly, if all the other towns reduced air pollution and this one town did not, it would enjoy the benefits without having to bear the cost. It provides a negative incentive for proactive and ambitious environmental measures.

Given the 'weak profile' of environmental priorities, it is not surprising that even leading municipalities and countries have a hard time putting into practice proactive and integrated approaches to the environment (see also Bouwman et al. 2005, Burström and Korhonen 2001, Creedy et al. 2007 and Jordan 2008). The prime signifier referred to by governments is sustainable development. But while sustainable development is widely referred to and used, its actual impact in policy actions and social, spatial and environmental governance is less convincing (CEC 2004, EEB 2006, Jordan & Lenschow 2010, Kamphorst 2006 and O'Riordan & Voisey 1998). Towns often fail to move beyond well-meant ambitions and plans. As such, initiatives such as Local Agenda 21 and the EU Sustainable Cities Campaign became considered to have met with only limited success (CEC 2004).

Searching for environmental quality 11

On the other hand, the access that local authorities and stakeholders have to instruments, resources and expertise is also a cause for doubt regarding their possible role in environmental policy (e.g. De Vries 2000, Fleurke & Hulst 2006). Time, funding, competent staff and legal instruments are among the resources needed to cope with the responsibility of developing and delivering locally proactive and integrated policies. Research findings suggest that the availability of these resources is certainly not evident at the local level (e.g. Burström & Korhonen 2001, Flynn 2000 and Walberg et al. 2000). As this can undermine the potential success of increasing the role of the local level (e.g. Eckersley 1992, Jordan 1999 and Prud'homme 1994), a degree of scepticism again seems appropriate.

Where this leaves us: central hypotheses

Despite widely accepted benefits, doubts have arisen with regard to increasing the role of the local level in environmental policy in order to develop and deliver proactive and integrated approaches to the environment. If local willingness or ability to develop and deliver proactive and integrated approaches are limited or constrained, local authorities will also fail to respond appropriately to newly decentralized tasks and responsibilities. If these partly replace existing national and international regulations, local environmental quality might even deteriorate, with consequences for public health. In addition, since various localities will perform differently in making and implementing their environmental policies, there will tend to be differences in environmental qualities and protection against health effects. Equity issues then come into play, pushing for moral debates as to what degree of difference is acceptable. Finally, given potential differences in policies and protection, the legal security experienced by companies or citizens can be undermined. To summarize, increasing the role of the local level in environmental policy is not necessarily producing the changes desired in local environmental quality.

What I instead assert is that an increase in the role of the local level is contingent on many circumstances (see also De Roo 2004, Fleurke & Hulst 2006). It involves the kind of capacities that are available locally and the local willingness to use them. It involves the kind of policy issues that are or are not appropriate for dealing with locally. And it involves the kind of national and international policies that local parties are faced with and can enable, stimulate or constrain local parties in the choices they make. Therefore, despite the benefits of extending the role of the local level, my hypothesis in this book is that central policies and regulations can provide a foundation of policies for employing and supporting decentralization so as to improve the governance capacity for developing and delivering proactive and integrated approaches to the environment.

In order to pursue and reflect on this hypothesis, I first aim to explore the contingent relations between adopting various approaches to governance, the circumstances and conditions existing when adopting them and the kind of consequences that can be expected. In doing so, I will explain how central policies and regulations can influence the occurrence of various practical consequences when increasing the role of the local level for developing and delivering proactive and integrated approaches to the local environment. Based on the contingencies

12 *Searching for environmental quality*

found, I subsequently aim to formulate arguments and conditions that can inform us about the kind of consequences we can predict when relying on various approaches to governance. In doing so, I hope to provide decision makers with arguments that enable them to make informed choices regarding governance renewal. It is an ambition that brings us right back to the debates on the plural governance landscape discussed above. This also brings us to the 'heart of the matter' that this study aims to address.

1.3 The heart of the matter

Our story was inspired by the observation that the local level is increasingly seen as a good place to develop and deliver environmental policies, specifically in relation to more proactive and integrated approaches to the environment. While this observation is supported by widely accepted benefits, there are also important doubts. Consequently, we are faced with a lack of clarity as to the merits of extending the role of the local level, also in environmental policy. One key aim of this book is to contribute to creating this clarity with a view to helping governments to make better informed choices when addressing the role of the local level in environmental policy.

The main *practical aim* for this book is to indicate the likely consequences of increasing the role of the local level in environmental policy for developing and delivering proactive and integrated approaches to the local environment, given the circumstances encountered. This practical aim, however, is closely related to a second key aim of this book: the *theoretical aim*. The theoretical aim relates to the problems encountered in navigating the plural governance landscape, which also results in theoretical difficulties to find arguments as to whether – and if so, when – to increase the role of the local level in governance. The aim is to find common ground comprising theoretical arguments that can help us navigate the plural governance landscape we face while focusing on the possible increase in the role of the local level of governance (i.e. decentralization). This should result in theoretical guidelines that can help decision makers make choices regarding governance renewal. These guidelines are subsequently applied in the context of environmental policy, specifically regarding the attempts to increase the role of the local level in developing and delivering proactive and integrated approaches to the environment. On the one hand, environmental policy is thus used as an empirical context in which to realize our theoretical aim. On the other hand, the arguments formulated through our theoretical inquiry should help realize the practical aim.

A theoretical inquiry

The theoretical inquiry focuses on developing guidelines for connecting various governance approaches to various issues and circumstances we face, while focusing on the possible increase in the role of the local level in governance (i.e. decentralization). Two main questions are central in guiding this inquiry.

Searching for environmental quality   13

The first question addresses the assumption that guidelines *can* be found to help us navigate the plural governance landscape. In this context, I take inspiration from contingency theory, i.e. the idea that the decision in favour of an approach or strategy in a given situation should be contingent upon the circumstances of the situation (e.g. Bryson & Delbecq 1979 and Lawrence & Lörsch 1967). Contingency theory, thus, assumes that the performance of different approaches or strategies is influenced by the circumstances encountered. Therefore, it offers us the possibility to formulate guidelines to choose between approaches or strategies, based on the circumstances encountered. Nevertheless, contingency theory arose in a period dominated by confidence in objective knowledge and universal rationality. As a consequence, early contingency theories, which I here coin as 'classical' contingency theory, are strongly object-oriented in trying to describe and objectify the contingencies between an approach or organization and its contextual circumstances. In addition, this object-oriented knowledge is then rationally translated into changes in governance approaches and organizational configurations, where rationality is equated with what is called instrumental or technical rationality, where all people subscribe to a similar rational logic (e.g. Berting 1996, De Roo 2003 and Healey 1997). In other words, these classic approaches tended towards the belief that it was possible to objectively establish which approaches should be chosen under different circumstances. Classic approaches to contingency, clearly, are thus positioned in a scientific tradition that is quite far removed from the scepticism we encounter when discussing the plural governance landscape towards objective truths and universal rationality.

In this book, contingency theory will be approached from a sceptical philosophical attitude, implying a denial of the possibility for objectively establishing which approaches should be chosen under different circumstances. Nevertheless, the ambition in this book is to formulate guidelines to help us choose between various approaches to governance. I therefore propose to reformulate contingency theory and, in doing so, answer a first key theoretical question (T1), which is: *how can contingency theory be reframed so as to help us navigate the plural governance landscape?* While answering this question, I will propose a new approach towards contingency, which I will coin as a *post-contingency approach.*

On the one hand, a post-contingency approach draws upon the object-oriented focus of classical approaches to contingency in arguing that differences in the contextual circumstances encountered provide us with arguments for choosing between different approaches to governance. On the other hand, however, a post-contingency approach also incorporates the argument that the way in which people establish what is 'real' and 'rational' is influenced by the intersubjective context in which they are embedded. Therefore, post-contingency also draws from an intersubjective-oriented focus in suggesting that how planners and stakeholders frame the circumstances encountered and subsequently respond to them is also a socially mediated matter of choice. Incorporating an intersubjective-oriented focus brings contingency thinking outside of a reliance on objective knowledge and notions of universal rationality. It does not, however, imply that choosing between approaches to governance is to be equated with 'whatever people believe

14   *Searching for environmental quality*

to be an appropriate approach'. Not only does a post-contingency approach argue that collecting object-oriented knowledge of the circumstances remains relevant for informing such a choice, it argues that any choice between various approaches to governance is constrained according to some basic logical relations, even if contextual circumstances are perceived and interpreted differently. To explain this argument, a post-contingency approach argues that not all organizational formats are equally well suited to performing certain functional ambitions, and vice versa. A decentralized format is argued to be ill suited for addressing generic policy ambitions that should performed equally in all jurisdictions. In contrast, a centralized format is argued to be ill suited for delivering specific policy approaches for addressing unique local circumstances. Post-contingency, therefore, incorporates the argument that there is only a limited number of configurations or organizational formats and functional ambitions that match. As a result, it becomes possible to differentiate between governance approaches with more or less matching configurations, allowing us to formulate guidelines that help us choose between different governance approaches.

Following an explanation of how a post-contingency approach can be adopted in navigating the plural governance landscape, I will continue in this book with the aim of formulating guidelines as to whether and how to increase the role of the local level. Consequently, the second theoretical question (T2) is: *which guidelines can be formulated that help to inform decision making on a possible increase in the role of the local level in governance?*

A practical inquiry

Following on from my practical aims, the main practical question for research (P0) is *what are the likely or possible consequences of increasing the role of the local level in environmental policy for developing and delivering proactive and integrated approaches to the local environment, given the circumstances encountered when pursuing this increase?* It is this question that will guide the practical line of inquiry.

I begin by discussing recent developments in environmental policy, which pursue the idea that the local level is considered a good place to develop and deliver proactive and integrated approaches to the environment. While renewing environmental governance, various central governments in Europe are encouraging the local level to become more engaged. In doing so, they assume that this will help local authorities and their partners to develop and deliver proactive and integrated approaches to the local environment. I am interested in the *consequences* of their attempts and *why* these consequences occur. First, therefore, I ask (sub-question P1): *what are the consequences of central government attempts at encouraging local authorities and their partners to develop their own proactive and integrated approaches to their environmental challenges?* Next, the outcomes of my theoretical inquiry can help to explain *why* these consequences occur. To this end, a second practical sub-question (P2) guides the research: *do central governments take account of the guidelines as formulated when promoting local proactive and integrated approaches to the environment?* Based on the answers to questions

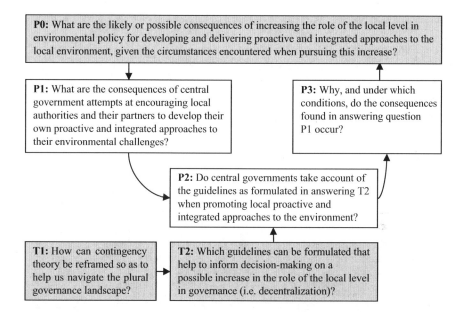

Figure 1.1 The structure of the inquiry

P1 and P2, I finally address the third practical sub-question (P3): *why, and under which conditions, do the consequences found in answering question P1 occur?*

The practical inquiry is not just a matter of analysing the performance of central governments, but is also used to reflect upon the theoretical insights developed. The intention is not to test the guidelines in detail, since this would require a large set of practical cases. The empirical data collected does, however, allow for a reflection on the applicability of the guidelines that are developed in practice and, if necessary, to add nuances, contextualize or alter them. The questions that structure the inquiry on which this book is based are summarized in Figure 1.1.

1.4 Structure of the study

This book consists of both a practical and theoretical inquiry to fulfil the defined aims. The two inquiries are expressed in the structure of this book. Before that, Chapter 2 will first explain why the local level is increasingly considered a good place for developing and delivering environmental policy, while addressing the main risks and doubts involved. In other words, Chapter 2 sets the stage for the two inquiries that follow.

Setting the stage

Chapter 2 begins with a discussion of the general shifts in governance that can be witnessed in recent decades, while focusing on the European context.

16 *Searching for environmental quality*

The chapter first explains why increasing the role of the local level is considered a means for renewing governance to bring it in line with changed theoretical perspectives and societal conditions. In addition, the chapter addresses the plural governance landscape that follows from these recent shifts. As such, Chapter 2 sets the stage for the later theoretical inquiry by expressing the kind of doubts as to where the shifts in governance are leading us, or should lead us.

Second, Chapter 2 links the recent changes in governance to the development of environmental policy in a western European realm, and especially the recent changes in this policy field. Related to the rise of sustainable development and more recently 'resilience', these changes are mostly evident during the past two decades and point towards more proactive and integrated approaches to the environment. The motivation for these changes will be discussed, while the potential role of the local level in delivering them is highlighted. The changes point towards more proactive and integrated approaches to the environment and sketch a picture of the possible implications for the activities of local authorities and parties. Also discussed in Chapter 2 are the main risks associated with increasing the role of the local level in environmental policy. Based on discussing both the motivations and risks associated with decentralization in environmental policy, Chapter 2 will also set the stage for the practical inquiry. It does so by both expressing the *envisioned* outcomes of increasing the role of the local level in environmental policy and by expressing the *possible* outcomes of the strategies used to increase this role.

Finally, Chapter 2 explains that the practical inquiry of this book cannot be separated from its theoretical inquiry. Instead, the difficulties we face in navigating the plural governance landscape are directly influencing the difficulties we have in assessing how the local level can be involved in environmental policy so as to develop and deliver proactive and integrated approaches to the environment. To conclude, then, the aim of Chapter 2 is to explain the fundamentals of the discussion that is central to this book.

A theoretical inquiry

The theoretical inquiry takes place in Chapters 3 and 4. Chapter 3 explains how contingency theory can be reframed to help navigate the plural governance landscape (question T1). It results in my proposal for a post-contingency approach. In this chapter, I begin by addressing the plurality we face in governance practices and in theories on governance. Furthermore, I will address the consequences of this plurality for finding guidelines for choosing between various approaches to governance. A post-contingency approach is then proposed as a tool for dealing with these consequences. Based on this, Chapter 3 sets the stage for finding guidelines for decision making on whether or not to decentralize power and responsibilities to the local level, based on the likely consequences.

This also brings us to Chapter 4 that draws upon the idea that guidelines can be formulated to inform us about the likely consequences of decentralization, given the approaches used and conditions encountered. Chapter 4 addresses research and discussions on decentralization and policy implementation as collected in

Searching for environmental quality 17

research reports and academic literature. Drawing upon a post-contingency approach, Chapter 4 uses this research and discussion for identifying the kind of consequences that can be expected from decentralization operations under different conditions. Chapter 4 ends by formulating a series of practical guidelines that can help inform choices between different governance approaches based on the contingencies identified between different circumstances, governance approaches and their consequences, in a context of diverging stakeholder interpretations and values.

A practical inquiry

The practical inquiry consists of two chapters based on the study of two government attempts to encourage local proactive and integrated approaches to the environment. These are the attempts made by the European Union (EU) and the Dutch national government. The choice for the EU was relatively simple. In the first place, it is currently (2016) the most influential government body in environmental governance in Europe. In the second place, the EU has a history of almost two decades of attempts to stimulate local proactive and integrated approaches to the environment. In the meantime, the EU is also interesting for other reasons. The EU has the dynamics of a supranational body, which strongly influences its style of policy-making (compare Barnes & Barnes 1999, Jordan 2002, Zito 2000). Hence, the EU is subject to different institutional mechanisms and traditions that influence policy development, compared to most nation states. The EU, for example, has a slight bias towards more centralized, uniform and regulatory policies, certainly in environmental policy. This bias, as I will discuss, also influences the EU's attitude towards the role of the local level in pursuing proactive and integrated approaches. Chapter 5 addresses the approach taken by the EU and focuses on the rise of the 'Thematic Strategy on the Urban Environment' in the period 1990–2008. It is this 'Thematic Strategy' that the EU planned to use for encouraging urban authorities to develop and deliver proactive and integrated approaches to the environment.

The other case addresses the Dutch national government and was chosen for two main reasons. First, the Netherlands is traditionally considered to be one of the leading nations in environmental governance (see Andersen & Liefferink 1997, Weale 1992). Developments in environmental governance in the Netherlands often take place relatively early in comparison to many other countries and, at the same time, are often rather sophisticated. The second reason is related to this: the recent increase in the role of the local level in Dutch environmental policy. While Dutch environmental policy used to depend on a strongly centralized governance system, decentralization and deregulation have been pursued for over a decade. Therefore, many consequences of decentralization have also already become visible (see also Baartmans & Van Geleuken 2004, De Roo 2004 and Kamphorst 2006). Chapter 6 addresses the approach taken by the Dutch national government, in particular since the start of decentralization operations in the mid-1990s.

Conclusions

Chapter 7 contains the conclusions of the study presented in this book, as well as a reflection on the results and study. A first main, and more practical, conclusion is that the generic shift away from a reliance on centralized and regulatory-based policies in environmental policies is uncalled for. The role of the local level can be important, especially in terms of adding value to central policies in the face of unique local problems and opportunities. I argue, however, that this role should remain within a context of central policies and regulations that provide a robust foundation for setting minimum safeguards and for supporting and stimulating local initiatives. A second key conclusion is more theoretical and refers to the added value of adopting a post-contingency approach to help us navigate the plural governance landscape. Illustrated by the cases studies, I will explain why a post-contingency approach can be a logical next step in advancing theories on policy and planning making. I will not stop there, though, as Chapter 7 is also the place for highlighting avenues for further research and debate that can help further elaborate a post-contingency approach.

2 Governing the environment in a world of change

Although used as far back as the 14th century, the word 'governance' has not been used frequently in social sciences until the late 1980s. During the last decades, however, 'governance' has grown into a commonly used concept within planning and policy sciences. While clarity regarding its detailed meaning has faded with its increased use in literature and debate, the use of 'governance' in planning and policy science is unprecedented. Spoken in general, governance is simply a synonym for 'steering', or what Pierre and Guy Peters (2000) call 'the process of governing'. Nevertheless, many authors describe governance more narrowly as a distinct style of steering, hinting towards a relatively adaptive and flexible style that is based on a fluid sharing of responsibilities between mutual dependent actors both within and outside formal governments. In recent decades, it is exactly this more narrow description of governance that has gained in popularity in both theoretical debates and practice. Hence, the popularity of the word governance is a clear signifier of a process of institutional change, both in thinking about how societies should be governed and in how they are actually being governed.

Changes in governance are visible in many policy fields, including environmental policy. In this chapter, I will address the recent renewal in environmental policy in the EU, while specifically focusing on the increased role of the local level in developing and delivering proactive and integrated approaches. The problematic relation between established governance structures (e.g. centralized, regulation-based government bureaucracies) and our current dynamic, fragmented and complex societies will be highlighted. As environmental policy is traditionally among the more centralized governance policies (e.g. Andersen & Liefferink 1997, Lemos & Agrawal 2006, Van Tatenhove, Arts & Leroy 2000 and Weale 1992), it is a worthwhile case to discuss how our existing governance structures try to respond to changing societal circumstances and needs.

The problematic relation between established governance structures and our societies both produces autonomous changes to policies and provides a call for deliberate renewal operations. While trying to respond to changing societal conditions, however, governance renewal is not always pursued with clarity regarding its consequences (see also Wätli 2004). Consequences of governance renewal can be quite relevant to respond to, though, especially when it comes to protecting public health and safety, a key ambition of environmental policy. Arguably

20 *Governing the environment*

environmental policy might thus be more susceptible to the risks of governance renewal, again confirming it as an interesting policy field to study. I will argue, therefore, that we should be well aware of the potential consequences of governance renewal so as to take them into account during decision making.

Section 2.1 introduces a conceptual framework that I will use for describing and discussing recent changes in environmental policy. The rise of the concept of governance will be discussed, as will the related shift away from a dominance of government controlled policies towards more hybrid forms of governing. The result is an overview of the current diversity of governance approaches and a conceptual framework in which the prime changes in governance can be positioned. With this framework as a background, I shift focus to the rise of the so-called 'governance through coordination' model (e.g. Martens 2007, see also Lemos & Agrawal 2006, Pierre & Guy Peters 2000). It is this model that describes the regulation-based government bureaucracies that also dominated the development of environmental policies in many EU countries during its rise in the 1960s and 1970s (section 2.2). Despite its benefits, there are also some problems associated with relying on the coordinative model. As is discussed in section 2.3, these problems have also become visible within the field of environmental policy and have fuelled some important renewal operations.

The renewal of environmental policy in the EU is not just fuelled by problems with relying on the 'governance through coordination' model. Also fuelling these renewal operations is the ambition of many EU governments to develop more proactive environmental policies that are integrated into other policy fields. These governments include the Dutch national government and the EU, both of which are addressed in this book. In section 2.4 I will explain that the ambition to pursue more proactive policies provides additional challenges to the governance through coordination model, reinforcing the need for governance renewal operations. The use of so-called 'new' environmental policy instruments (e.g. Jordan et al. 2005), centralization due to the increased role of the EU, and decentralization in domestic environmental policy in various EU member states are all examples of such renewal in environmental policy (section 2.5). I will specifically focus on the increase in the role of the local level such as through decentralization. Not only is decentralization considered a response to current coordinative problems, but it is also seen as a means for supporting the development and delivery of proactive and integrated approaches.

Despite the benefits associated with increasing the role of the local level in environmental policy, there are also some consequences that might not be welcomed. Section 2.6 discusses some of the possible consequences of increasing the role of the local level in developing and delivering proactive and integrated approaches to the environment. In section 2.7, I continue by explaining that anticipating these possible consequences is of pivotal importance when renewing governance. Section 2.8, therefore, ends with arguing for guidelines for deciding both upon the pursuit of governance renewal operations and upon setting the kind of conditions for anticipating undesired consequences. That is, this chapter ends with setting an agenda for the remainder of this book.

2.1 The diversity of governance

For most of the 20th century "government enjoyed an unrivalled position in society in that it was the obvious locus of political power and authority" (Pierre & Guy Peters 2000; p. 4). With the United States as a notable exception (but see Fiorino 2006), it was mostly the *central state* that would exercise control over lower tiers of government. In the last decades, however, the exercise and organization of power and authority in our Western society has been undergoing important changes.

Summarizing the main changes we are witnessing, Jessop (1994) refers to the 'hollowing out' of the nation state. This hollowing out implies that power and responsibility are reallocated from the central state 'upward' to supranational bodies, 'sideways' to non-government, market and civil organizations and 'downwards' to lower tiers of government. Supranational bodies such as the European Union have taken over many competences and responsibilities, whilst decentralization and deregulation have increased the role of regional and local governments. In the field of environmental policy the EU has also taken over the role of individual nation states as the dominant force in developing environmental policies, while decentralization has affected environmental policy in various EU member states, including the Netherlands.

While centralization and decentralization are changing the allocation of responsibilities in a 'vertical' sense, power and responsibility are also spreading in a 'horizontal' sense. In other words, governments at all levels of authority are increasingly dependent on non-government organizations, business and the 'civil society' to make decisions and implement policies. The related 'dispersal of authority' (Hooghe & Marks 2001) undermines the ability of governments to make decisions and their capacity to carry them out (Stoker 1998; p. 18). As Pierre and Guy Peters conclude, this "weakened position of government then forces consideration of how their role can be strengthened, and of alternative modes of political governance" (2000; p. 1). Indeed, many governments are accepting that the governance landscape is changing and are redefining their place in this landscape. This process of redefining can also be witnessed in the realm of environmental policy, for example, in the rise of market-based instruments, participative approaches and the increased influence of non-government organizations, business and the 'civil society' (e.g. Jänicke & Jörgens 2006, Jordan et al. 2005, Lemos & Agrawal 2006, Mol et al. 2000 and Vig & Kraft 2013).

The variety of governance

In considering the kind of changes following the relative demise of government control in governing societies, many scholars refer to them as a so-called 'shift from government to governance' (e.g. Hajer et al. 2004, Healey 1997, Hooghe & Marks 2001, Kooiman 1993, Pierre & Guy Peters 2000, Rhodes 2000, 2007 and Stoker 1998). For these scholars a *government* style of governing relies on formal government control, on top-down regulatory steering and on institutional

22   *Governing the environment*

procedures (compare Hajer et al. 2004, Healey 1997, Kooiman 1993, Pierre 1999 and Stoker 1998). *Governance* is then used in contrast, as a style of governing relying on a more fluid sharing of competences between formal governments and what is called the 'civil society' (the public, businesses and non-government organizations). As power and responsibility are also shifting to both supranational and local levels, governing is also increasingly described in terms of multilevel governance (e.g. Bache & Flinders 2004, Bernard 2002, Enderlein et al. 2010, Héritier 2010, Kooiman 2003 and Marks & Hooghe 2001). Multilevel governance not only is related to a sharing of competences between formal governments and non-government actors (governance), but also explicitly involves a sharing of competences and responsibilities between various levels of authority (multilevel).

In discussing the more general shift from government to governance, many scholars are attempting to identify the exact changes we are witnessing in our Western governance landscape and also trying to assess where these changes could lead us (e.g. Arts & Van Tatenhove 2005, Brandsen & Holzer 2009, Hajer et al. 2004, Martens 2007, Nelissen 2002 and Pierre & Guy Peters 2000). To understand where recent shifts in governance are taking us, I argue we should first recognize where we are coming from; i.e. what is considered a 'government style of steering'. Theoretically, a 'government style of steering' is widely accepted as the most influential model of governance in the 20th century, and without a doubt in the organization of environmental policy in Europe. Inspired by the writings of, for example, Fredrick Taylor (1911), Lyndall Urwick (1929, 1953) and Max Weber (1922), it is a style of steering (i.e. 'governance') that draws on the effectiveness and efficiency of bureaucratic organizations and the coordinative capacity of the central state. In drawing from Martens (2007), we will refer to this style of governance not as 'government', but as the 'governance through coordination' model (compare Knill & Lenschow 2004, Lemos & Agrawal 2006 and Pierre & Guy Peters 2000).

Although strongly dominating the governance landscape in at least the EU during much of the 20th century, the 'coordinative model' has been affected by recent changes in governance. These changes have made it increasingly problematic to define governance in most of the EU just in terms of the 'coordinative model'. Instead, alternative modes of governance have arisen, which are not necessarily relying on the coordinative capacity of governments. Many authors tried to categorize and contrast these alternative modes (e.g. Lemos & Agrawal 2006, Martens 2007 and Nelissen 2002). In doing so, two key trends stand out in how governance can be organized. On the one hand, recent decades have seen a shift in thinking and practice inspired by *neoliberalist* ideas (see Allmendinger 2002a, Harvey 2005, Taylor 1994). Celebrating the merits of competition and market processes, these ideas are often popular among right-wing political movements. Especially during the 1980s, they impacted governance practices, most notably in the US (Reagan regime) and the UK (Thatcher regime). Although subject to shifts in popularity, recent and current renewal operations in governance in many Western countries are influenced by neoliberal ideas (see also Allmendinger

Governing the environment   23

2002a; pp. 92–94). Examples are decentralization to increase competition among localities and privatization and deregulation for reducing government control over market processes (e.g. Castree 2008a, 2008b, Mol et al. 2000 and Oates 2001).

On the other hand, the last decades have also seen what Healey (1992) calls the *communicative turn* in planning theory (e.g. Dryzek 1990, Healey 1997, Innes 1996 and Sager 1994, see also Giddens 1998). This 'turn' is not confined to planning alone but also related to the sociological works of, for example, Habermas, and pragmatic philosophies (see Forester 1989, Hoch 2007). The communicative turn has manifested itself in an increased popularity of participative or collaborative governance approaches, also in practice. This turn is supported by the idea that knowledge and rationality are not 'out there to be discovered' by scientific inquiry, but rather, are constructed in debate (e.g. Berger & Luckman 1967, Healey 1997). Active participation of multiple societal groups and stakeholder bargaining are means for coming to decision making from this perspective (e.g. Jordan 2002, Lemos & Agrawal 2006).

Three ideal types

The neoliberal and communicative turn both point towards extreme alternatives to the organization of power and authority based on the coordinative capacity of the central state. Consequently, they help to demarcate the boundaries of the governance landscape. In drawing from academic literature on governance it is also possible to structure the landscape of governance by identifying three *ideal type* models for organizing governance. In the literature, these ideal types can, for example, be found as 'structures of governance' (Pierre & Guy Peters 2000), 'patterns of governance' (Knill & Lenschow 2004) or 'models of governance' (Martens 2007, Pierre 1999). These ideal type models highlight several extreme manifestations of governance; i.e. they are caricatures of 'the real thing'.

Again drawing from Martens (2007) (compare Knill & Lenschow 2004, Lemos & Agrawal 2006, Pierre & Guy Peters 2000), we distinguish two alternative models of governance to the coordinative model: the 'competitive model' (neoliberal turn) and the 'argumentative model' (communicative turn). As each of these three models is presented as an ideal type, they can hardly be expected to exist in reality in their pure manifestation. The added value of using these ideal types lies mostly in the fact that they "demarcate the boundaries within which real-life governance processes can be positioned" (Martens 2007; p. 48). Hence, they also help to highlight the differences in how governance can be organized. In Figure 2.1 all three are summarized in what can be called the 'governance triangle' (Lemos & Agrawal 2006, Martens 2007). It is within this triangle that various theories and practices of governance can be positioned. Similarly, it is within this triangle that recent 'shifts in governance' (Van Kersbergen & Van Waarden 2004) can be positioned. The neoliberalist turn inspired a trend towards more market steering and competition and can be seen as a shift from the top to the left bottom corner. Similarly, the communicative turn can be seen as a shift from the top to the bottom right corner.

24 *Governing the environment*

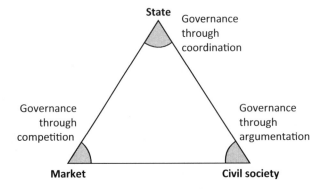

Figure 2.1 The governance triangle (based on Martens [2007] and Lemos and Agrawal [2006])

Hybrid manifestations and networks

The emergence and active adoption of approaches built on either or both the governance through competition and the governance through argumentation models, complement and partly replace existing coordinative instruments in a variety of ways. In practice, we can see processes of linking existing coordinative organizations and approaches to new means of organizing and governing. This is resulting in the emergence of new policy instruments (e.g. Golub 1998, Jänicke & Jörgens 2006, Jordan et al. 2005 and Vig & Kraft 2013), and mixed or 'hybrid' networks in which governance occurs (e.g. Kenis & Schneider 1991, Kickert et al. 1997, Kooiman 1993, Marsh & Rhodes 1992 and Sørensen & Torfing 2007).

New policy instruments are, for example, market-based instruments such as taxes, charges and subsidies that are based on combining the coordinative capacity of governments with the competitive model (see also Jänicke & Jörgens 2006, Jordan et al. 2005, Lemos & Agrawal 2006, Olmstead 2013 and Stavins 2003). The argumentative model comes forward in yet another group of instruments, such as public participation, voluntary agreements and informational policy devices such as eco-labels and campaigns (e.g. Arimura et al. 2008, Jänicke & Jörgens 2006, Jordan et al. 2005 and Mol et al. 2000). Often, new policy instruments are used by governments or are based on government-issued legislation. These new policy instruments, therefore, are typically inspired by both the coordinative model and by alternative models of governance, implying that these alternative models complement rather than replace the coordinative model.

As new policy instruments have been developed that cross the boundaries between coordinative, competitive and argumentative models of governance, the process of governance is changing. The blurring of boundaries between state, market and society can be seen with the rise of 'governance networks' that facilitate processes such as co-governance and negotiated coordination (e.g. Kooiman

Governing the environment 25

1993). As Jessop (1994) explains, these networks involve both the linking of actors and organizations in a horizontal and vertical sense. Horizontally, they follow the dispersal of power in society and involve the linking of state, civil society and market parties. Consequently, as Torfing explains, we see the emergence of "networks of actors, such as politicians, administrators, interest organisations, private firms, social movements and citizen groups involved in public governance" (2005; p. 305). Vertically, linkages are pursued between levels of authority in sharing power and responsibilities, notably in response to multi-scalar policy issues. Consequently, as Bressers and Kuks (2003; p. 65) suggest,

> Sectors in society are not governed on one level, or on a number of separate levels, but through interaction between these levels . . . one reason for this is a growing recognition that the problem situation itself often contains various interacting levels (such as environmental problems).

Multilevel governance, which emerged out of studies on the role of the EU in relation to domestic governance, has even emerged as a distinct field of governance studies focussed on vertical linkages between levels of authority (e.g. Bache & Flinders 2004, Bernard 2002, Enderlein et al. 2010, Héritier 2010, Jordan & Fairbrass 2002, Kooiman 2003 and Marks & Hooghe 2001).

Already during the 1970s and 1980s various studies would focus on the emergence of networks of actors that would, for example, surround specific issues (e.g. Heclo 1978), policy agendas (e.g. Jordan 1990), policy implementation (e.g. Hjern & Porter 1981) or that would constitute so-called 'advocacy coalitions' (e.g. Sabatier 1986, Sabatier & Jenkins-Smith 1993). Within these studies the notion of what constitutes a 'network' differs, partly depending on the empirical phenomena these studies address. Consequently, as Hanf and Scharpf (1978) explain, "the term 'network' merely denotes, in a suggestive manner, the fact that policymaking includes a large number of public and private actors from different levels and functional areas of government and society" (1978; p. 12). Research on governance and policy making in networks expanded from the 1980s onwards, resonating with the ongoing 'shifts in governance' that reinforced the emergence and importance of such networks. Literature on 'network governance', as Klijn explains, expanded 'dramatically', "making it increasingly difficult to draw clear outlines of the literature and its findings without making inappropriate generalizations" (2008; pp. 514–515). Rather, we are witnessing a vast variety of literature now trying to capture and analyze the many different manifestations of networks in which governance and policy making take place (e.g. Blanco et al. 2011, Börzel 1997, Carlsson 2000, Jessop 2002, Jones et al. 1997, Kenis & Schneider 1991, Klijn & Skelcher 2007, March & Olsen 1995, Marsh & Rhodes 1992, Newig et al. 2010, Powell 1990, Provan & Kenis 2007, Rhodes 2000, Sørensen & Torfing 2007 and Torfing 2005).

The rising importance of network governance has triggered some scholars to consider 'governance through networks' as a fourth ideal type model, in addition to the coordinative, competitive and argumentative models (e.g. Pierre & Guy Peters

26   *Governing the environment*

2000, for discussion; see also Börzel 1997, Powel 1990). Governance through networks, as a specific form of governance, would then point towards governance, "based on non-hierarchical coordination, opposed to hierarchy and market as two inherently distinct modes of governance" (Börzel 1997; p. 1, see also Rhodes 2000, Van Kersbergen & Van Waarden 2004). Here I consider networks not as the basis for a specific form of 'model' of governance, mostly based on accepting that network governance is itself subject to a large variety of manifestations that make developing the kind of 'clear outlines' and generalisations on network governance that Klijn (2008) opts for so problematic.

The diverse manifestations of networks in which governance takes place has urged various scholars to try to create some clarity by developing classifications (e.g. Klijn & Skelcher 2007, Powell 1990 and Provan & Kenis 2007). While it has not provided a common typology or classification, this work has helped clarify that 'network governance' is a wide and overarching concept covering different empirical phenomena and theoretical assumptions. As, for example, Blanco et al. (2011) and Klijn (2008) express, it is especially relevant to express 'policy networks' and 'governance networks' as distinct, albeit complementary concepts that the literature on networks and governance sometimes conflate for being similar. Policy networks tend to be more focused on the relation between the state and interest groups, while governance networks are considered focused especially on innovative and novel combinations of actors, not necessarily government related (e.g. Blanco et al. 2011; p. 299). To try to come to closure, Torfing (2005; p. 305) concludes that the notion of a network can best be regarded as an 'umbrella concept' with diverging 'conceptual constellations' (see also Carlsson 2000; p. 504). In a similar stance, Klijn and Skelcher (2007) argue it is best to consider 'network governance' as "a higher level concept associated with a particular mode of societal organization" (p. 587), with various categorizations of how governance is manifested in networks underneath it.

More important than coming to an exact classification of the variety of manifestations of network governance, I propose, is the acceptance that network governance is not pointing at a distinct 'style' or 'model' of governance but is an expression of exactly this variety of manifestations. It is also well supported by scholars showing us the many different mixtures that exist in networks of, for example, degrees of self-regulation in networks (e.g. Torfing 2005), top-down or bottom-up control (e.g. Sabatier 1986), mixtures between state and civil society initiatives (Bogason & Musso 2006) or the degree of influence of state, market and civil society partners (e.g. Powel 1990). While its 'fuzziness' might be considered an analytical limitation in discussing 'network governance', I argue that this broad description of network governance also allows us to use it in descripting what we are witnessing over the past decades in western Europe: an increased plural manifestation of governance (see also Nelissen 2002; p. 20). That is, to use the words of Sørensen and Torfing, the "plurality in the forms of governance networks attests to the broad relevance of the concept for describing and analysing contemporary forms of governance" (2009; p. 237).

A plural picture

The 'shifts in governance' affecting western Europe are united in pushing for a move away from traditional monistic forms of governance, "where a sharp distinction between state, civil society and market exists" and where "the position of the political domain will be mainly defined in terms of the rationalities of the state" (Arts & van Tatenhove 2005; p. 8). In line with the emergence of networks in which governance takes place, Arts and van Tatenhove express that "contemporary societies show increasing encroachment, interweaving and interference of the three subsystems [state, market and civil society] and demarcation lines are rather vague" (see also Hajer et al. 2004, Jessop 1994, Nelissen 2002). The result, as Sørensen and Torfing (2009; p. 236) explain, is that "although traditional forms of top-down government remain in place, public governance increasingly proceeds in and through pluricentric negotiations among relevant and affected actors interacting on the basis of interdependency, trust, and jointly developed rules, norms and discourses". Governance practices, as is the conclusion, are in a state of transition pointing away from central state control and the dominance of the coordinative model of governance.

While shifts in governance unite in their relative shift away from the coordinative model, they do not seem to point to the emergence of a new dominant mode of governance. As Martens (2007) explains, we should also hardly expect such answers. Instead of expecting "that a new, dominant, mode of governance will emerge" (2007; p. 53), we should expect to end up with more 'fuzzy' notions of governance where the roles and responsibilities of state, authorities and non-governmental actors are both spread and variable according to the detailed governing activity they relate to. Instead, the early 21st century seems to be characterized by what Offe (1977) predicted as a 'restless search' for adequate governance approaches.

The emergence of a plural governance landscape can in the first place be found when assessing the practice of governance renewal. This practice is showing extensive administrative experimentation (e.g. Hajer et al. 2004), the reliance on 'pick and mix' theories (Allmendinger 2002b) and 'trial and error' operations (e.g. De Roo 2002, see also John 2001, Martens 2007, Nelissen et al. 1999, Stoker 1998 and Van Tatenhove et al. 2000). These changes are not just time- and space-specific, but also "each policy area has its specific dynamics and 'creates' its own arrangement" (Nelissen 2002; p. 20). The result is a wide variety of governance approaches and initiatives, as the literature on network governance vividly illustrates (e.g. Carlsson 2000, Sørensen & Torfing 2007a). Also serving as an illustration are the several hundreds of organizational formats on developing local integrated policies that De Roo and Schwartz (2001) found in the Netherlands and Belgium alone. Similarly, participative approaches are thriving all around Europe and the US, while the growing abundance of public and private partnerships and of voluntary agreements shows a similar pattern of growth.

28 *Governing the environment*

While experimentation seems to be the norm in practice, theoretical debates are also showing a wide variety of theoretical perspectives and arguments. As was also discussed in Chapter 1, there is currently little theoretical agreement on the kind of new governance approaches to pursue. Instead, we face what Allmendinger (2002b) calls a post-positivist planning landscape that is subject to a continuous condition of what Healey et al. (1979) refer to as 'theoretical plurality'. What it means is that we are facing a wide plurality of theoretical positions that each draw upon different socio-cultural and philosophical values and beliefs (see also Bevir & Rhodes 2001). This theoretical plurality manifests itself in fundamental disagreements between the supporters of different theories and perspectives on the organization and process of governance. So instead of providing us with ready-made answers regarding the organization of governance, theoretical debates also confront us with a plurality of perspectives that is hard to navigate (compare Bevir 2010).

Illustrated by both practice and theory, we are lacking clear-cut answers regarding the merits and risks associated with governance renewal. Rather, we face ongoing processes of experimentation with different modes of governance that are each supported by alternative and possibly conflicting theoretical positions. In the midst of the resulting plurality of approaches and theoretical arguments, however, we should be well aware of the fact that renewing governance does not only offer new and inspirational pathways, but also involves risks we might want to avoid. That is, we should be well aware of the kind of consequences we might encounter when renewing governance. As is illustrated by the experimentation in practice, this awareness is certainly not omnipresent.

My ambition

Changes in governance follow both deliberate choices and autonomous societal and institutional processes such as power dispersal and the related increased impact of market parties, citizen groups and non-government organizations on governance. Deliberate changes are inspired by a variety of motives, but are often dominated by a general dissatisfaction with the coordinative model's capacity to cope with modern societal challenges (e.g. Lemos & Agrawal 2006, Martens 2007 and Nelissen 2002). In response, governments try to improve their capacity to govern by means such as improved policy coherence and deregulation to *improve* the functioning of the traditional coordinative model and its regulations, while they also aim for incorporating approaches based on either the competitive or argumentative model (i.e. market-based instruments, covenants, participation, etc.). Finally, through decentralization, privatization and deregulation, governments hand over power and authority to the market (competitive model) and civil society (argumentative model). In other words, next to trying to improve its functioning, governments are also *partly dismantling* their coordinative apparatus.

Still, both theory and practice are inconclusive of the kind of governance changes that should be pursued. Given these doubts, the head start that governments are taking in renewing governance is subject to risks, such as a reduction of the capacity to govern or the production of unforeseen and unwanted policy consequences. Such risks are certainty relevant to be aware of when it comes to environmental policies. Focused on the protection of ecosystems and especially human health and safety, governance renewal should be treated with care and prudence. Criteria such as legal security, social equity, minimal levels of protection against pollution and preventing risks are highly valued within the realm of environmental policy and are traditionally protected through government control; i.e. based on the coordinative capacity of mostly central governments (see also Oates 2001). Environmental policy, therefore, is one of the policy fields where changes in the coordinative capacity of governments or the dismantling of the coordinative model can have important consequences. Decentralization, deregulation and privatization usually result in greater autonomy for local governments, individual groups or people. They might prioritize, for example, development-oriented interests such as economic growth or financial gain at the cost of environmental qualities (e.g. Eckersley 1992, section 2.7).

The possibility of undesirable consequences with a serious impact should urge us all to be well aware of the consequences of governance renewal, especially in the realm of environmental policies. Revealing these possible consequences, as was discussed in Chapter 1, is one of the key aims of this book. This chapter contributes by discussing the current changes in the field of environmental policy, by explaining why they take place and by expressing the kind of doubts and questions surrounding these changes. After first discussing the rise of the coordinative model as the main benchmark for environmental policy in the EU, I will proceed with explaining both why this model was so widely confided in, criticized and altered and, finally, why moving away from it can be risky.

2.2 The coordinative model in perspective

The coordinative model rose within a context of high confidence in the ability and responsibility of the central state to dominate governance, while it was also associated with efficient, effective and democratically legitimate policymaking. When environmental policies emerged in much of the Western world during the 1960s and 1970s, it was also the coordinative model that provided the main inspiration. In this section I will both address the arguments behind the emergence and popularity of the coordinative model and explain how it impacted early environmental policies during the late 20th century. It is only in the following section (2.3) that I will continue with addressing the more recent doubts and challenges to its dominant position.

The rise of the coordinative model

A key trigger for the rise of the coordinative model of governance can be found in the Enlightenment project. The Enlightenment produced the idea that human reason and knowledge would allow humans to solve all key societal issues (e.g. Allmendinger 2002b, Harvey 1989, Parsons 1995 and Rosenthal et al. 1988). During

30 *Governing the environment*

the 19th and early 20th century, this idea was gradually translated into theories upon how to best organize the governance of societies and businesses. Max Weber (1922) was among the prime scholars to do so. Among other things, he argued for more rational forms of organization and decision making, and for a separation between policy-making as a political function and the use of administrative bureaucracies for carrying out policies (see also Parsons 1995).

To Weber (1922), rational-legal forms of authority were more stable and stronger than authority that followed the succession of elite leaders or of charismatic leaders (see also Allmendinger 2002b, Parsons 1995). Derived from laws, contracts, public consent and rules, rational-legal forms of authority were to Weber considered 'technically superior' to alternative forms of control. Furthermore, facts would be separated from values in order to prevent highly emotional or intuitive decision making. During the 20th century Weber's ideas were increasingly adopted. Among other things, governments gained administrative capacities for making decisions and carrying them out. In the meantime, widespread and successful implementation of rational ordered organizations occurred in businesses and industries (i.e. Fordism), while theoretical developments continued to support rationality as a guiding principle in governance and governance organizations (compare Harvey 1989). Keynes (1936), for example, argued for more informed and scientifically driven decision making. In addition, Fredrick Taylor (1911), brought his ideas on specialization forward (see also Parsons 1995, Rosenthal et al. 1988). Taylor, staying close to Weber's (1922) ideas on bureaucracy, used the metaphor of a machine to describe how human organizations could best be structured. Relying on specialization, central and hierarchical control and rational behaviour, Taylor's (1911) ideas became influential in policy sciences alongside Weber's ideas (e.g. Rosenthal et al. 1988).

Central in the ideas of Weber and Taylor was that rationally ordered organizations would produce more rational decisions and would also implement these decisions more effectively and efficiently. The main concern within the theories of Weber and Taylor, therefore, was with achieving pre-given ends as effectively and efficiently as possible (see also Allmendinger 2002b). To do so, specialization would accommodate skilled administrators that would work on clear and well-defined tasks. Specialization proved an appealing strategy, especially for policy fields where specialist expertise proved valuable. Environmental policy is among those fields, as it addresses issues such as air quality, soil remediation, risk assessments or exposure to toxic substances. The kind of scientific data and expertise needed for developing policies and implementing prevention or mediation strategies have given environmental policy a rather technocratic character. Consequently, environmental policy fits in well with the idea of organizing administrations based on sectoral policy agendas, programmes and regulations, all of which influenced and were implemented by specialists (e.g. Andersen & Liefferink 1997).

Effectiveness was, to Weber and Taylor, linked to the idea that strong lines of control, often hierarchically organized, would increase not only the reliability of organizational outputs, but would also result in predictable outputs (e.g. Heywood

2002). The clear lines of command in such 'machine-like' organizations would make it easier for the 'top' of the organization to exercise control and coordinate the activities of the various specialized departments. A hierarchical organization facilitated that the top would only have to manage the subsequent level, allowing it to remain within a realistic 'span of control'; i.e. the number of organizational units and tasks a manager can be expected to manage (Urwick 1956). If organized to allow for such a realistic span of control, direct hierarchical exercise of power could be installed over different levels of authority so as to ensure the delivery of centrally decided policy objectives. Hierarchical control also proved an appealing strategy in the development of environmental policy (e.g. Lemos & Agrawal 2006). In the first place, early Western environmental policies were developed during the 1960s and 1970s in the face of a high urgency that called for swift policy delivery and predictable results. Second, environmental policy focuses on the protection of ecosystems and public health. Public health especially begs for equal treatment between people and also begs for little error. The result is that there is a desire for relatively uniform policies that have relatively predictable outcomes. The high priority assigned to uniformity and the predictability of policy outcomes make it attractive for central governments to exercise control through a strong hierarchical organization, further reinforcing the appeal of the coordinative model in the realm of environmental policy.

The confidence in the effectiveness of the coordinative model was especially supported in the years following the Second World War. In planning theory, this is most vividly illustrated by the seminal writings of Meyerson and Banfield (1955) and their elaboration of the 'rational planning model', now often described as an instrumental, functional or technical rational approach (see also Dryzek 1990, Schön 1983). While their work acknowledged diversities in planning practices to which their 'rational planning model' counter responded, it became a point of reference for dominant theoretical assumptions held by planning theorists in the mid-20th century. These theoretical assumptions included the appreciation of a single true world to be fully understood when carefully studied. Hence, a technical rationale, expressing strong confidence in scientific inquiry and human reasoning being capable of producing certain knowledge and solve key societal issues (e.g. Allmendinger 2002b, Harvey 1989, Parsons 1995, Rosenthal et al. 1988). Through a careful and detailed analysis it was assumed possible to correctly model the social and physical world in which to intervene and hence, to predict the consequences of different policies and plans. Based on such models and predictions, planners would subsequently be able to design policy strategies that would produce outcomes considered best by the standards society or politicians had decided upon (see also Alexander 1992; p. 39). That is, planning would be about finding the most effective and efficient 'means' to reach the 'ends' given.

The ideas of Weber and Taylor also addressed the political question of legitimacy regarding the choice of policy ends. The idea was that if ends would be democratically chosen, they would be in line with the protection and advancement of the 'public good'. Furthermore, executive administrators were assumed to obey the commands made by those in charge. In other words, administrators

32   *Governing the environment*

were essentially 'value-free' and were assumed to secure that the envisioned policy outputs would be produced without alterations. It was also this 'value-free' nature of administrators that would allow them to advise policymakers objectively and hence, would improve political capacity to select the right strategies. Again, a link with environmental policy can be made. Many forms of environmental stress are not sensitive to individual preferences. Exposure to, for example, toxic substances, small airborne particles, radiation or explosions has rather similar consequences for all people. In other words, these consequences show little variation between individuals. A limited variation can even be seen for environmental stress where individual perception plays a more prominent role, such as with odour or noise nuisance. Given such limited variations, environmental policies are based largely on scientific inquiry expressing levels of exposure that are harmful (see also Lemos & Agrawal 2006). Emotional or ideological differences tend to become less prompt in the face of the predictability associated with such scientific data. In addition, moral arguments push for equal protection for all humans and thus further fuel a call for uniform and centrally controlled standards. The organization of governance such as advanced with the coordinative model is intended to do exactly that: ensure that centrally decided protection levels are implemented and maintained without exceptions.

The rise of environmental policies

While the influence of the coordinative model was widespread in the decades following the Second World War, environmental policies were also slowly emerging. During the 20th century ongoing industrialization of the Western world had created increasingly serious environmental problems. Although many localized problems existed from the late 18th century onwards, the post–Second World War 'economic boom' resulted in a rapid increase of polluting emissions and environmental stress. Ongoing and swift industrialization reinforced by a growing dissemination of technology eventually intensified and expanded the environmental stress. Increasingly, most of the Western world started suffering from the externalities (i.e. the side effects) of economic development. At first externalities were addressed ad hoc and incidentally. Key issues such as sanitation, waste, acute health problems, etc. were under the control of specifically designed policy programmes. From the 1960s onwards, however, environmental problems became increasingly urgent and featured more prominently on the policy agendas in most Western countries.

In addition to rapid economic growth, the 1960s showed an increased knowledge of the impact of environmental stress on human health and wellbeing. This caused a serious sense of urgency, confirmed by often visible effects of environmental pollution. Events such as the Great London Smog (1952), the Santa Barbara Oil Spill (1969), the fire on the Cuyahoga River in Cleveland (1969) and the overall levels of smog, forest destruction by acid rain and pollution of rivers and lakes made it hard to ignore these effects. Similarly, books such as Rachel Carson's *Silent Spring* (1962) and articles such as Garret Hardin's 'Tragedy of

Governing the environment 33

the Commons' (1968) helped to fuel a growing social pressure to act against these effects. Against this background the Club of Rome commissioned the production of 'The Limits to Growth' (Meadows et al. 1972), the report that put the environment on political agendas all around the world. Especially following the 1972 UN conference in Stockholm, most Western countries responded by swiftly creating an impressive body of environmental policies and supporting bureaucracies.

Inspired by the wide confidence in the coordinative capacity of governments and most notably the central state during this time, most environmental policies in the Western world were modelled in the fashion of the coordinative model. This is certainly true for the EU and the Netherlands, which have my prime interests in this book (see also Andersen & Liefferink 1997, Van Tatenhove et al. 2000, De Roo 2003 and Weale 1992). In both the EU and the Netherlands, environmental policies were dominated by environmental (emission and immisson) standards. In being general policy guidelines and often formulated as quantitative values, these standards could be applied uniformly, with much force, while being unambiguous to those implementing them. Developing and implementing these standards relied solely on the regulatory capacity of government bureaucracies and the technical solutions available for those implementing them. In the meantime, basing policies on standards and central control was also surrounded with much optimism. The confidence in the predictability of coordinative policies such as standards was supported by the idea "that laws and other legal regulations were considered the best answer to the problem: once a law was passed, the problem would be solved immediately" (Nelissen et al. 1997; p. 167). Given this confidence and the need for swift action, standards became rather popular instruments (see also De Roo 2003, Fiorino 2006 and Lemos & Agrawal 2006).

During these early years, environmental regulations were on the one hand used to reduce the environmental loads produced by human activities; i.e. they focussed on the emissions of environmental stress. Standards included obligatory end-of-pipe measures such as filters, catalysts and silencers. In addition, they included regulations imposing limits on the amount of environmental stress caused by engines or installations. For example, noise levels of engines were regulated, safety risks of industrial processes were calculated and reduced and the amounts of toxic or carcinogenic chemicals used and emitted by industry, agriculture and traffic were regulated. Existing production processes were altered and the materials used or produced by manufacturers were changed to reduce emissions. Environmental regulations were also used for setting limit values dictating the absolute maximum of environmental stress tolerated in the environment; i.e. the maximum level of pollution tolerated in soil, water and air. These *environmental quality standards* were not focused on regulating the emissions themselves, but at the exposure of humans and ecosystems to environmental stress. Finally, *immission standards* were used for regulating the maximum amount of pollutants tolerated to be immitted into an area or object.

Within 10 years after the beginning of the 'environmental revolution' most Western nations had well-developed bodies of environmental policies that, among other things, dictated both minimum quality and maximum emission and

34 *Governing the environment*

immission levels (e.g. Andersen & Liefferink 1997, Jänicke & Weidner 2002 and Van Tatenhove et al. 2000). During the 1980s, environmental policy was eventually established as a mature policy field in most EU member states (see also Andersen & Liefferink 1997, Weale 1992). Within the EU itself a specific Directorate General for the Environment was established in 1981, while in member states such as the Netherlands a specific ministry for the environment was established in 1982. The reliance on the coordinative model in the realm of environmental policy was high during these days, manifesting itself in rather centralistic and legalistic policies. The Netherlands was one of the key examples of this (e.g. De Roo 2003, Liefferink 1997), while other examples include most western European countries, such as Austria, Denmark, Finland, Germany, Sweden (see Andersen & Liefferink 1997), France (Lowe & Ward 1998) and also the early developments in the US (Fiorino 2006, but as exception, see Lowe & Ward 1998 on the United Kingdom).

Gradually, the role of the EU also intensified, starting with the first Environmental Action Programme of the EU (OJEC 1973). This programme was first based on political pressure of more 'green' EU member states to develop international environmental policies. However, within the EU the motive of environmental protection always has to be accompanied by the protection of the EU common market (e.g. Barnes & Barnes 1999, Jordan & Fairbrass 2002, McCormick 2001 and Wallace & Wallace 1997). The idea of a common market involves the movement of people, goods, services and capital between EU members without barriers. Differences in environmental regulations between member states potentially cause competitive (dis)advantages for industries or companies between countries and can thus constitute barriers that frustrate the functioning of a common market. The common response of the EU is to harmonize domestic environmental policies and regulations. Consequently, the EU has played a prime role in the last 30 plus years in domestic environmental policy, with a focus on creating similar policies and levels of protection all over the EU (policy convergence). A key result is that many environmental policies in EU member states were dominated by centrally issued EU standards and principles, reinforcing the dominance of the coordinative model, albeit with the addition of a supranational centre. Although these tendencies have changed somewhat in recent years (see section 2.5), the coordinative model has clearly been an essential benchmark for the development of environmental policies in the EU.

2.3 Coping with coordination

The reliance on the coordinative model produced some impressive results. Several large-scale environmental issues have been largely resolved or are showing key improvements. These issues include famous examples such as the reduction of acidification in Europe, the cleaning up of the river Rhine, reversing the depletion of the ozone layer and the reduced use of pollutants such as lead or DDT. Similarly, coordinative instruments remain important in setting safeguards against environmental pollution and degradation. But despite the successes, relying on

Governing the environment 35

the coordinative model also has its limitations. Many of these limitations would gradually become visible from the 1970s onwards. Although not unique for the field of environmental policy, these limitations would also affect the functioning of environmental policies. In this section I discuss these limitations in relation to environmental policies based on dividing them into three main categories: the occurrence of policy incoherencies and conflicting policies in relation to specialist and fragmented government organizations, the limited predictability of policy outcomes especially in the face of interrelated problems, complexity and unique local circumstances, and finally, varying beliefs and preferences of different societal groups. I will then continue in the next section (2.4) with yet another key challenge to the coordinative model that is more specific to the field of environmental policy: the desire to develop more proactive and integrated policy approaches also linked to the rise of sustainable development as a policy guideline.

Fragmentation and specialization

The coordinative model relies on the assumption that rationally organizing governance through specialist departments in a hierarchical organization is effective and efficient. Taken into its extremes, the coordinative model relies on 'perfect bureaucracies' that are governed from the top by well informed and perfectly rational decision makers. Weber (1922) himself acknowledged that real life bureaucracies cannot be expected to work as fluent, effective and optimal as the ideal type he promoted. It is, however, especially during the last decades that critiques have increased on the potential for government bureaucracies to produce incoherent policies, coordination deficits and finally, for being rigid, inefficient and inflexible (e.g. De Leeuw 1984, Pierre & Guy Peters 2000). The differences between environmental and economic policies serve to indicate the potential for conflict. Within single policy fields, incoherencies and inefficiencies also occur, and environmental policy is no exception (e.g. Andersen & Liefferink 1997, De Roo 2003, Dovers 2003, Jordan & Lenschow 2010, Lachman et al. 2001, Van Tatenhove et al. 2000 and Weale 1992).

Within the coordinative model, government bureaucracies are divided into units that each specialize in a separate government function. In environmental policy, separation is, for example, expressed by addressing different environmental stressors by separate agencies. Such a separation means that individual departments and units can specialize in areas such as noise abatement, air quality, soil remediation, water quality, etc. Specialization, however, is accompanied by difficulties to ensure that these separated policies are developed and implemented in a coherent way (e.g. Breheny & Rockwood 1993, CEC 2006, Miller & De Roo 2004 and Van Tatenhove et al. 2000). The coordinative model relies on a hierarchical organization where the top of the organization – such as the central state – will coordinate separate policies in order to protect coherency in formulation and implementation. Coordinating all the different sectoral and sub-sectoral policies can easily be outside the top management's 'span of control'; i.e. the number of subordinates or areas a manager can safely be assumed to oversee and control.

36 *Governing the environment*

Not only are there many different policies, they are also constantly changing. Hence, as De Roo (2004; p. 8) suggests "practice forces us to acknowledge that policies are to such a degree dynamic that full coherency and consistency of policies is a fiction" (translation CZ).

Although coordination deficits and incoherencies have existed for a long time, they have mostly become more visible in recent decades. This follows a growing recognition that many social and economic processes and flows of resources are interrelated (e.g. Lemos & Agrawal 2006), while there is also "the emergence of new, cross cutting policy problems such as the environment or sustainability that demand much more cooperative solutions" (Jordan et al. 2005; p. 482, see also O'Riordan & Voisey 1998). Hence, it is nowadays widely accepted that many "contemporary problems are complex and interrelated, defying treatment by means either of narrow, sectoral policies or of all-encompassing, super-policies" (Briassoulis 2005; p. 2, compare De Roo 2003, Persson 2002). Coping with interrelated policies and issues requires coordination between various sectoral interests and policies (see also Jodan & Lenschow 2010, Lafferty & Hovden 2003 and Persson 2002). Such coordination is not easy to deliver by the central state alone as it is problematic for central governments to be fully aware of the kind of interrelatedness of policies that is often manifesting itself in specific and unique ways. Acquiring the desired time and space-specific knowledge is again easily outside of the central government's 'span of control'. The consequence: sectoral and specialist policies are often not based on an understanding of the interrelatedness of policy issues. Hence, next to having the potential to be conflicting and incoherent, they can also cut through integrated themes and issues.

Limited predictability

Following the logic of the coordinative model, many policies in the decades following the Second World War can be characterized as command and control policies; i.e. the central state issues policies that administrative units need to carry out. Command and control was based on two main assumptions: 1) the applicability of the technical rational approach where the policies issued are based on correct information and rational decision making and 2) that the implementation of policies is a natural consequence of the formulation of well-designed plans and policies; i.e. policy outcomes are highly predictable (see also Parsons 1995). Practice, however, has shown that both assumptions do not necessarily apply.

In the first place, correct information and rational decision making are less obvious than considered within the technical rational tradition. The technical rational approach assumes that it is possible to correctly model the social and physical world and, based on the information gathered, to make fully rational decisions. Following the works of scholars such as Simon (1957), Lindblom (1959) and Davidoff (1965), the ideas of both certainty in knowing the social and physical world and of choosing rationally were seriously nuanced. Rather, planners would be confronted with a lack of information (Simon), cognitive limitations and uncertainties (Lindblom) and also political conflict (Davidoff). Knowledge,

Governing the environment 37

in other words, is at least *bounded* by these factors, meaning that actors can be rational only within the boundaries of what they know, understand, influence and believe.

Recent decades have indeed seen an increased and also more fundamental rejection of the possibility of certain knowledge and uniform rationality. On the one hand, this rejection follows general changes in philosophic and scientific thinking, which can abstractly be related to the shift from *modernism* to *postmodernism* (compare Allmendinger 2001, Beauregard 1989, Boelens 2010, Harvey 1989, Healey 1997 and Portugali 2012). Where modernism tends towards a strong belief in the human capacity to understand and control reality (i.e. the technical rational approach), postmodernism has a sceptical or even deconstructive attitude towards this belief. Following a gradual demise of modernism as dominant scientific discourse, there is a decreasing support for the 'modernistic myth' that "the world [can] be controlled and rationally ordered if we could only picture and represent it rightly" (Harvey 1989; p. 27). Rather, degrees of uncertainty and disagreement are to remain when gaining knowledge and responding rationally to this knowledge. Therefore, the context in which reality and truth are considered, knowledge is created and in which rationality is applied is highlighted as crucial to how they are interpreted and framed (compare Hillier 2005); i.e. what is correct knowledge and rational behaviour is at least influenced by the observer and his or her context.

The importance of uncertainty for planning also gained theoretical and empirical support from the rise of theoretical and empirical findings related to what can shortly be summarized as complexity sciences (e.g. Gell-Man 1994, Holland 1995 and Kauffman 1995). Complexity sciences show us a picture of physical and social systems showing non-linear behaviour, and those systems continuously evolve due to internal mechanisms of adaptation in relation to changes in the system's environment. They are ideas that have in recent years also increasingly been applied in the realm of governance and planning (e.g. Batty 2005, Byrne 2003, De Roo & Silva 2010, De Roo et al. 2012, Hillier 2008, Innes & Booher 2010 and Portugali 2000). The result is that planners and policy makers face issues that display fundamentally unpredictable changes. Uncertainty, then, is not just a feature of incomplete information or cognitive limitation to overview a situation, but is considered a fundamental feature of systems displaying non-linear behaviour. With the realization that uncertainties are fundamental characteristics of the world planners aim to intervene in, it becomes even more pressing to understand that reality and truth are concepts open to multiple interpretations that are likely to change once new information or ideas come forward.

Second, but related, the predictability of policy actions has proven quite more limited than first assumed. Even though relying on the 'modernistic myth' proved fairly correct in dealing with many issues in the 20th century (e.g. industrial development and 'Fordism', military operations, natural sciences), many societal issues proved less easy to represent and control. Instead, implementation of policies often proved difficult, whilst failure to implement was not at all uncommon. Vividly recorded by, for example, Pressman and Wildavsky (1973), Bardach (1977) and Derthick (1972), serious policy failures were encountered in

38   Governing the environment

the implementation of many government programs (see also Hall 1980). These failures made that the illusion that success would simply follow the installation of new top-down regulations, so typical for the 1960s and 1970s, was increasingly nuanced (see also Barret & Fudge 1981, Ingram & Mann 1980 and Sabatier 1986). This nuancing also occurred within environmental policy. Although practice showed the overall improvement of the environmental conditions on many terrains and in many places, it could not hide the lack of success on many other occasions (e.g. Andersen & Liefferink 1997, Nelissen 1997).

Policy failures were (are) on the one hand attributed to the failure of administrators to respond to central policies (compare Mazamanian and Sabatier 1983, Pressman & Wildavsky 1973 and Van Meter & Van Horn 1975). Policies are often skewed when administrators interpret policies within a context of the interest, beliefs and priorities they hold (e.g. deLeon & deLeon 2002, Yanow 2000). These distortions can be both deliberate and undeliberate. Or as deLeon & deLeon argue, "when multiple players are involved (and they almost always are . . .), implementation becomes a battle to determine a correct reading of the mandate and its accurate execution" (2002; pp. 464–465). Struggles between multiple levels of authority are then bound to occur, as is reported by Jordan et al. (2002) on EU environmental policies (see also Barret & Fudge 1981, Elmore 1979, Schofield 2001). Such struggles need not only be based on differences of opinion between levels of authority, but can also follow from difference in information and knowledge upon the circumstances under which implementation needs to occur.

On the other hand, then, policy failures can be attributed to the failure of decision-making units such as the central state to be sufficiently informed on the contextual circumstances that affect implementation (e.g. Elmore 1979, Hjern & Porter 1981 and Lipsky 1980). When developing policies, policymakers typically rely on simplified models (i.e. assumptions) of the reality in which these policies ought to intervene. As was noted earlier, it can easily be outside of the central government's 'span of control' to acquire relevant time- and space-specific knowledge. Consequently, contextual influences on the policy issues faced are deliberately excluded when developing (central) policies so as to keep them manageable. However, "because an issue cannot usually be considered out of context, that context will also affect the end result" (De Roo 2003; p. 111). If such contextual influences are predictable and stable, they can be included in the models made. Also, if contextual influences are limited, discarding them is of little consequence. In both cases, simplifications made will prove useful and there are also many occasions that the technical rational approach associated with the coordinative model can be successful.

Contextual influences become more significant once issues, causes and effects are interrelated; a larger number of disagreeing and influential stakeholders are involved; and when contextual circumstances are not stable and predictable (see also Emery & Trist 1967). In those cases, discarding contextual influences can have serious unanticipated and possibly adverse consequences. These are what, for example, De Roo (2003) calls more complex issues, where central agencies are left without the capacity to sufficiently model and predict policies and consequences or are constrained in their control over other stakeholders. Now the acceptance of unpredictable behaviour and dynamically changing circumstances, further aggravated if systems also display non-linear behaviour, urge planners to be flexible and adaptive in dealing with them. While this might be an argument for lower levels of authority to consciously alter central policies so as to better match them with the contextual circumstances and stakeholder interests, these lower levels of authority are limited in their capacity to do so due to the predefined objectives and policy formats given to them by central authorities. Relying on the coordinative model and its predefined and common policy formats is based both on simplified models of reality and a lack of leverage for lower levels of authority to alter them. As a result, policy failures, local resistance and unanticipated consequences of implementation are among the likely results of relying on the coordinative model under more complex circumstances.

Environmental policy is certainly not exempt from problems during implementation. To begin with, extensive scientific research cannot hide the fact that many effects of human activities on our environment are still unknown or subject to debate. Knowledge, therefore, is necessarily limited and uncertain. Popular examples are global warming and predicting the exact consequences of, for example, small airborne particles on human health. Second, and next to uncertainties in scientific knowledge, many issues are also remarkably difficult to model. Issues such as urban sprawl and increased car use are among the examples where exact effects, causes and possible strategies for dealing with them are hard to map. After decades of work on reducing car dependency, for example, we still face an increase of car traffic in Europe and, faced with limits in predicting the effects of measures, we also face different opinions and stakeholder preferences.

Third, the impact of unique contextual circumstances is also important in the realm of environmental policy. Many environmental quality standards are set to guarantee that certain exposure levels are not exceeded and hence, apply generically. However, local circumstances influencing the implementation of these standards can make this difficult if not unrealistic. To illustrate: meeting noise quality standards might be feasible or even easy in a rural village or a suburban town. In a busy city centre this can be quite different. Meeting standards thus requires different investments in different localities. Under complex circumstances, such investments might not be considered to outweigh the benefits (see also Oates 2001). In the Netherlands the so-called Integrated Environmental Zoning (IEZ) project showed that meeting strict environmental standards could even result in excessive costs, such as demolishing large city parts or closing down most of the local industry (Borst et al. 1995, see also Chapter 6). More recently, meeting EU air policy regulations on small airborne particles caused similar problems in many urban areas, such as the Dutch Randstad, Flanders and many German cities (e.g. Backes et al. 2005, Zandvoort & Zuidema 2007). Implementing central policies under such circumstances can easily be considered unacceptable by key societal stakeholders, suggesting these circumstances are too complex to be dealt with by relying on the coordinative model.

40 *Governing the environment*

Fragmented societies, diverging beliefs

Another limitation related to the coordinative model that was already implicitly touched upon follows from societal changes that have undermined the legitimacy of representative democracies. The coordinative model is supported by the idea that central governments are able to protect and advance the 'public good'. In other words, governments and most notably the central state would make decisions that are also considered rational and correct to others. In recent decades these assumptions are challenged by what Martens (2007) calls 'the legitimacy crisis of representative democracy'; which is based on a decreased belief in, and support for, the central governments to protect and advance the 'public good' (see also Alonso et al. 2011). At least two key forces are contributing to this crisis.

A first force contributing to the legitimacy crisis of representative democracy is an increased *dispersal of power in society*. This dispersal is already a well-documented phenomenon that in a rather abstract sense is related to changes in the relation between state, civil society and market (e.g. Bevir & Rhodes 2003, Hooghe & Marks 2001, Kickert et al. 1997, Kooiman 1993, Newman 2001, Stoker 1998 and Van Tatenhove et al. 2000), or what Torfing (2005; p. 3005) describes as "the pluralisation of economic, social and political actors, sites and processes". 'Power' is increasingly spread among a diverse range of actors (see also Booher & Innes 2001, Kearns & Paddison 2000, Stoker 1998 and Torfing 2005). Each actor and stakeholder group wields power, exercises influence and seeks to set the agenda for the decision-making process. Landowners, trade unions, pressure groups and various government agencies all have their own recourses, be it land, property, knowledge, legal competences or funding. On the one hand, these stakeholders can use their resources to lobby for policy outcomes that are in their interest. This is, for example, well documented when it comes to the development of EU environmental policies (e.g. Kohler-Koch & Buth 2009, Mazey & Richardson 2002, McCormick 2001 and Zito 2000). This also goes for civil society actors, which are increasingly empowered by the liberalization of access to information, (e.g. Castells 1996, Winsemius 2001). On the other hand, citizens and stakeholders also claim their place in the process of governing. Central governments need to share their powers for developing policies and controlling their delivery with non-government actors at various levels of authority. Furthermore, central governments also need to share power and responsibility with local governments and related local stakeholders claiming their place in policy development and delivery; i.e. resulting in the aforementioned governance networks. By making the control over decision making and implementation no longer predominantly a government matter, power dispersal further challenges the coordinative model.

A second force contributing to the legitimacy crisis of representative democracy is caused by processes such as individualization, the rise of the 'information age' (Manuel Castells) and globalization (e.g. Arts & Van Tatenhove 2005, Beck & Beck-Gernsheim 2002, De Vries 2000 and Hajer et al. 2004). As Kearns and Paddison note, during the last decades "new lifestyles are being tried and adopted: people live in different ways; travel in different directions; shop at different times"

(2000; p. 846; compare Dryzek 1990, Healey 1997, Winsemius 2001). It results in increased fragmentation and diversity in societies, its processes and activities (e.g. Andersen & van Kempen 2001, Healey 1997, Koppejan et al. 2011, Papado-poulos 2000, Sassen 2001, 2002, Stoker and Mossberger 2001, Torfing & Trian-tafillou 2011 and Vranken 2001). As Martens (2007; p. 51) states, this "growing diversity of lifestyles reflects the increasing number of social groupings defined around ethnicity, gender and life-style choices" (Martens 2007; p. 51). Increased social fragmentation and diversity of preferences make it more cumbersome to define the 'public good'. Elected politicians can hardly be expected to respond to this diversity, undermining their capacity to 'define' the public good without alienating large parts of the electorate. Environmental policy certainly also has to face debate, starting with highly different beliefs and convictions regarding how to interpret and respond to environmental stress and pollution. Even global issues such as climate change or excessive resource consumption are among the issues that are interpreted and judged quite differently between societal groups. Consequently, central governments are facing a plurality that cannot necessarily be united. Such problems are possibly even more intense when it comes to envi-ronmental issues that are subject to large variations in perception among societal groups such as noise or odour nuisance and the perception of safety risks will be judged differently by different people and societal groups. What is considered 'lively' by one is considered a 'nuisance' by another. Finally, given highly dif-ferent values, lifestyles and routines developed by different groups and people, reaching 'the public' through policies and campaigns will also be difficult without expecting highly different preferences and responses. Representative democratic bodies, therefore, are increasingly less able to make decisions that the vast major-ity of the public considers as legitimate.

Improving on the coordinative model

A first step in responding to the limitations of the coordinative model is to try to improve its functioning. Starting mostly during the 1980s, many EU member states and the EU itself also pursued this strategy, also for trying to improve their environmental policies (e.g. Andersen & Liefferink 1997, Barnes & Barnes 1999, De Roo 2003 and McCormick 2001).

First, governments can pursue the strategy of introducing additional coordina-tive instruments to correct limitations of existing instruments or their undesired side effects. The underlying assumption of adding more regulations is that they help to increase a government's coordinative capacity to govern and to produce more control. Additional policies and regulations can indeed create improved gov-ernance capacities. However, adding new policies and regulations is not always producing more capacity to govern. It can even be counterproductive where gov-ernments run the risk of falling into a trap which Joseph Heller already hinted towards in his 1961 bestseller *Catch-22*: bureaucracy creates more rules, more procedures and becomes more complex, which eventually diminishes its effec-tiveness and efficiency, resulting in even more regulations (i.e. what is called a

42   *Governing the environment*

'catch-22' situation). It is a condition De Leeuw (1984) describes with his 'wet van de bestuurlijke drukte' ('law of management overload'). De Leeuw witnesses that an increase in governance efforts and additional governmental capacity to act at first indeed tends to increase the effect of this governing activity. However, when continuing to push for more governmental capacity and efforts, the relative additional effect caused by this additional effort typically decreases as it is less helpful to add people and regulations to increase the effect of the governing activity. When capacity and effort continue to be added, De Leeuw witnesses that this can even limit the ability to deliver the outcomes and hence, result in a decreased effect of the governing activity. There are simply too many rules, regulations and agencies involved to coordinate and communicate properly, whilst implementation becomes increasingly difficult in the face of the multitude of agencies and regulations to deal with. Common consequences are increasingly specialist rules and regulations, increased policy fragmentation, unanticipated incoherencies, conflicts between regulations and agencies and, overall, high workloads for implementers to cope with the mosaic of policies and requirements they face (see also Osborne & Geabler1993, Toffler 1970).

A second strategy is for governments to improve the implementation and enforcement of existing policies. Within the realm of environmental policies this strategy also became a key theme during the 1980s in the EU and various EU member states (e.g. Andersen 1997, Barnes & Barnes 1999, Kronsell 1997, Liefferink 1997 and Nellisen 1997). Clear 'implementation deficits' were found to exist, urging for improving the process of ensuring that implementation takes place (see also Knill & Lenschow 2000). This nuanced the desire to add more regulations, while it also made people aware that "more legislation did not necessarily mean a cleaner environment; it meant that craftier methods for dodging environmental laws were developed" (Nelissen 1997; p. 258). In response, intensifying control and guidance on implementation was considered an important strategy. For example, in Denmark (Andersen 1997) and Norway (Reitan 1997) enforcement already became a key focus in the 1980s. In the EU of the 1980s and 1990s, a gradual shift also occurred from developing new environmental policies towards implementing those policies already in place (McCormick 2001, see also Glachant 2003).

Third and finally, governments can also streamline existing fragmented policies. The swift and often rather ad hoc development of environmental policies in the 1970s and early 1980s had indeed caused serious problems with fragmentation and the associated coordination deficits. During the 1980s this increasingly became a policy priority in various European countries and for the EU itself (e.g. Andersen & Liefferink 1997, De Roo 2003, Lachman et al. 2001, Van Tatenhove et al. 2000, Weale 1992). During the 1980s, the streamlining of existing sectoral policies became a priority in countries such as Austria (Lauber 1997), Denmark (Andersen 1997), the Netherlands (e.g. De Roo 2003, 2004) and Sweden (Kronsell 1997). The idea was to produce a more coherent and manageable environmental policy field (e.g. Andersen & Liefferink 1997, De Roo 2003, Lachman et al. 2001, Van Tatenhove et al. 2000 and Weale 1992). Better coordinated permit systems and framework directives were used for removing some problems, whilst deregulation was used to simplify and streamline regulations (see also in relation to the symptoms of 'management overload' encountered). Within the EU, efforts were also made to better coordinative environmental regulations (e.g. McCormick 2001). The EU, for example, increasingly began working with so-called framework directives (i.e. overarching laws) for coordinating daughter directives that contain detailed prescriptions.

Despite the efforts to improve the functioning of the coordinative model it continued to face limitations and critiques. It had become increasingly evident that a fully coherent, streamlined and effective set of policies was indeed little more than an ideal picture. Furthermore, challenges related to the legitimacy crisis of representative democracies and power dispersal remained. Finally, even if environmental policies would become better integrated, more efficient and effective, improving its integration with other policies remained a challenge not dealt with. The coordinative model, despite its advantages, thus continued to be challenged. Most notably, it was the attempt to pursue the integration of environmental policies with other policy fields that came to signify a more radical critique on the coordinative model within the realm of environmental policy: its reactive nature and focus on uniform policy outcomes. It was this critique that from the late 1980s onwards united many of the previous critiques on a sole or strong reliance on the coordinative model of governance in improving environmental conditions. That is, it is a critique that addressed the desire to respond to interrelated policy issues, power dispersal, social fragmentation and the desire to proactively respond to risks and uncertainties.

2.4 Towards proactive and integrated policies

The rapid development of environmental policies in the 1970s and 1980s resulted in an impressive body of policies. Although this body of policies was gradually improved on and became more coherent, it remained a distinct set of policies that meant to respond to the possible harmful consequences of other policies and societal developments. Dominated by restrictive environmental standards, permit systems and clean-up operations, environmental policies had remained largely *reactive;* i.e. designed for setting restrictions on development-oriented policies (see also Andersen & Liefferink 1997, Butler & Olouch-Kosura 2006, Jänicke & Jörgens 2006 and Milbrath 1989). Despite their positive contribution to improving overall environmental conditions and to provide safeguards against intolerable damage to public health and ecosystems, they were increasingly considered to be *only a part* of the puzzle of environmental policies (see also CEC 1990, 2004, De Roo 2003 and Van Tatenhove et al. 2000). After all, instead of functioning as one of the underlying criteria for making decisions in policy fields such as agriculture, spatial planning, economic development or transportation, environmental quality was essentially translated into a set of legal standards that developments had to be checked against. Environmental quality was more an administrative check on development than a criterion for making choices in itself. Furthermore,

44 *Governing the environment*

these reactive policies had no explicit intention to proactively respond to environmental risks and uncertainties.

Doubts about reactive policies

During the 1980s the drawbacks of the reactive nature of early environmental policies became increasingly clear. As Simonis, for example, explains

> . . . it is expensive, the measures it employs take effect so late that the ecosystems involved can no longer be saved, it is focused on only a part of the relations between environment and society [and] it is pursued as a media-specific policy; i.e. controlling air and water, noise, and waste . . . [E]xpenditures for environmental protection are made when damage to the natural environment has occurred. They are belated; they are repairs to the process of economic growth, signs of a post-fact policy that reacts to damages (and must react to them) but does not, or cannot, prevent them.
>
> (Simonis 1988; p. 193)

In other words, "reactive forms of environmental policy have been and are perhaps increasingly unable to respond effectively or efficiently to many of the negative impacts of economic development" (Gouldson & Roberts 1999; p. 4).

Given their drawbacks, relying on reactive policies has risks (Barrow 1995). This is also what Butler and Olouch-Kosura (2006) state when they consider that "the reactive attitude to ecosystem management leaves society vulnerable to adverse surprises, which tend to be belatedly recognized and poorly managed. It is conceivable that some of these adverse ecological surprises (e.g. pollution, erosion, coastal eutrophication, or runaway climate change) could surpass a threshold, overwhelming social capacity and, consequently, damaging human well-being" (see also Walker & Meyers 2004). They thus argue for a more proactive and anticipatory approach. Gouldson and Roberts (2000) also addresses the risks of reactive policies. While drawing on Beck (1992), they consider the increased complexity of many societal processes to make it hard to oversee the risks associated, and these risks are potentially extremely high. It is an argument that closely resonates with the uncertainties related to systems displaying non-linear behaviour as debates within complexity sciences. Also referred to as 'complex adaptive systems', ideas of non-linearity are also used in the realm of planning, ecology and environmental studies. As Duit and Galaz (2008) explain, radical changes can occur in complex adaptive systems once a certain 'threshold' or 'tipping point' is reached (see also Folke 2006, Gunderson & Holling 2002 and Walker & Meyers 2004). Consequently, rather than showing a predictable linear process of change, systems can quite suddenly collapse or cause a sequence of cascading effects, exacerbating the problem far beyond what is initially expected (see also Duit & Galaz 2008). Overseeing these risks and identifying critical thresholds can be highly problematic if not beyond our reach when it comes to systems as complex as our climate, aquatic ecosystems or, to give another example, the energy market.

Reactive policies, in the modernistic tradition of the coordinative model, however do rely on societal capacities to oversee these risks (see also O'Riordan & Voisey 1998).

Finally, many reactive policies focus on maintaining or achieving *minimum quality levels*, either related to emissions allowed or ambient environmental qualities. Although these policies provide stimuli for meeting these minimum quality levels, they are no impetus for doing more (e.g. Oates 2001). Ideally, governance would be about finding a balance between the costs and benefits of additional environmental measures in comparison to, for example, economic or social costs and benefits; i.e. policies should urge decision makers to find such a balance. Reactive policies often fail to provide such stimuli.

The shortcomings associated with reactive environmental policies have led to calls for a more proactive, anticipatory and ambitious approach to environmental issues. It is what Barrow (1995) calls a shift from a 'react and mend' approach to an 'anticipate-and-avoid' approach. In addition, Simonis (1988) argues that "the path that future national and international environmental policy should take is clear: to transform from react-and-cure strategies to anticipate-and-prevent strategies". Proactive policies, therefore, try to avoid future damages before they are done by constraining damaging activities, by trying to anticipate future problems and, if this is too challenging, to take precautionary action. Especially from the late 1980s onwards, the call for more proactive and anticipatory policies has become visible in environmental policies. Most notably, this call became associated with the concept that best characterizes environmental governance in the late 20th and early 21st century: sustainable development.

Towards a proactive attitude: sustainable development

Sustainable development is most specifically promoted through the report 'Our Common Future' of the 1987 UN Brundtland commission. The definition used in this report is also the most commonly referred to when discussing sustainable development: i.e. "development that meets the needs of the present without compromising the ability of future generations to meet their own needs" (WCED 1987; p. 43). As such, sustainable development is focused on both the needs of current generations and those of future generations. Although certainly not without critiques, sustainable development has over the past decades made a major impact on policy agendas across the world, ranging from global to local levels (e.g. Jordan 2008).

Sustainable development assumes that delivering and maintaining prosperity and wellbeing in the future requires environmental protection and economic and social development to go together. Hence, sustainable development supports the idea that environmental quality and protection need not be at odds with economic and social development. Rather, sustainable development aims to move beyond the idea that environmental policies should *react* to economic and social development. Instead of being advanced through developing as a distinct set of reactive regulations or policies, sustainable development considers environmental quality

46   *Governing the environment*

and protection key criteria that all other policies and developments should promote. Consequently, sustainable development unites a proactive policy approach with that of an *integrated* or *holistic* policy approach.

As Gouldson and Roberts explain, "environmental objectives could be pursued more proactively if they were better integrated into the design and implementation of the various policies that seek to facilitate and promote economic growth and development" (2000; p. 5, also Barrow 1995, Simonis 1988). Better integrated policies, therefore, are seen as a positive condition for the development of more proactive policy approaches. Unsurprisingly, among the 'key defining features' of sustainable development is the notion of 'environmental policy integration', referring to the "integration of environmental objectives into non-environmental policy-sectors" (ibid, also Briassoulis 2005, Jordan & Lenschow 2010, Persson 2002). Environmental policy integration is supported by the "recognition that the environmental sector alone will not be able to secure environmental objectives, and that each sector must therefore take on board environmental policy objectives if these are to be achieved" (Lafferty & Hovden 2003; p. 1). Furthermore, policy integration can also help "bind together currently departmentalized, disparate and uncoordinated policies that fail to tackle contemporary, cross-cutting, complex socio-environmental problems, sometimes being among the forces producing these problems" (Briassoulis 2005; p. 351, also Jordan et al. 2005, Lafferty & Hovden 2003 and Persson 2002).

Although used before the widespread introduction of sustainable development, the popularity of environmental policy integration increased with the rise of sustainable development as a policy guideline from the late 1980s onwards. The integration of environmental objectives in other policy fields was also actively pursued in most EU countries and within the EU itself during especially the 1990s (for examples, see Andersen & Liefferink 1997, De Roo 2003, McCormick 2001 and Van Tatenhove et al. 2000). Such integration can be seen in the widened use of environmental impact assessments (EIA) and the EUs EIA directive (85/337/EEC), making an EIA obligatory in all EU member states (e.g. Arts 1998, Glasson et al. 2005). But while environmental policy integration supports the pursuit of sustainable development, it is arguably only a first step in doing so. Essentially, it is still about improving the functioning of the coordinative model, now focussing on increasing the coherency and interaction between different policy fields and with a focus on the central government (see also Lafferty & Hovden 2003). Increased coherence and integration of central government policies are certainly important in the pursuit of sustainable development, but pursuing proactive and integrated approaches associated with sustainable development also demands more than what the coordinative model seems to offer. It is on doing so that the call for more proactive and integrated approaches also relates to the main challenges for the coordinative model just discussed.

To begin, producing integrated policies that proactively embed environmental criteria in different policy sectors and decisions already provides a key challenge for the coordinative model. Integrated policies mean to explicitly respond to the

Governing the environment   47

interrelatedness of issues and ambitions on various spatial and temporal scales. As Martens also explains,

> Even if sustainability is robbed of its radical edge and controlled by established government bodies, it still poses challenges to the coordinative model. Ironically, this is precisely because sustainability underscores the importance of a more comprehensive, holistic and integrated approach and thus pushes the need for coordination to its limits.
>
> (2007; p. 52)

Furthermore, it is also well understood that interrelated policy issues and ambitions are often associated with different societal behaviours, practices and interests that are represented by different societal groups and stakeholders. Proactive and integrated approaches, therefore, need to respond not just to interrelatedness, but also to social fragmentation and power dispersal. It again illustrates that proactive and integrated approaches require exactly what the coordinative model has trouble delivering.

Finally, its tendency to be anticipatory shifts the perspective of proactive policies to an acceptance of dynamics, risks and uncertainties. Avoiding risks within a realm of high complexity as Beck (1992) addresses, implies that proactive policies mean to deal with uncertainties and events we cannot yet fully comprehend or even recognize (see also Berkes & Folke 1998, De Roo & Silva 2010, De Roo et al. 2012, Duit et al. 2010, Folke et al. 2010, Gunderson & Holling 2002 and Simmie & Martin 2010). After a focus on trying to better understand the idea of sustainable development and translating such an understanding into policies and action during mostly the 1990s and early 2000s, the past decade has shown an increased recognition that proactiveness should also embrace ideas of non-linearity and complexity. Hence, recent debates on sustainability and the governance of environmental challenges are exploring new concepts and approaches that might help them better understand these risks. Especially dominant among these is the notion of resilience and, related, that of societal transformations and transitions. It is the most recent chapter in the debate we as planners and policy scientists are engaged in while we try to move beyond a sole reliance on the coordinative model.

Towards proactive policies in a dynamic world: resilience

Complexity sciences have helped planners and policy sciences to understand the world they aim to intervene within as being in a constant state of flux. Within such an understanding of the world, conceptions of stability and linearity are being replaced by those of dynamically interacting social and physical systems that are producing non-linear, emergent and co-evolutionary behaviour. As Davoudi (2012) explains, "among the prescribed remedies for dealing with [such] a state of flux, the one that is rapidly gaining currency is 'resilience'" (p. 299). Swiftly

48 *Governing the environment*

gaining in popularity and use, resilience even seems to partly replace 'sustainability' as the 'buzzword' in academic debates on planning and policy sciences in the early 21st century (e.g. Porter & Davoudi 2012). Following its introduction in the realm of ecological studies by 'Buzz' Holling in 1973, resilience initially had only limited impact in academic debates. This limited impact was also partly due to the interpretation of resilience in the first decades after Holling's initial studies. Although Holling hinted towards the role of dynamics, unpredictability and non-linearity, resilience was first adopted in relation to the capacity of ecological systems to withstand shocks and bounce back to their original states. It is a view on resilience that focusses on systems in stable equilibria, on degrees of disturbance systems can handle and on their return times; i.e. it is a view that "applies only to behaviour of a linear system, or of a non-linear system in the immediate vicinity of a stable equilibrium where a linear approximation is valid" (Folke 2006; p. 256). Hence, it is also a view that fits in well with notions of control, certainty and the coordinative model. It is a view, however, that with the growing impact of complexity sciences in academic debates has increasingly been replaced with a more dynamic perspective on resilience, also in the literature on planning and policy making.

Within the literature on environmental planning and policy making resilience is especially used in studying the behaviour and governance of so-called 'socio-ecological systems'; i.e. the integration of social and ecological systems that relate to each other in a dynamic fashion (e.g. Berkes & Folke 1998). While often related to the interaction between human systems, such as cities, towns or economies, with ecological systems, urban areas can also be seen as socio-ecological systems. Now, it is the human activity that not just affects ecological systems through the air, water and soil, but also affects human health itself. Resilience, then, has to do with how we might respond to changing trends of human behaviour, economic growth, urban development, pollution and even large-scale trends such as climate change, globalization and technological development.

While taking a more dynamic perspective, resilience is linked to systems that use mechanisms of self-organization to adapt and respond to both internal and external processes of change (see also Folke et al. 2010, Gunderson & Holling 2002 and Simmie & Martin 2010). It is a view Davoudi calls 'evolutionary resilience' and that fits in with the rise of complexity sciences and more specifically with studies on complex adaptive systems (e.g. Berkes & Folke 1998, Gell-Man 1994, Holland 1995 and Kauffman 1995). Resilience, from such a perspective, is not necessarily about a system's capacity to return back to its original or 'normal' state of equilibrium. Instead, an evolutionary perspective on resilience "challenges the whole idea of equilibrium and advocates that the very nature of systems may change over time" (Davoudi 2012; p. 302). Therefore, taking an 'evolutionary' perspective means that "resilience is not conceived of as return to normality, but rather as the ability of complex socio-ecological systems to change, adapt and crucially, transform in response to stresses and strains" (ibid, p. 302). It is a perspective that again provides some key challenges to the coordinative model, while it also has allowed resilience to partly complement and partly replace sustainability as an inspirational planning guideline in at least two ways.

Governing the environment 49

Resilience and sustainability

Resilience first complements and adjust notions of sustainability by highlighting the idea that socio-ecological systems need not be subject to a single stable state of equilibrium for them to be sustainable. Resilience, therefore, challenges the idea that sustainability should refer to a socio-ecological system's capacity to *sustain* its current state or behaviour in the long run. Sustaining existing states or behaviour, after all, can be risky and potentially undesirable when faced with new and unforeseen disturbances and pressures. Such disturbances and pressures are not at all unlikely considering the changeable nature of our societies and environments that complexity sciences have shown us. If we are to adapt to climate change in western Europe, just to illustrate, we are best to also allow our ecosystems to gradually migrate, our cities to gain in capacity to cope with heat, smog or heavy rainfall and our rivers to accommodate higher fluctuations of water flows. Resilience responds by suggesting that what matters more than a socio-ecological system's capacity to sustain its current state or behaviour is its capacity to change, adapt and transform within a world that is in a constant state of 'flux' (see also Davoudi 2012, Folke 2006 and Gunderson & Holling 2002). Disturbances and pressures, therefore, might well call for new responses and behaviour to aid the system's survival. Resilience responds by implying that "rather than assuming a social system characterized by stability and equilibrium, analytical focus is placed on understanding processes of change and surprises and on how governance arrangements try to *cope with* and *adapt to* a constantly dynamic and changing environment" (Duit et al. 2010; p. 364).

With its focus on socio-ecological systems, change and adaptation, a resilience perspective provides some key challenges for the coordinative model. On the one hand, resilience does so by highlighting the need for more integrated policy approaches that respect the dynamic interactions between both social and ecological systems and between alternative social, economical and socio-ecological systems. This includes linkages between alternative policy fields such as spatial planning, water management, agriculture, infrastructure planning or housing, each of which aims to cope with parts of the dynamic social and physical systems we aim to influence through governing. Hence, resilience confirms the need for integrated policy approaches that take account of the interrelatedness of different problems, their causes and effects, and related stakeholder interests. Furthermore, in trying to respond to such interrelatedness, a resilience perspective also shifts perspective to local conditions influencing how problems, causes, effects and interests actually relate. Therefore, resilience also highlights the importance of using local knowledge to benefit from available social capital and more participatory forms of planning for governing socio-ecological systems (e.g. Adger 2010, Berkes et al. 2000, Brown 2011, Folke et al. 2010, Galaz 2005, Olsson et al. 2006, Ostrom et al. 2003, Shaw 2012 and Walker et al. 2002). Locally sensitive and more participatory approaches will be difficult to accommodate by the coordinative model and its reliance on central state control. Instead, resilience challenges the coordinative model by urging for a more significant role for locally based governance approaches (see also section 2.6).

50   *Governing the environment*

On the other hand, resilience challenges the coordinative model since, as Folke (2006; p. 253) explains, "the resilience perspective shifts policies from those that aspire to control change in systems assumed to be stable, to managing the capacity of social–ecological systems to cope with, adapt to, and shape change" (Berkes et al. 2003, Smit and Wandel 2006). Taking a resilience perspective, thus, provides a key challenge for the coordinative model and its focus on control and predictability. Consequently, instead of having proactive policies that aim for sustaining the current states and behaviour of systems, proactive policies should also respect change and encourage and allow for systems to also adapt in response. This implies that policies should allow for some degree of flexibility so as to leave room to shift from one development path to another.

Resilience, transformability and transitions

Second, the evolutionary perspective on resilience has helped scholars reframe the difficult quest for a sustainable society. An evolutionary perspective on resilience implies that pressures and disturbances are not just forces we should to try to anticipate and canalize. They are also forces that can foster creativity, development and, hence, transformation (e.g. Davoudi 2012, Folke 2006, Folke et al. 2010 and Gunderson & Holling 2002). Transformability, then, can be seen as "the capacity to transform the stability landscape itself in order to become a different kind of system, to create a fundamentally new system when ecological, economic, or social structures make the existing system untenable" (Walker et al. 2004; p. 5). It is a concept closely related to what others describe as socio-technical transitions (e.g. Foxon et al. 2008, Grin et al. 2009, Loorbach 2007, Meadowcraft 2009, Rip & Kemp 1998 and Rotmans et al. 2001).

Transformational change can be a desirable process planners and policy makers aim to stimulate and foster, and hence that can be seen as a proactive policy ambition (e.g. Folke et al. 2010, Walker et al. 2004). Over the past two decades the shift to a more sustainable society is being increasingly conceptualized as a desirable transformation or transition of our global socio-ecological system (e.g. Folke et al. 2010, Foxon et al 2008, Grin et al. 2009, Loorbach 2007, Meadowcraft 2009, Rip & Kemp 1998, Rotmans et al. 2001 and Walker et al. 2004). A conceptualization of this shift as a transformation or transition is inspired by the idea that existing proactive policies have not been able to stimulate such a transformation or transition to take place; a point well illustrated by ongoing unsustainable trends in relation to, for example, climate change and energy use, environmental threats to public health, poverty and social exclusion, poor management of natural resources, biodiversity loss, etc. (for a discussion see Global Footprint Network 2010). A focus on transformations, transitions and resilience suggests that our current socio-economic system can be seen as a complex web of interrelated actors and networks, in a physical, economic, social and institutional sense (De Boer & Zuidema 2015). Apart from limitations to fully oversee and grasp such a complex web, ownership and power are fragmented, limiting the capacity of any actor to alter the system. Also, this complex web develops routines of interaction,

Governing the environment 51

expectations about each other's behaviour, muddling-through behaviour and other kinds of self-reinforcing mechanisms that make the system resistant to change as well as path dependent (Martin & Simmie 2008). Traditional planning and policy approaches seem to fail as they do focus on the capacity of a single or group of stakeholders to come to decisions regarding desired 'end-states' and the approaches to achieve them (Allmendinger 2009). In response, a resilience perspective highlights that failure to respond to the persistence of our unsustainable practices is at least partly due to a focus on incremental improvements to our environment, rather that pushing for a wider societal transformation.

Central concepts for understanding societal transformation are innovation in 'niches', learning-by-doing, adaptation and co-evolution. As Folke et al. explain,

> A resilience perspective emphasizes an adaptive approach, facilitating different transformative experiments at small scales and allowing cross learning and new initiatives to emerge, constrained only by avoiding trajectories that the SES [social ecological system] does not wish to follow, especially those with known or suspected thresholds.
>
> (2010; p. 5)

Much similar to the academic literature on socio-technical transitions, this is a perspective that urges for social, technical, economic and administrative innovation to take place. Those innovations that prove promising can inspire learning and might lead to upscaling – increases in size, span of activities and social or political influence (e.g. Gillespie 2004). Subsequently, changes taking place within one societal or technological domain might well trigger changes in other societal domains or resonate with them; i.e. a process also called co-evolution (e.g. Foxon 2010, Hadfield & Seaton 1999, Kemp et al. 2007 and Norgaard 1984). It is, for example, when technical innovations in using solar, biomass and wind potentials can trigger local business and households to increasingly use them. Energy companies can subsequently be triggered to co-invest; grid companies are urged to adapt their infrastructure making it even easier to invest; and governments have to reflect both on existing tax regimes, energy legislation and possibly, further pave the road for investments and behavioural change. It is the kind of positive feedback showing co-evolution between innovations and changes occurring in different social, economic, institutional and technological sub-systems and that combined can create new development paths. Or as, for example, Kemp et al. explain, "in transition terms we speak of co-evolution if the interaction between different societal subsystems influences the dynamics of the individual societal subsystems, leading to irreversible patterns of change" (2007; p. 80).

Based on complex systems thinking, co-evolution helps to explain how new physical, socio-economic or institutional structures can emerge out of the interaction between existing societal processes. Ideally then, we do not just aim for policies that highlight the need for systems to be able to adapt to change. We also aim for policies that have the capacity to stimulate societal transformations by creating the right conditions for processes of innovation, upscaling, learning

52   *Governing the environment*

and co-evolution to take place. But suggestions on how to trigger and govern such societal transformations or transitions are not necessarily very detailed or supported by extensive academic backup yet (see also Shove and Walker 2007). What is clear, though, is that governing societal transformations or transitions implies a move away from the kind of top-down policy approaches associated with the coordinative model (e.g. Loorbach 2010). Rather, there is a desire to enable and stimulate experimentation and innovations in localized 'niches', so as to allow for a bottom-up development of possible new technologic, administrative or socio-economic practices. In the meantime, processes of learning and adaptation should aid in disseminating good practices and holding back those that are less desirable (e.g. Folke et al. 2010, Geels 2011, Kemp 2006, Rotmans et al. 2001). Therefore, and resonating with the literature on network governance, governing societal transformations or transitions is also about "the complex interaction patterns between individuals, organizations, networks, and regimes within a societal context, and how over time, these can lead to nonlinear change in seemingly stable regimes" (Loorbach 2010; p. 167). Finally, once co-evolution kicks in unpredictable developments are likely to take place while feedback loops can create rapid and unforeseen changes. Such complexity and unpredictability can make it difficult to exert much control over societal transformations or transitions, or even to know the outcomes in advance. It again urges planners to move beyond traditional planning approaches and their focus on stating a final envisioned 'end-state' and controlling the process towards such 'end-states'. So, when trying to foster societal transformations, complexity and unpredictability are motives for shifting towards more adaptive and flexible planning approaches that draw on a constant process of learning-by-doing (Geels 2011, Kemp 2006 and Rotmans et al. 2001).

Beyond the coordinative model?

The challenge of sustainable development is already a great one for the coordinative model; i.e. where "the very radicallity of the ideas underpinning sustainability questions the traditional organisation of government in sectors, hierarchies and levels and the established modes of governance that are related to them" (Martens 2007; p. 52). Jordan comes to a similar conclusion and suggests that "it is significant that sustainable development is being pursued using new rather than older modes of governance" (2008; p. 29). With the rising focus on unpredictable change and dynamics even an additional dimension is added. Loorbach (2010; p. 164), for example, explains that "new modes of governance are sought that reduce the lack of direction and coordination associated with governance networks in general, and increase the effect of existing forms of government and planning in the context of long-term change in society". It is these lessons and conclusions that have made sustainable development, resilience and, more generally, their focus on proactive and integrated approaches key inspirations to move beyond the coordinative model in pursuing improved environmental conditions.

2.5 New directions in environmental policy

Resonating with more general 'shifts in governance', environmental policies in the Western world and also many EU countries began to change somewhere during the 1980s and most significantly the 1990s (see also Andersen & Liefferink 1997, Jänicke & Jörgens 2006, Jordan et al. 2005, Lemos & Agrawal 2006, Mol et al. 2000 and Vig & Kraft 2013). These changes would begin with processes to improve the functioning of the coordinative model and with processes of environmental policy integration. Gradually, though, more significant changes were made. On the one hand, the state reallocated power and responsibility to nongovernmental groups in line with the neoliberal and communicative turns. On the other hand, the dominance of the central government was also reduced through both centralization following the increased role of the EU, international organizations and decentralization in many EU member states. It is especially the process of decentralization catching my attention in this book as the development of more proactive, adaptive and integrated approaches to the environment is among the prime motives supporting decentralization. Before I will address decentralization in more detail in relation to proactive and integrated approaches, I will first discuss the role of the neoliberal and communicative turns that are not disconnected from the process of decentralization either.

Moving beyond the coordinative model: the neoliberal and communicative turns

During the 1980s many EU countries began with the introduction of new policy instruments such as market-based instruments, voluntary agreements and informational devices while empowering market and civil society actors (e.g. Andersen & Liefferink 1997, Jordan et al. 2005). These instruments were added to existing governmental regulations, sometimes replacing them. They helped 'internalize' environmental costs and benefits in both market processes and public debate. Such internalization stimulates more proactive environmental behaviour as environmental costs and benefits are actively taken into account by both market and civil society actors. Therefore, these new policy instruments also became tools used to support the pursuit of sustainable development. Indeed, as Jordan states, "the adoption of new environmental policy instruments such as eco-labels . . ., taxes, and voluntary agreements grew massively after Rio [the 1992 United Nations Earth Summit]" (2008; p. 27, see also Biermann 2007, Levy & Newell 2004).

To begin with market-based instruments, governance renewal follows the logic of the neoliberal turn and is inspired by the competitive model of governance. The main motive underlying the use of market-based instruments is their supposed higher efficiency than regulations to internalize the costs of environmental externalities in products and services. Examples of market-based instruments include eco-taxes, pollution charges, tradable permits and government subsidies

54 *Governing the environment*

(e.g. Jordan 2002, Lemos & Agrawal 2006, Sonnenfeld & Mol 2002 and Stavins 2003). Typically, "market-based instruments are regulations that encourage behavior through market signals rather than through explicit directives regarding pollution control levels or methods" (Stavins 2003; p. 358). Market-based instruments have been used as early as the 1960s, but as Ekins (1999) shows, their use in most Western nations has increased, especially since the 1980s. They are often used for reducing the use of certain products or pollution by target groups such as industries, agriculture or households. They are also used to ensure that emissions do not exceed certain maximum amounts such as with tradable emission rights.

Market-based instruments "are founded upon the bedrock of individual preferences and assumptions about self-interested behaviour by economic agents" and the "claim advanced in their favor is their superiority in terms of economic efficiency related to implementation" (Lemos & Agrawal 2006; p. 8). Those products and services that are more harmful to the environment are made more expensive and less attractive as compared to those that are less harmful. The ideology is that "market-based instruments provide for a cost-effective allocation of the pollution control burden among sources" (Stavins 2003; p. 359), also as compared to the bureaucracy that is typical for regulatory policies. Command-and-control policies often force companies, industries or households to take similar measures and meet similar standards, regardless of the costs of doing so. The practical consequence is that some actors have to make disproportionate investments for reaching only marginal environmental benefits, whilst other actors can reach serious environmental benefits with only marginal investments. The answer provided by market-based instruments is to distribute investments for improved environmental performance in such a way that those who can create benefits most efficiently are also induced to do so. Thus, some actors will aim for higher ambition levels than others, inspired by the kind of investments they require to reach those benefits. Finally, many command and control measures provide no stimuli for polluters to do more than meet minimum requirements. Market-based instruments do urge polluters to reduce the costs of pollution to at least the degree that the marginal costs of prevention are equal to the marginal costs of pollution.

Most eco-taxes, charges, tradable permits or subsidies are installed at an (inter) national level, whilst their local applicability is limited. They contribute to lower background levels of pollution by limiting the emission of pollution and the use of more environmental friendly products and services. Market-based instruments, however, also have disadvantages. For example, they do not necessarily take the distributional effects of pollution into account. Concentrations of polluting activities can still exist, causing poor local environmental conditions. Factories can emit much locally if they have acquired sufficient emission rights. In addition, cars might be cleaner due to financial constraints and stimuli, but urban areas can still be flooded with cars causing, for example, air quality or noise issues. These distributional effects mean that local environmental quality can be quite poor, urging for remaining coordinative policies to provide safeguards for excessive pollution or stimuli for improving conditions. In the meantime, coordinative policies also remain important for local governments, as the use of market-based

Governing the environment 55

instruments in a local realm is itself often quite limited. Local authorities do not always have the adequate legal competences for issuing such tools, nor are they always inclined to do so. After all, financial constraints on economic activities can cause competitive disadvantages for a locality in comparison to neighbouring localities (see also Oates 2001).

Second, governance renewal was also inspired by the argumentative model and the associated communicative turn. New instruments in environmental policy especially included instruments such as voluntary agreements and informational devices. Voluntary agreements are deals made between governments and a group of stakeholders or other (lower level) governments regarding a certain set of objectives and/or regarding how certain objectives should be achieved (e.g. Jordan 2002, Mol et al. 2000). They come in very different forms, varying from legally binding contracts to fully voluntary statements of intention (Jordan 2002). Voluntary agreements were first adopted in few countries (e.g. France, Germany and the Netherlands) and later became widespread in the EU (e.g. CEC 2001b, Mol et al. 2000, Sonnenfeld & Mol 2002). Voluntary agreements are, for example, used by governments who want to make arrangements with a specific target group on possible environmental measures; e.g. groups such as motorists, large companies or specific economic sectors such as fishing, petrochemical industries or waste collection. Voluntary agreements are often pursued at an EU-wide or national level and can also involve agreements between local authorities and individual industries or local groups (Patton 2000). Although voluntary agreements can be important to push environmental policies forward, especially where legal standards are considered unambitious, they do hold a risk. Only if they are based on solid contracts between the key partners will they provide safeguards to excessive pollution or ensure proactive action. This is not always the case, as many covenants and agreements are statements of intent, rather than clear agreements on the objectives to be reached (e.g. Ashford 2002).

Finally, and subject to the same limitations as voluntary agreements, are informational devices such as eco-labels and campaigns (see also Siebert 1987, compare Visser & Zuidema 2007). Campaigns on, for example, television, radio or in newspapers are used to stimulate environmentally friendly behaviour. They are based on the argumentative power of the message portrayed. When it comes to companies, campaigns can be linked to semi-voluntary labels such as the ISO 14001 or the EU Environmental Management and Audit Scheme (see also Jordan 2002). Finally, eco-labels are used for informing consumers of the environmental impact of a product or service they mean to purchase. Cars, houses and refrigerators are among the examples of products that can be labelled.

Despite their influence on renewing environmental governance, the neoliberal and communicative turns have not eclipsed the role of regulations or of the coordinative model in environmental policy. They have merely resulted in a relative shift that means to help adapt the societal capacity to govern to societal challenges such as power dispersal, interrelated policy issues and social fragmentation. They are assumed to stimulate effectiveness and efficiency as individual people and stakeholders are stimulated by market-based instruments or voluntary actions

56   *Governing the environment*

to take their own measures in pursuing a better environmental quality. Environmental objectives, consequently, can be integrated in the behaviour and decisions made by market parties, citizens and government agencies, resulting in a more proactive approach to the environment. Despite their potential and promise, however, Jordan (2008) explains that the actual impact of these new instruments on the delivery of proactive environmental policies and sustainable development is not at all evident. Also, the promise of using new environmental policy instruments, he concludes, should not yet be considered an open call for denying the importance of existing governmental regulations (see also Ashford 2002, Jordan et al. 2005).

Instead of replacing regulatory tools, market-based instruments, voluntary agreements and informational devices can actually be supported by (central) government legislation; i.e. they function under the 'shadow of the law' (Jordan et al. 2005; p. 489). They are thus examples of the kind of tools resulting in the aforementioned hybrid governance networks where the coordinative, competitive and argumentative models of governance are connected. A key example of this is the rise of ecological modernization as a dominant approach in environmental policy from the 1980s onwards. Ecological modernization highlights the role of technological and societal innovations as responses to many pressing problems that undermine our society's sustainability (e.g. Buttel 2000, Jänicke 2008, Mol et al. 2009, Mol & Spaargaarden 1992). Examples include new means for energy production, increased efficiency in using and recycling resources and altered societal behaviour. Ecological modernization tries to use regulations to push for ongoing innovations in both the economic and societal domain so as to gradually make a social transformation to more sustainable practices; i.e. through what Jänicke and Jörgens call 'smart regulation' (2006). In doing so, the hope is of "a new environmental rationality arising from failures and crisis within industrial societal institutions" (Warner 2010; p. 546). This view is not without critiques, which, for example, Buttel (2000) and Warner (2010) note in stating that it is doubtful whether ecological modernization goes far enough to push for a full societal transformation. In addition, claims are made upon ensuring that a reliance on the innovative capacity of markets and companies is combined and not a substitute for traditional command and control regulations (e.g. Ashford 2002). Nevertheless, ecological modernization is exactly the kind of approach that tries to combine the merits of regulatory pressure with that of the competitive and argumentative models of governance. That is, it is an example of governance renewal going 'outwards' from the central state, so as to include civil society and market parties.

Moving beyond the coordinative model: centralization and decentralization

While an increased reliance on the competitive and argumentative models helps to reallocate power in a 'horizontal' way from governments to non-governmental actors, changes are also made regarding the relative allocation of responsibilities in a 'vertical' sense. Partly, environmental governance in the EU is being characterized by a long-term process of centralization following the increased role of the EU itself on domestic environmental policies. The shifting upwards of power and responsibility from localities and member states towards the EU is closely related to the EU ambition to harmonize domestic environmental policies and regulations (see also section 2.2).

In the meantime, we are also witnessing an increased role of global institutional powers within the field of environmental governance (e.g. Ashford 2002, Gehring & Oberthür 2006 and Sonnenfeld & Mol 2002). As Sonnenfeld and Mol (2002) explain, for example, ongoing processes of globalization have challenged national governments to increase forms of cooperative international governance. Similarly, the rise of civil society actors, non-governmental organizations and the ongoing diffusion of policies and technologies are causing the creation of international environmental institutions and governance practices. Key examples of clear international regulation also exist. The 1987 Montreal agreement on the protection of the ozone-layer, the 1989 Basel Convention on hazardous waste, the 1997 Kyoto Protocol or the sustainable development summits in 2002 (Johannesburg), 2012 (Rio de Janeiro) and 2015 (Paris) have gained follow-up or at least international political attention. In the meantime, institutions such as the United Nations Environmental Programme, the International Panel for Climate Change and the World Health Organization have gained in exposure and impact. Efforts to create and institutionalize international environment governance vary in rates of success. Isolated successes such as the follow-up on the Montreal agreements are counterbalanced to relative modest successes as related to, for example, the Kyoto protocol. It shows the intricate nature of global environmental governance, where no single authority has the ability to enforce follow-up and where leadership is hard to organize. Consequently, in 2002 Sonnenfeld and Mol saw mostly a rapid growth of building blocks for international environmental governance, but still limited impact. Axelrod and Vandeveer follow these lines in 2014, where they do witness an 'emerging global governance system', but also still problems of implementation and enforcement.

The rise of international and global environmental governance need not simply be a process of copying the bureaucratic apparatus of nation states. As Sonnenfeld and Mol (2002) recognize, the dynamics in an international arena beg for alternative modes of governing. On the one hand, large multinationals, non-governmental organizations, existing institutions such as the World Trade Organization and the media all play a role. Intensified by the internet and swift communication technologies, as Axelrod and Vandeveer (2014) note, attempts at international cooperation can easily become highly politically charged. On the other hand, the world is not a homogenous place. Countries differ substantially, not only in their administrative and economic means, but also in the priorities and issues they face. The result is that common denominators are hard to find and, instead, more flexible and tailor-made implementation paths are often needed. Most notably, then, this begs for careful deliberation and interaction between multiple levels. As Gehring and Oberthür (2006) note, for example, interaction between alternative institutions across levels and interests is often a positive

58　*Governing the environment*

condition for effectiveness. It is an argument also recognized in a European context, where the ongoing process of centralization has been balanced by the rise of the notion of *subsidiarity* (e.g. Barnes & Barnes 1999, Jeppesen 2000, Jordan 1999, McCormick 2001 and Zito 2000).

Subsidiarity asserts that central authorities should only perform the functions that they can perform better than other (lower) levels of authority. The underlying logic is that there are many functions that can well or even better be performed by lower levels of authority (i.e. nations, regions, provinces and municipalities). Then, depending on the service or function to be performed, an assessment should be made of 'which level should do what' (Flynn 2000, Jordan 1999). While making such an assessment, subsidiarity suggests that decisions should be made at the lowest possible level of authority so as to be closest to the actual problem; it is an argument that often, but not necessarily, implies a support for decentralization. Consequently, it is a concept signifying how processes of decentralization fit in with a growing body of international policies, regulations and institutions.

Although established in original EU Treaties, subsidiarity only became widely used in EU debates since the early 1990s (e.g. Jordan 1999). Fuelled by fairly broad support of EU member states, centralization could progress swiftly in the 1970s and 1980s. This progress was manifested in an astonishing growth of EU environmental policies and regulations (Hildebrand 2002). After 1990, subsidiarity gained in importance as an argument in EU policymaking. This had several important consequences. Mostly, the pace of centralization following new EU environmental legislation slowed down (e.g. Jeppesen 2000, Jordan 1999). In addition, central policies issued by the EU from the 1990s onwards are in general more sensitive to domestic or stakeholder interests than those before this date (see also Jordan 1999). In other words, these central policies allow for interpreting policies within a national or local realm by lower levels of authority. The resulting sharing of power and responsibilities is an important contributor to the rise of multilevel governance. Hence, subsidiarity is indeed not equated with decentralization, but merely provides a context in which assessments are made upon the possibilities to pursue decentralization.

In a general sense, decentralization can be described as

> the transfer of responsibility for planning, management and the raising and allocation of resources from the central government and its agencies to field units of government agencies, subordinate units or levels of government, semi-autonomous public authorities or corporations, area wide, regional or functional authorities, or non-government private or voluntary organizations.
>
> (Rondinelli & Nellis 1986; p. 5)

Following this definition, decentralization can involve both horizontal and vertical shifts in the allocation of power and responsibility. Hence, decentralization can occur in different ways and thus, has different envisioned and actual impacts on the organization and practice of governance (Prud'homme 1994).

Horizontal changes involve the deliberate dispersal of power and authority from government agencies to (semi)autonomous agencies or actors. These actors can be businesses, non-government organizations or even citizens that take over responsibilities for producing (semi)public goods and services. Vertical decentralization can be seen as "the devolution of power and responsibility over policies from the national level to the local level" (De Vries 2000; p. 493). Taken strictly, vertical decentralization seems to only involve a shift of power and responsibility within government bureaucracies. Thus, there is still a reliance on the coordinative capacity of government agencies. The main difference is that it is less the central state and more the local or regional authorities that are taking the lead. In practice, vertical decentralization typically involves more than just a change in the comparative role of various levels of authority. Instead, it can also result in horizontal decentralization of power and authority at the local level; i.e. local governments share power and authority with local stakeholders. In fact, horizontal decentralization is often a consequence of vertical decentralization and can be among its envisioned outcomes. That is, vertical decentralization is a possible strategy for increasing the role of non-government actors in governance and hence, for increasing the role of the argumentative or competitive model of governance. As a result, decentralization, even when instigated in a vertical sense, indeed resonates with both the neoliberal and communicative turns.

Although authors such as Prud'homme (1994), De Vries (2000) and Fleurke and Hulst (2006) warn for the 'dangers of decentralization' it has since, especially since the late 1980s, become a popular strategy to renew environmental policies in quite a number of European states (e.g. Lemos & Agrawal 2006, Van Tatenhove et al. 2000 and Wätli 2004). Examples include countries such as Sweden (Bergström & Dobers 2000), Norway (Hovik & Reitan 2004), the United Kingdom (Gibbs & Jonas 1999) and the Netherlands (de Roo 2004). Among the dominant arguments for such decentralization processes are not just possibilities to respond to the limits of the coordinative model, but also, to further increase the societal capacity to pursue sustainable development and its associated proactive and integrated approaches. I will now first address the main benefits and arguments associated with decentralization, while returning to its risks and the desire to reassess the role of centralized governance formats in the final part of this chapter.

2.6 Decentralization and area-based approaches

Decentralization is pursued for a variety of reasons, few of which are actually undisputed (see also De Vries 2000, Prud'homme 1994). De Vries summarizes these as

> the possibility of tailor-made policies, short lines between the allocating agency and the receivers thereof, service delivery based on greater knowledge of the actors at the local level, with regard for local circumstances, greater possibilities for civil participation and, in general, more effective and efficient allocation of public goods and services.

(2000; p. 493)

60   *Governing the environment*

Lemos and Agrawal (2006) and Oates (2001) add to this that it can also promote the competition among subnational units and, as such, increase efficiencies. It is furthermore the "proximity to the every-day life of citizens and companies in the municipal territory or region" (Burström & Korhonen 2001; p. 37) which is considered to help local authorities to accommodate local partnership working or bargaining processes. Hence, in allowing for tailor-made policies, decentralization allows policies to be adapted to specific local circumstances, changes and stakeholder interests. Decentralization, therefore, has some important benefits, which I will summarize in four key arguments all related to the desire to push for more proactive and integrated approaches to the environment.

Why decentralize? Interrelatedness and mutual dependency

Adapting policies to local circumstances can be relevant for issues that are deeply interwoven with their local contexts and that are surrounded by mutually dependent stakeholders each having their own opinions and interests. On the one hand, decentralized governments are in a good position to respond to interrelated policy issues. Issues that are interrelated often have causes and effects that not only cross spatial scales or areas, but that also cross policy fields and relate to different stakeholder interests and policy objectives. How different causes and effects interact, produce changes and effects over time and also touch upon stakeholder interests and policy objectives is, however, also dependent on the detailed local circumstances. Influenced by, for example, geographical, political, economic or cultural circumstances, different localities will have different priorities and different opportunities and hopes for the future. For coping with the local character of policy problems, area-based approaches that thrive on time- and place-specific knowledge and a keen awareness of changing local needs and circumstances have important benefits over standardized centralized policy solutions. Fuelled by time- and place-specific knowledge, the idea is that the interrelatedness of issues and interests can be translated *in situ* into integrated strategies (compare De Roo 2004, Dryzek 1987, Jordan 1999 and Liefferink et al. 2002). Doing so can be difficult for central governments as identifying time- and place-specific relations between causes, effects and policy issues can easily be outside their span of control. In contrast, as De Vries (2000), Fleurke et al. (1997) and Rydin (1998) also emphasize, decentralization is often meant to create policies tailored to the local circumstances, a more informed and cautious balancing of competing interests and eventually in 'integrated governance' to cut through existing segmented policies (i.e. to advance proactive and integrated approaches).

On the other hand, but related, decentralized governments are also in a good position to respond to local power dispersal and social fragmentation. The proximity to local circumstance can also be beneficial for entering into collaborative 'governance networks' with local and regional stakeholders and for crossing the public–private divide (e.g. Sørensen & Torfing 2009). Area-specific approaches mean to bring together collaborating and competing stakeholders in a network of bargaining, collaboration and balancing of various interests and objectives

Governing the environment 61

(see also Cameron et al. 2004, De Roo 2003 and Turok 2004; compare Steward 2001). Or, in the words of Franke et al. (2007), area-specific approaches can then be "the foundation of communication and cooperation between all professional and non-professional stakeholders" (p. 17). To Hajer et al. (2004), this is even central to area-specific approaches, which they consider

> a form of policy where various governments are not organized in a hierarchical organization so as to try to govern various societal actors, but a form of policy where public and private actors are organized in a horizontal network within which bargaining and collaboration are the means to come to suitable solutions.
>
> (Hajer et al. 2004; p. 27, translation CZ)

In other words, area-based approaches often rely on localized networks in which governance takes place.

Common examples of localized networks are so-called public–private partnerships between government and non-government parties. They resemble (or can be) voluntary agreements, typically inspired by a specific project or issue. A second common example is involving those people and groups that are involved in contributing to environmental issues or that suffer from the consequences. In doing so, representative democracy can be supplemented or even partly be replaced by forms of direct democratic involvement (compare Woltjer 2000). As the proximity of local governments to local stakeholders and citizens put local authorities at an advantage over the central state in engaging in participative approaches (see also Hooghe & Marks 2001), decentralization is again considered beneficial for the development of area-specific approaches that can bring collaborating and competing stakeholders together in a locally grounded governance network. As such, it is to stimulate the development of more dynamic policy approaches that are tailored to the local circumstances.

Why decentralize? Resilience and local knowledge

Second, the increased popularity of using resilience as an inspiration for environmental governance is also reinforcing the call for area-based and locally sensitive policy approaches. In section 2.4 I discussed that a more dynamic perspective upon resilience challenges the centralized and top-down nature of the coordinative model. Resilience urges for more adaptive forms of governance so as to better respond to conditions of uncertainty. Informing these responses is a process of learning upon both the changing circumstances and upon how the socio-ecological systems we aim to govern dynamically respond to these changing circumstances. While such learning can certainly be partly accommodated by central governments, agencies and experts, there is also reason to address the role of more local communities in facilitating such learning. While focussing strongly on the resilience of socio-economic systems, Folke et al. also explains that it "comes as no surprise that knowledge of ecosystem dynamics and associated management

62 *Governing the environment*

practices exists among people of communities that, on a daily basis and over long periods of time, interact for their benefit and livelihood with ecosystems" (2005; p. 445, see also Berkes et al. 2000). Folke et al. continue by also suggesting that "it is important to address the social dimension and contexts for adaptive governance in relation to ecosystem management, including processes of participation, collective action, and learning" (p. 447). Consequently, aspects such as social links, social capital, trust, sense making and participation are among the key resources identified for societies and communities to cope with their own (localized) environmental problems and socio-economic challenges (see also Adger 2010, Brown 2011, Galaz 2005, Olsson et al. 2006, Ostrom et al. 2003, Shaw 2012 and Walker et al. 2002).

Taking a resilience perspective is more than just a call for adaptiveness to changing circumstances; it is also a call for adaptiveness to specific local circumstances, pressures, knowledge, and debates. It is a suggestion that does not only apply for governing socio-ecological systems, where often the ecological dimension of these systems is a central focus point. It also applies for urban environmental challenges, where rather than a focus on the ecological dimension of cities, environmental stress is related to human health and wellbeing; e.g. noise nuisance, air pollution, climate change, safety risks. The idea is now that out of sensitivity to local circumstances and stakeholders, well-informed and dynamic policy approaches can be developed that are supported for making sense to those involved. They are the kind of policies that integrate local knowledge, interests and issues, whilst proactively seeking for improvements based on experimentation, monitoring and learning. Consequently, as Adger, for example, expresses in relation to climate change policies, resilience calls for "active and empowered local government able to promote social capital and social learning between civil society and government" (2010; p. 5). Although calls for sensitivity to local conditions are no calls for a full shift towards decentralized governance formats, a resilience perspective does urge for room for more decentralized and bottom-up governance practices than the strong reliance on centralized policies that the coordinative model tends to offer (see also Davoudi 2012, Holling & Gunderson 2002).

Why decentralize? Implementing sustainable development

Third, scholars also increasingly address the need for policies that are adapted to specific contextual conditions and dynamics when it comes to implementing sustainable development. Related to difficulties to implement sustainable development, many critiques have emerged upon alternative routes towards governing the pursuit of sustainable development. One of these routes is to try to put sustainable development into practice by noting that implementing sustainable development is constrained as it is not accompanied by a very clear and univocal definition. Sustainable development is a fairly abstract or even 'fuzzy' concept (e.g. Connelly 2007, De Roo & Porter 2007 and Jordan 2008). In remaining relatively abstract, sustainable development has the capacity to 'hide' existing

Governing the environment 63

tensions between the three main policy objectives it tries to achieve; i.e. economic growth, social welfare and environmental quality and protection (e.g. De Roo & Porter 2007, Mitlin & Satterthwaite 1996). Hiding these tensions is especially beneficial when the ambition is to unite seemingly conflicting objectives, interests and societal groups within one single policy agenda. When translating sustainable development into action-oriented policies, however, problems emerge. As Creedy and Zuidema (2007) also explain, the abstract notion of sustainable development only covers a basic and mutually accepted 'core of understanding'. Once this core gets taken apart, they continue, a whole array of different interests and interpretations gets uncovered (see also Campbell 1996, Connelly 2007, Jordan 2008, Richardson 1997 and Rydin 1997). The result is that struggles emerge between parties concerning what sustainable development actually means, reducing its practical value as a policy guideline. Richardson, therefore, simply considers sustainable development as "a political fudge: a convenient form of words, promoted, though not invented, by the Brundtland Commission, which is sufficiently vague to allow conflicting parties, factions and interests to adhere to it without losing credibility" (1997; p. 43).

With its vague and 'fuzzy' nature hiding possible conflicting interests, sustainable development does help to unite stakeholders in a single discourse. Clearly defining sustainable development would be difficult without alienating key societal actors and perspectives. Furthermore, it would undermine its appealing character and instead, uncover a range of different interests and interpretations. Consequently, clearly defining sustainable development is not necessarily helpful and, quite possibly, counterproductive. Alternatively, as Kates et al. (2005; p. 20) put it, pursuing sustainable development can be implemented based on "the creative tension between a few core principles and the openness to re-interpretation and adaptation to different social and ecological contexts". So instead of searching for detailed definitions, scholars have shifted focus to suggesting that sustainable development should be 'given meaning' in direct relation to its specific context and, more specifically, the local circumstances in which sustainable development is pursued (see also Baker et al. 1997, Connelly 2007).

In reflecting on the pursuit of sustainable development, Kemp et al. (2005; p. 15) conclude that

> What is most needed, appropriate and workable always depends heavily on the context. One could say that sustainability is about locally suited options that are globally sustainable. But it is also about local awareness and behaviour that shares the larger agenda.

Rydin, in a more practical fashion, also explains that "the notion of top-down implementation of a sustainable development strategy at the urban level is inappropriate" (1997; p. 153). Rather, Rydin opts for debating sustainable development on a local scale where interactive processes based on (following Lindblom) 'mutual adjustment' can result in linking sustainability to the "distinctive characteristics of the local area, . . . [and] the ambiguities of local policy to sustainable

64   *Governing the environment*

development" (1997; p. 154, see also Jordan 2008). It is based on the idea that "the social, economic and political contexts are very much local in nature" (Dhakal & Imura 2003; p. 116). So instead of uncovering a 'layer' of conflicting interpretations and interests when trying to make sustainability more concrete through a more detailed definition, Rydin urges for the concept to be given more substance through connecting its abstract claims to the specific context (see also Healey 2007). Consequently, when pursuing proactive and integrated policies, centralized policies would allow for some flexibility so as to be adapted to contextual circumstances. Subsequently, it is up to local governments and stakeholders to use this flexibility and develop action-oriented policies that are based on time- and place-specific knowledge. The hope is then that "the local scale of projects allows for a development process based on an understanding of local needs, conditions, dynamics and potentials, and that includes local residents and stakeholders in a collaborative planning process" (Cameron et al. 2004; p. 311).

Why decentralize? Innovation in localized 'niches'

Fourth, and finally, I draw from the earlier discussion on complexity science and its call for framing the pursuit of sustainable development as social transformations and transitions. Rather than relying on the coordinative capacity of a central government for instigating and fostering societal transformations or transitions, these approaches highlight processes of innovation, learning-by-doing, adaptation and co-evolution for doing so. As Loorbach (2010) explains, for example, these approaches thus involve a move away from the kind of top-down policy approaches associated with the coordinative model. Instead, central to these approaches is to enable and stimulate experimentation and innovations in localized 'niches' (e.g. Kemp & Van den Bosch 2006, Smith et al. 2005). Although innovation should not be constrained to niches (see also Berkhout et al. 2004), it is clear that they play a key role in especially the start-up phase of societal transformations or transitions by allowing for innovations (see also Geels & Schot 2007).

Abstractly spoken, a 'niche' is a place relatively isolated from the larger socio-economic system where innovations and learning can take place in relative isolation from regressive societal and economic forces (e.g. Kemp & Loorbach 2006). It is in such a context that individual actors and local practices can benefit from freedom to experiment and improvise. Therefore, it is in these localized 'niches' where, for example, innovation takes place in engineering or developing new technologies, products and markets. Nevertheless, equating 'niche' developments with the picture of bright and creative minds exploring new technologies is too simplistic (e.g. Berkhout et al. 2004, Dóci et al. 2014 and Smith et al. 2005). A 'niche' is also the playing ground of local governance, where new societal practices and organizational formats are invented, tried and tested. Local governance arenas, thus, are among the 'niches' for experimenting with and adopting social innovations, new governance practices and organizational formats also in the realm of environmental governance (for examples, see Bulkeley & Broto 2012, MacCallum et al. 2012 and Nevens et al. 2013). Examples related to environmental governance could be novel strategies to reduce air pollution, respond to noise nuisance or promote biking, but also new management strategies to better integrate environmental interests in decision making, create partnerships with public or private parties or to create energy cooperatives. They are the kind of innovations that, if successful, might well be among the tools and strategies other localities will also start using. Successful practices can subsequently challenge the existing system of institutions, practices and regulations – what is also called the socio-technical regime – and show new development pathways (e.g. Hekkert & Ossebaard 2010, Kemp et al. 2007 and Simmie 2012b). Local innovations in relation to the integration of environmental objectives into spatial planning, for example, were among the main drivers for what eventually became a serious transition in Dutch environmental policies (De Roo 2003). Similarly, changes to the Dutch energy law were also largely induced by the increased understanding that the many emerging local energy practices urge for different national policy frames (see also De Boer & Zuidema 2015).

In terms of innovations in governance, the local level can be defined as a 'niche' where both governmental and non-governmental actors can pursue new governance approaches and organizational formats to deal with their environmental challenges. As compared to the centralized practices of the coordinative model, this requires a more decentralized mode of governance. Enabling and stimulating experimentation and innovations in localized niches, after all, is not just about the activities of local governments and stakeholders. It also requires them to be allowed sufficient regulatory, financial and political room to manoeuvre and innovate (e.g. Hekkert & Ossebaard 2010, Kemp et al. 2007). Existing centralized governance formats and generic policies can be among the regressive forces that might constrain innovation in local governance arenas – for example, when certain innovations are not tolerated by existing legislation or policies. But while some degree of decentralization is thus needed to allow for novel practices to develop, the learning from these practices and translating such learning into the development or more fundamental societal (regime) changes also requires some degree of coordination. Furthermore, local innovation might also well benefit from centralized governance formats being among the driving forces for innovation (e.g. Ashford 2002, Ashford & Hall 2011, Jänicke & Jörgens 2006). Instead of a simple call for decentralization, the idea of governing social transformations and transitions is associated with more hybrid and multilevel governance approaches (e.g. Geels & Kemp 2000, Loorbach 2010 and Verbong & Geels 2007). Or, as Kemp summarizes, "transition management relies on the interaction between processes at three levels [macro, meso, micro]. Transition management tries to align these processes through a combination of network governance, self-organization and process management leading to modulation of ongoing dynamics", making that "transition management can be considered as a specific form of multi-level governance" (2009; p. 82).

What about centralized governance?

While discussing four main categories of arguments in support of decentralization, I already proposed some important nuances with regards to how these

66 *Governing the environment*

arguments should be interpreted. In practice, creating more proximity of decision making arenas to local circumstances is used as an argument for decentralization in various EU member states. Often, this is also done to pursue more integrated approaches and also to capitalize local knowledge, social capital and governance innovations in pursuit of proactive policy agendas. The benefits associated with centralization, however, do not deny the need or relevance of central policies.

To begin with, decentralization is not always needed or desirable. Not all issues are strongly embedded in their local context and in need of area-based approaches, while not all issues should be framed as if they are part of larger societal transformations or linked to ideas of creating more resilient socio-ecological systems. In fact, many environmental issues have a rather common manifestation throughout countries and localities. They can be dealt with by routinely using many tried and tested policy instruments and approaches, as I will explain in the following section (2.7). In addition, there are important risks involved in relying on decentralized governance formats, also in relation to environmental policy. Section 2.1 noted the deterioration of public health conditions, large differences in environmental qualities between localities and reduced legal security experienced by companies or citizens. In response, central policies and regulations can be used to guarantee minimum protection levels that apply equally throughout various local jurisdictions. It is again an argument that I will pursue in the following section (2.7).

Second, even if decentralization is considered desirable, this does not imply a 'black or white' choice between *either* centralized *or* decentralized governance formats. Reality is way more subtle and typically involves varying degrees of decentralization given the many ways in which central policies and decentralized governance can interact. Central policies can play a role in, for example, supporting local performance, providing checks and balances against excesses, stimulating and enabling local policy innovations for setting the boundaries within which local area-specific approaches are to take place, etc. (e.g. De Roo 2004, Fleurke et al. 1997). When discussing decentralization operations, the question should not just be about the question whether decentralization is desirable. Rather, it should also be about questions related to the *degree* to which power and authority can be decentralized and *within which frames* of central policies decentralization might be pursued so as to prevent undesirable consequences. Hence, in the following section (2.7) I will address possible consequences of relying on decentralized governance in environmental policy. Based on that, I will argue that holding onto central governance formats for issuing uniform and generic policies is indeed important in environmental policy, also when we hope to increase the role of proactive and integrated approaches to the environment.

2.7 The relevance of central policies

During the last two decades the local level is gaining in importance in the development and delivery of environmental policies (e.g. Andersen & Liefferink 1997, EEAC 2003, Evans et al. 2005, Feitelson et al. 2004, Lemos & Agrawal 2006, Selman 1996 and Wätli 2004). Many local authorities and stakeholders started

Governing the environment 67

their own initiatives, proving themselves localized 'niches' for innovation (e.g. Bulkeley & Broto 2012, Nevens et al. 2013). Some, for example, developed tools and instruments to formulate and implement their own environmental policies (e.g. Creedy et al. 2007, De Roo et al. 2011, Miller & de Roo 2004, 2005 and Moore & Scott 2005) and policies to ensure coherency between environmental policies and policy fields such as spatial planning, economic growth, social development, etc. (e.g. Creedy et al. 2007, De Roo et al. 2010 and Franke et al. 2007).

While pioneering towns and cities are taking initiative, (inter)national governments also developed structural policies and tools to stimulate local initiatives (e.g. CEC 1990, 2004, Miller & De Roo 2004 and VROM 2001). On the one hand, various national governments are involved in deregulation and decentralization operations (e.g. Lemos & Agrawal 2006, Van Tatenhove et al. 2000 and Wätli 2004). The idea is to reduce the impact of top-down policy imperatives that potentially constrain local authorities and stakeholders in setting their own priorities and targets. Instead, environmental priorities and targets should increasingly be formulated at the local level; i.e. on an area-specific basis. On the other hand, (inter)national governments are also persuading local authorities to take initiative in not only setting targets, but also developing their own strategies, approaches and environmental policies. A quick scan made in 2005 by the EU Expert Group on the Urban Environment (EGUE 2005) indicated that at least 11 of the (then) 25 EU members had national policies in place to stimulate local policy initiatives. Also, existing (inter)national programmes include well-known examples such as Local Agenda 21 initiatives (O'Riordan & Voisey 1998), the European Sustainable Cities and Towns Campaign (CEC 1996), the Aalborg Commitments and the use of management schemes such as the Environmental Management and Auditing Scheme (EMAS) at the local level. As this illustrates, (inter)national governments are attempting to invite, enable and stimulate local authorities to develop proactive and integrated strategies (see also Franke et al. 2007).

While we are witnessing moves towards the local level, practical experiences are not yet convincingly showing that this move is also necessarily successful. Rather, the beginning of the 21st century is showing that success in pursuing more proactive and integrated approaches to the environment is still difficult. Despite their commitment to deliver proactive and integrated approaches, even leading municipalities and countries are reportedly finding it difficult to implement new policies (e.g. Creedy et al. 2007, Kamphorst 2006; compare Baker et al. 1997, Campbell 1996, Godschalk 2004, Richardson 1997 and Rydin 1997). Consequently, initiatives such as Local Agenda 21 and the European Sustainable Cities and Towns Campaign have only met limited successes (e.g. CEC 2004). While these initiatives have received follow up in debates and policy plans and programs, their real impact in policy actions and social, spatial and environmental governance is less clear (e.g. CEC 2004, EEB 2006, Kamphorst 2006 and O'Riordan & Voisey 1998).

The problems of local authorities to develop and deliver ambitious, proactive and integrated environmental policies already provide a warning against a reckless pursuit of decentralization. On the one hand, therefore, decentralization

68 *Governing the environment*

should be accompanied by a clear understanding of its possible consequences and the conditions that influence the occurrence of these consequences. Although this is true for all policy domains, I argue that it is especially relevant when it comes to environmental policy. After all, environmental policy has some important characteristics that can make decentralization risky and instead provide arguments in support of more centralized approaches. On the other hand, and related, centralized governance formats, despite their disadvantages, also have important benefits that might prevent undesirable consequences from occurring. They might become arguments not to decentralize at all, or to decentralize within a context of such centralized governance formats.

Environmental policies: protection and social dilemmas

Environmental policies have some distinct characteristics that might limit the potential for decentralization of the responsibility for developing these policies and controlling their implementation. The first of these is the focus of environmental policy on protecting people and ecosystems against harmful environmental stressors and risks. On the one hand, this can limit the potential for decentralization as motives such as legal security and equity push for similar levels of protection for all people and ecosystems. In response, central governments, such as nation states or the EU, typically install uniform and generic policy ensuring that all lower levels of authority implement similar levels of protection. Alternatively, decentralization tends to result in different levels of protection that are set between alternative jurisdictions, while levels of success in implementation might also differ. Consequently, motives in support of decentralization have to be carefully assessed in contrast with motives related to legal security and equity.

On the other hand, the protection of people and ecosystems against environmental stress also calls for predictable and reliable policy outcomes. Such predictability and reliability are far from evident, though, especially if the task is left to lower levels of authority. Environmental interests are obviously not the only interests to defend and pursue. Central governments respond by putting environmental objectives and levels of protection into a legislative context, giving much legal and political pressure for them to be implemented. Local governments, typically, have only limited options to create a similar degree of legal and political pressure, while they might also be unwilling to do so. After all, environmental interests might well have to compete with other – often more development oriented – interests (e.g. economic growth, financial gain). Environmental policies, however, tend to have a relatively 'weak profile' within local governance. Environmental issues often have a relatively technical nature, making them hard to understand for local politicians, administrators or other stakeholders. Furthermore, their impact is not necessarily tangible (ozone layer), easy to express in financial terms (such as noise nuisance), rather invisible (as with safety risks), diffuse (as with air pollution), subjective (odour) or very much long term (global warming). Consequently, translating environmental policies into clear and tangible benefits in the here and now is much more difficult than it is with, for example, new urban developments,

Governing the environment 69

expansion of industries or businesses, urban mobility or economic growth. Without legal backup, therefore, it is questionable whether environmental interests are strong enough to gain the necessary societal, political and administrative support to compete with especially financial and economic interests. Hence, as Eckersley (1992) also concludes, devolving power to local arenas can easily cause problems if local authorities adopt development-oriented paths that degrade the environment (see also Bouwman et al. 2005, De Roo 2004, Jordan 1999, Milbrath 1989 and Oates 2001). In the meantime, though, also on a national level, political preferences can shift, resulting in changing policy ambitions, legal frameworks and regulations. Consequently, it cannot just be assumed that having a centralized governance framework would indeed ensure that 'weak' environmental interests are fully represented or protected.

A second distinct characteristic of environmental issues is that they can easily be manifested as so-called 'social dilemmas'. Social dilemmas concern situations where individual interests do not correspond with common interests (e.g. Dawes 1980, Sager 2002, Van Lange et al. 1992 and Vlek & Steg 2007). As a consequence, selfish behaviour is prone to occur: outcomes that are desirable from a collective perspective (i.e. good for the group as a whole) will not be produced through the accumulation of individual choices based on individual interests (i.e. good for the individual agent). Social dilemmas are often associated with three main metaphorical stories: the prisoner dilemma, the public good dilemma and the commoners dilemma. The prisoner dilemma focuses on a situation with only two agents. The public good and commoners dilemmas are more relevant when addressing environmental and sustainability issues, as they concern multiple agents. The public good dilemma addresses the *production* of non-rival and non-exclusive goods, where the question is who will pay for this production. The commoners dilemma (e.g. Hardin 1968) addresses the *consumption* of non-rival and non-exclusive goods, where the question is who will be responsible for the adverse effects of consumptions (e.g. depletion, pollution). Obviously, both are strongly related. And also, both suffer from a similar problem; i.e. individual interests do not coincide with collective interests.

Social dilemmas are common societal phenomena that are often linked to environmental problems (e.g. Hardin 1968, Vlek & Steg 2007, see also Oates 2001, Wätli 2004). Strong examples are environmental issues that have external effects, implying that environmental pollution spills over into other jurisdictions than the one in which the pollution is created. Therefore they are environmental issues that manifest themselves on more spatial scales than just the local. As Lemos and Agrawal state, such a "multiscalar character of environmental problems – spatially, socio-politically, and temporally – adds significant complexity to their governance" (2006; p. 12). This is well illustrated by the example of air pollution. Air pollution is caused by a wide variety of sources that are spread throughout multiple jurisdictions. The pollution itself is easily distributed by the wind and hence, typically manifesting itself in regional, national or even international patterns. Under normal conditions, all jurisdictions are interested in reducing air pollution. If only one or two single jurisdictions invest in reducing pollution, the

70 *Governing the environment*

effect on the ambient air quality will be modest as these few jurisdictions control a small fraction of all polluting activities. They bear the full costs of the investment, which can be quite serious; e.g. reduce mobility, cleaner technology, reducing factory production, economic disadvantages of stringent environmental policies compared to other municipalities. They will only receive a modest benefit (i.e. slightly improved air quality), which they share with surrounding jurisdictions. The locally confined cost (investment) thus easily exceeds the local benefit; i.e. there is little reason to invest. Alternatively, if most jurisdictions do act, social dilemmas also make it easy for the last few jurisdictions to become 'free riders' (e.g. Pugh 1996). If multiple jurisdictions invest in improving air quality, 'costs' are spread and the overall 'benefits' (i.e. the improvement of air quality) will increase. This can make it more attractive for a single jurisdiction to invest, as its neighbours also share the costs. However, a single jurisdiction can also decide to become a 'free rider' and let other jurisdictions pay the 'costs' while equally enjoying the benefits without bearing these costs. Again, there is a mechanism making it attractive for a single jurisdiction not to invest.

The example of air quality can easily be replaced by, for example, climate change, the depletion of the ozone layer, acid rain or loss of biodiversity. Essentially, it applies to all environmental issues where there are large external effects involved; i.e. the effects of pollution or of environmental improvement 'spillover' the jurisdictional borders (e.g. Oates 2001, Wätli 2004). 'Free rider' behaviour or a general apathy in taking initiative is in fact a real risk associated with many environmental issues. And as Prud'homme (1994) argues, this is even truer if local environmental action results in the loss of local business. Devolution of power and authority towards the local level is therefore a risk when it comes to environmental issues with external effects. Although action on a local level is still possible and desirable, the idea is now that centralized governance formats are in place to put pressure on both avoiding free riders and to prevent a general apathy in taking initiative. As Lemos and Agrawal indicate "a common prescription to address the multilevel character of environmental problems is to design governance mechanisms across levels of social and institutional aggregation" (2006; p. 12). Thus, multiscalar issues should not be addressed at one level of authority or on separate levels, but based on interaction between these levels (Bressers & Kuks 2003; p. 1); i.e. what is called multilevel governance (e.g. Bache & Flinders 2004, Bernard 2002, Jordan 2002, Kooiman 2003 and Marks & Hooghe 2001).

Centralized formats: economies of scale, routine and efficiency

Apart from their possible contribution to correct for the 'weak profile' of the environment and to respond to social dilemmas, centralized governance formats also remain to have some important general benefits. Especially, there are important economies of scale involved in dealing with many environmental issues. Many environmental issues call for high levels of *technical expertise*. Examples include the management of safety risks, calculating air pollution, coping with polluted soil or water, etc. As many of these issues are quite common in most localities, most local authorities deal with these issues. It can be problematic for them to hire experts and specialists on all of these themes. Central agencies can employ experts more easily than local agencies. Especially small agencies, such as small municipal offices, can have problems attracting experts on all different themes (see also Flynn 2000). Therefore, developing and delivering policies and regulations for dealing with these issues is problematic for many localities. In response, it is more efficient to install central policies and regulations. Specialist knowledge can be united on a higher scale and be translated into common procedures and regulations that apply locally.

Economies of scale are also involved in issues that have a rather *common manifestation* throughout countries and localities in a generic way. Earlier I noted that local policies and area-based approaches mostly make sense in dealing with locally embedded issues that defy generic policies. The majority of environmental issues that localities face, however, are rather common. Examples of issues that most localities face are noise and air pollution caused by traffic, sewage, safety risks of hazardous transport, etc. Often, these issues are to such a degree similar that there are common solutions available (e.g. Winsemius 1986). Examples include sound screens or double glass for noise abatement, the isolation and cover up of modest soil pollution and the issuing of common permits and licences based on uniform regulations. As many issues can be managed based on such common formats, it is inefficient to reinvent the wheel in each country or locality. Instead, it is more efficient to issue common procedures and regulations for encouraging a routine implementation by national, regional and local administrators. Hence, there is much reason for centrally issued common formats to deal with local environmental issues.

2.8 Subsidiarity and multilevel governance

In this chapter I not only discussed the benefits associated with increasing the role of the local level in the pursuit of proactive and integrated approaches to the environment. I have also argued that increasing the role of the local level should not go without awareness of the role and benefits of centrally formulated policies and regulations. These help not only by responding to uniform issues, but also to issues with external effects, to the desire to guarantee minimum levels of protection of public health and safety and to the 'weak profile' of the environment (compare Lemos & Agrawal 2006, Liverman 2004, Oates 2001). The challenge, therefore, is to organize environmental policy in such a way that it accommodates area-based approaches, without decreasing its capacity to effectively and efficiently solve uniform issues and to protect human health and safety.

For guiding us on our way in finding hybrid combinations of central coordinative approaches and more dynamic, tailor-made and decentralized approaches we encountered two key concepts in this chapter that might inspire our search The first of these is subsidiarity, suggesting that decisions should be made at the lowest possible level of authority so as to be closest to the actual problem. While doing so, subsidiarity also suggests that not all governance functions should be

72 *Governing the environment*

decentralized; it merely provides a context in which assessments are made upon the possibilities to pursue decentralization. Subsidiarity, thus, shifts our attention to linking government functions and spatial and organizational scales; i.e. which organizational structures can best perform certain government functions.

While subsidiarity raises questions about the link between government functions and structures that go along with it, multilevel governance adds the idea that governance can take place in a context of mutual dependency (see also Benz & Eberlein 1999, Lyall and Tait 2004, Svedin et al. 2001). It does not rely on hierarchic and centralized regulative structures (i.e. governance through coordination), but rather on flexible and empowering types of steering (Lyall and Tait 2004). Often, multilevel governance involves a 'blurring' of boundaries where organizations and levels in organizations overlap with regard to such things as responsibility, power, knowledge or available means such as money or manpower (e.g. Svedin et al. 2001). Multilevel governance, therefore, also relates to cooperation between various levels of authority. To this end, multilevel governance produces governance mechanisms *where the local level is neither fully constrained by central policies and regulations, nor fully autonomous.* It implies that higher-level authorities maintain a degree of control over the behaviour of lower levels of authority. Examples of such control are structures of monitoring, framework legislation or oversight to ensure a degree of local accountability.

Multilevel governance reveals a key message, which is that it is not necessary to choose between a full reliance on centralized governance or decentralized governance. Instead, multilevel governance points at an interaction between levels of authority, where pursuing dynamic policies in a decentralized setting takes place within a frame of central policies and regulations. Then, decentralization is merely a *relative* shift of power and authority towards the local level within a frame of central policies (e.g. Ashford 2002, Flynn 2000, Prud'homme 1994 and Zuidema 2004). Central policies and requirements then play a role in creating the kind of conditions that reduce the risks of decentralization and increase the likelihood of effective and efficient problem solving at a local level. That is, the coordinative model of governance remains to play an important role in setting the conditions required for decentralization resulting in its envisioned outcomes.

Towards guidelines

The key conclusion of this chapter is that governance renewal should not be equated with a generic shift away from a reliance on the coordinative model. Not only does the coordinative model have its own benefits that the concept of subsidiary also helps us remember, it also can set the conditions required for governance renewal resulting in its envisioned outcomes, even if such renewal means to celebrate the merits of alternative approaches to governance based on, for example, the competitive or argumentative model. The coordinative model of governance, therefore, might be challenged for good reasons, but is not fully obsolete or outdated. So when discussing decentralization, it pays to know in more detail how centralized and coordinative policies might condition or influence the outcomes of decentralized policy approaches. Finding guidelines so as to inform choices regarding a possible increase in the role of the local level in environmental governance is, therefore, also my main ambition in this book, not only because of the risks associated with environmental policies, but also given the theoretical doubts already existing upon the benefits of decentralization. While focussing on pursuing proactive and integrated approaches to the environment I thus aim to find arguments for choosing when, to what degree and within which frames of central policies we should pursue decentralization in environmental policy. It is an ambition, I argue, that brings us back into the realm of the plural governance landscape discussed in Chapter 1 and earlier in this chapter.

In section 2.1 I discussed how the emergence of a plural and fuzzy policy field confronts us with a wide variety of possible governance approaches and organizational formats to choose from. While this plurality increases our options, it is not accompanied with much clarity as to how and when to choose between certain approaches or formats. Rather, we are faced with a plurality of theoretical positions that each draw upon different socio-cultural and philosophical values and beliefs. Finding guidelines that can help in choosing when, to what degree and how we should pursue decentralization is, however, an attempt to try to navigate this 'theoretical plurality'. It is an ambitious attempt, which I will start with in the next chapter by arguing that guidelines can be formulated to choose between various governance approaches despite the current theoretical plurality and related different socio-cultural and philosophical values and arguments that exist. Although not meant to result in a definite closure to the 'theoretical plurality' we face, these guidelines can provide us with *arguments* for choosing between various governance approaches and the various theoretical perspectives they rely on. In doing so, I will in the remainder of this book try to formulate and practically assess these arguments with regards to increasing the role of the local level in pursuing proactive and integrated approaches to the environment.

3 Navigating the plural

The increased plurality of governance practices and the associated rise of multiple theories have created a wide range of possible governance approaches to choose from. While a wide range of approaches poses interesting opportunities for coping with different and often changing societal conditions, it also reduces clarity when choosing between these practices and theories. Rather, as we noted in Chapter 2, we seem to be faced with a 'restless search' (Offe 1977), 'trial and error operations' (De Roo 2002) and 'pick and mix' approaches (Allmendinger 2002b) while we try to navigate the 'theoretical plurality' (Healey et al. 1979). The objective of this chapter is not to 'resolve' this theoretical plurality, as I argue that there are viable arguments supporting different theories that suggest that there simply is no 'single best approach'. Instead, the objective of this chapter is to propose a series of arguments that can help navigate this plurality. To this end, I will propose what I call a 'post-contingency approach'.

Before I develop a 'post-contingency approach', section 3.1 will first explain the notion of plurality in more detail. I will explain that the current theoretical plurality is caused by a plurality of philosophical positions. Various theoretical approaches are based on different and sometimes conflicting philosophical positions regarding both how we can arrive at claims about what we know and how we might act rationally based upon this knowledge. This makes it difficult to find a common ground of either knowledge or reasoning that allows us to choose between these philosophical positions and, hence, the theoretical approaches to governance that are based on them. After all, whose perspective should we rely on if we are to develop such a common ground? In section 3.2 I will, however, argue that although plurality might *seem* to leave us without a common ground, a post-contingency approach does enable us to find a common ground for navigating the plural philosophical and governance landscape.

I will introduce my proposal for a post-contingency approach in section 3.3. A post-contingency approach draws upon a long tradition in planning and policy sciences, starting with the contingency theorists of the 1960s and 1970s (e.g. Fiedler 1967, Lawrence & Lörsch 1967 and Woodward 1958). Contingency theory posits that the performance of organizational functions and structures and the performance of governance approaches are influenced by the circumstances encountered; these are, so to say, contextual. Contingency theory has also produced guidelines that indicate which organizational functions and structures are appropriate for different circumstances. I will address what I call 'classical approaches' to contingency theory in section 3.4.

As I will explain in sections 3.5 to 3.7 these approaches typically draw upon a positivist perspective. They are thus built on the assumption that it is possible to acquire objective knowledge about the relationship between various contextual circumstances and the kind of approaches that are appropriate for these circumstances. Hence, classical approaches to contingency are largely incompatible with the post-positivist perspective that I rely on in this chapter. I will therefore explain the key criticism of these approaches (section 3.5) and will subsequently respond to it (sections 3.6 to 3.9).

My first step in responding to this post-positivist criticism is to embrace a more nuanced perspective on how different circumstances can inform a choice between different approaches to governance. I will do this in section 3.6, where I stay close to classical approaches to contingency. There I will focus on finding arguments for choosing between various governance approaches, based on the circumstances encountered, i.e. what I will call an 'object-oriented approach'. I will discuss how differing circumstances, defined in terms of their *degree of complexity*, provide us with an argument for choosing between various governance approaches.

My second step in responding to the post-positivist criticism on classical approaches to contingency is to move beyond classical approaches' singular reliance on an object-oriented approach. Instead, as I will explain in section 3.7, I will bring contingency theory into an intersubjective realm. On the one hand, an intersubjective-oriented focus comes forward in accepting that people often adhere to different preferences and objectives. Therefore, different decision makers might choose different responses to the circumstances they encounter, as they simply value and judge these circumstances differently. In other words, although the degree of complexity might provide an argument for choosing between various governance approaches, the ways in which people respond to this argument remains *a matter of choice*. On the other hand, an intersubjective-oriented focus also comes forward in accepting that issues and the circumstances surrounding them cannot be objectively known. Rather, defining these issues and circumstances will also be relative to the different frames of reference from which people perceive, interpret and judge what they experience. Relying on an object-oriented focus might help us understand the issues and circumstances we are faced with, defined here in terms of their degree of complexity; a sole reliance on an object-oriented focus will not, however, produce definite and certain knowledge. Knowledge might remain highly uncertain and disputed. This also shows that relying on an object-oriented approach has clear limits in providing us with arguments for choosing between various governance approaches. In response, in sections 3.8 and 3.9 I will discuss whether we can still find such arguments without relying on an object-oriented approach. I will explain that this is possible, based on connecting choices that decision makers make regarding their aims (functional focus) to how they intend to organize their approach (structural focus). I will conclude that the contingency between function and structure also allows us to formulate arguments for choosing between various governance approaches.

76 *Navigating the plural*

Finally, in section 3.10 I set the stage for Chapter 4, where I will translate a post-contingency approach into arguments that help to decide between central and local decision making in the field of environmental policy.

3.1 Plurality uncovered

The plurality of our governance landscape is not confined to theoretical debates on governance. It has its roots in philosophical differences and the increased fragmentation of social beliefs and preferences that characterizes the late 20th and early 21st centuries. Different and competing ideas on governance are founded on these social and philosophical differences. Plurality, in other words, is not confined to the governance landscape. Therefore, our aim is to better understand this plurality, before I continue to explain how we might cope with it.

Understanding plurality

To begin with, consulting the Oxford English Dictionary can help us on our way. It first shows that plurality has to do with diversity and difference. Second, it attributes four separate meanings to plurality. Three of these are important to us here, as they help us see how plurality can be understood within the realm of planning and policy-making.

The first of these three meanings is also the most fundamental as it takes a *philosophical* point of view. In this case, plurality refers to 'the theory that the world is made up of more than one kind of substance or thing; (more in general) any theory or system of thought which recognizes more than one irreducible basic principle. Also: the theory that the knowable world consists of a plurality of interacting entities. Opposed to *monism*' (Oxford English Dictionary 2009). Hence, plurality expresses a condition in which multiple perspectives exist of what is or is understood as 'true' and 'real'. Clearly, for such plurality to exist, a single and objectifiable truth is no longer the baseline assumption. Instead, it is based on the assumption that various people construct different ideas of what is real and rational.

The second meaning attributed to plurality in the Oxford English Dictionary is a sociological one. In this context, plurality refers to 'the presence or tolerance of a diversity of ethnic or cultural groups within a society or state; (the advocacy of) toleration or acceptance of the coexistence of differing views, values, cultures, etc.'. In other words, it refers to both a condition of social fragmentation and diversity *and* the acceptance of how this manifests itself in different perspectives held. Arguably, this social plurality relates to philosophical plurality; i.e. it manifests itself in the ideologies, values and beliefs people refer to when interpreting and constructing what they consider to be real and rational.

Third and finally, the Oxford English Dictionary points towards a political perspective on plurality, in which plurality refers to 'a theory or system of devolution and autonomy for organizations and individuals in preference to monolithic state power' and also to 'a political system within which many parties or organizations have access to power'. Hence, plurality refers to a spread of power among either various levels of organization or various groups. This is the same kind of dispersal I discussed in Chapter 2 to describe the increased plurality of our governance landscape. In practice, it is this kind of plurality that thus manifests itself in both the dispersal of power and governance among many groups and levels in society *and* in different competing theories and practices used by these groups or levels to address societal issues, i.e. a plural governance landscape. Importantly, this dispersal of power and mosaic of theories and practices relates to the variations in philosophical thought and the sociological differences regarding what and how to govern.

Hardly surprisingly, the three meanings attributed to plurality are related when looking at governance theory and practice. The plural governance landscape resonates with both social and philosophical plurality. First, societal changes such as increased social fragmentation and power dispersal contribute to social *and* political plurality. Social fragmentation manifests itself in an increased diversity of groups that adhere to different values and beliefs. Power dispersal also gives many of these groups the means to actively propagate and defend these values and beliefs. Second, even among those who address the issue of governance from a mere scientific or theoretical perspective, there are important differences in the values and beliefs they hold. Also inspired by recent societal changes, there have been many philosophical debates in recent decades (e.g. Harvey 1989). Often highlighted with the concept of 'postmodernism', the latter part of the 20th century has seen an increased diversity of the values and beliefs people refer to when they interpret and define concepts such as rationality and reality. This philosophical plurality also manifests itself in the different ideas and conceptions regarding how to approach planning and, in general, 'governance' (see also Allmendinger 2002b, Beauregard 1989, Healey 2008, Hillier 2008 and Hoch 1996). As such, the plural governance landscape is rooted in both societal and philosophical differences and involves a move away from what can be called a 'modernist' or 'positivist' tradition in governance theories (i.e. associated with the coordinative model of governance).

A post-positivistic landscape

Similar to other social sciences, planning and policy sciences are rooted in a tradition of modernist thinking or, more correctly, in a 'positivist modernist' tradition (Mumby 1997; compare Allmendinger 2001, De Roo & Voogd 2004 and Harvey 1989). Although modernism can be considered to encompass considerably more than this, it is often associated with strong confidence in technological development, science and the possibility of discovering objective knowledge about the true nature of the world. Underlying this confidence, from a philosophical perspective, is realist ontology and a foundational epistemology. Realist ontology assumes that our experiences and observations are representations of a reality that is 'out there' and exists independent of human experience. A statement is then considered true if it corresponds to the reality 'out there'. Possible differences

78   *Navigating the plural*

between people's beliefs of what is true and false are thus considered to be caused by distorted information and interpretations of this reality. The answer is to remove these distortions by representing this reality as accurately as possible through empirical inquiry or, in more detail, the scientific method. Rationality is then seen as a means for predicting and explaining phenomena, i.e. to develop testable hypotheses and as a procedure for establishing whether knowledge is true or false, based on empirical data. This is where the foundational attitude comes to the fore, in assuming that there is a universal rationality that is not affected by irrationalities such as emotions, opinions, values, etc., i.e. what is expressed in the social sciences as instrumental or technical rationality.

In 20th-century social sciences, modernist influences are especially strong in the positivist tradition and, mostly, the influence of logical positivism. Logical positivism combines the search for objective truths and an instrumental rational approach (e.g. De Pater & Van der Wusten 1996 and Mumby 1997). Positivism, more generally, is based on the idea that knowledge can only be considered true if it is based on what is perceived by the senses and on positive verification through empirical inquiry. *Logic* positivism adds to this the dominance of instrumental or technical rational reasoning. 'Positivist modernism' (Mumby 1997) has fuelled confidence in the coordinative model of governance, which is most vividly represented in the theories and practices in planning of the 1950s and 1960s (e.g. Meyerson & Banfield 1955). The latter decades of the 20th century have, however, seen changes in the philosophical landscape, manifesting themselves in multiple and sometimes incommensurable philosophical positions. Many of these distance themselves from modernism, while others can best be considered as attempts at improving or reconstructing modernism. These positions and attempts share a rejection of the radical or naïve realism associated with logical positivism, where we believe that we can objectively describe and explain a reality that is 'out there'. To Allmendinger (2002b), these attempts can be summarized by what he calls a 'post-positivistic typology' (compare Mumby 1997).

As Allmendinger (2002b; p. 83) expresses, at the end of the 20th century we were faced with "the breakdown of transcendental meaning (Lyotard), the discursively created subject (Foucault), the role of cultural influences in ordering society (Baudrillard) and a new appreciation of the pernicious role of power as a form of societal control (Foucault)". A variety of philosophical schools and positions has emerged in response to this. Allmendinger sees post-positivism as a signifier for this development, which has manifested itself, for example, in the rising popularity of philosophical and theoretical schools such as constructivism, poststructuralism and neo-pragmatism. Post-positivism is now not seen as a school of thought in its own right, but as an umbrella concept (see also Biersteker 1989, Patomäki & Wight 2000). Post-positivist perspectives unite in their sceptical attitude towards realist ontology and a foundational epistemology. While this sceptical attitude is arguably also present in earlier 20th-century positivist thinking (see Hunt 1991), post-positivism signifies an even more sceptical attitude. Consequently, as Allmendinger argues, "we have a post-positivist recognition of indeterminacy, incommensurability, variance, diversity, complexity and intentionality

Navigating the plural 79

in some routes of theoretical development – traits that question the very notion of 'planning'" (2002b; p. 88).

Central to what Allmendinger signifies with post-positivism is the notion that 'reality' and 'rationality' as people understand and know these are human constructions. That is, what is 'real' and 'rational' is established by humans through interpretation and by sharing meanings through language. Therefore, the role of language as a medium for structuring human experiences, perception and beliefs is common to most post-positivist ideas (see also Baert 1998, Mumby 1997). A key consequence of accepting the constructed nature of 'reality' and 'rationality' is their situatedness within different social and cultural settings. Instead of assuming that objective truths and (universal) instrumental rationality are possible, that which is considered 'real' and 'rational' is constructed by language and hence influenced by cultural beliefs and values. It is a view resonating with the writings of Friedrich Nietzsche and his approach of 'perspectivism' (e.g. Westin 2014). Perspectivism, as Small explains "means that the world is always understood within the perspective of some point of view; all knowledge is thus an interpretation of reality in accordance with the set of assumptions that makes one perspective different from another" (1983; p. 99). A perspectivist epistemology, or more generally accepting the situated nature of how reality and rationality are understood and interpreted, implies that any form of knowledge is established through processes of interpretation and is influenced by the interpreter's perspective, while similarly, claims to what is considered rational are also influenced by the distinct perspective taken.

Although debates are ongoing on the ontological consequences of accepting the constructed nature of what is 'real' and 'rational', this acceptance is in itself widespread among the dominant philosophical positions within social sciences. Examples of common philosophical positions that unite in this acceptance are constructivism, (post)structuralism and neo-pragmatism. Each recognizes the structuring role of language and culture in the development and use of constructions, while sharing the post-positivist scepticism towards realist ontology and a foundational epistemology. This is most explicit within (post)structuralism, resulting in the (post-)structuralist claim that reality and rationality should be understood as being constructed by subjects that are embedded in larger social (i.e. linguistic and cultural) structures. Then, based on their social contexts, historically manifested in what Bevir (2004) calls 'traditions', individual subjects construct their own beliefs based on their experiences, interpretations of these and, hence, considerations of what is 'real' and 'rational'. A point of debate in post-positivist thinking is the degree to which social structures, such as dominant discourses, influence or determine individual or group beliefs. Despite wide differences, post-positivist perspectives unite in the agreement that social structures at least influence beliefs of what is real and rational; i.e. the world is understood within the perspective of some point of view (e.g. Bevir 2004, Hillier 2005).

Using post-positivism as an umbrella concept is a simplification that should not deny the diversity of philosophical schools and positions that Allmendinger aims to signify with it. Rather, in the wake of its scepticism, post-positivism has

80 *Navigating the plural*

opened up the debate on planning, and most social sciences in general, to the plurality of beliefs that I discussed earlier. As social and cultural contexts differ, so will people's beliefs, i.e. reality and rationality are 'situated' and thus debatable concepts. This helps explain the plurality of theoretical positions and approaches to governance. Thus, as Allmendinger (2002b; p. 84) notes: "there is now a much more eclectic 'pick and mix' basis to theory development and planning practice that is better seen as relating to issues, time and space in a linear and non-linear manner". For Allmendinger, the process of understanding the rise of these different governance theories and practices should indeed begin with tracing them back to the social, cultural, spatial and temporal contexts in which they have arisen (i.e. 'situatedness'). After all, he explains, it is especially within this specific context that these theories and practices 'make sense'; political, sociological and philosophical plurality are not disconnected but, rather, resonate with each other.

The problem of relativism

While post-positivist perspectives and the associated plurality help us see connections between governance approaches and their different philosophical and social roots, they also involve a challenge in terms of coping with a plural governance landscape. After all, post-positivist perspectives hint towards the structuring effect of the (social) context with regards to claims to reality and rationality. As we are all situated within our own context, we evaluate claims to reality and rationality from *within* this context. What is considered real within one context (i.e. for some), might be incommensurable with what is considered real within another context (i.e. for others). Thus, what is considered 'real' or 'rational' is relative to the context in which claims to truth and rationality are made. Its scepticism towards a foundational epistemology, furthermore, contributes to post-positivism also not embracing the idea that there are any absolute or objective criteria to decide upon which of these claims is 'better' or 'true'. Consequently, as Allmendinger, for example, also notes, "post-positivism hints at a form of relativism", which he sees as "a tolerance and acceptance of different values and opinions to the point of being unwilling to judge others" (2002b; p. 84, see also Biersteker 1989). Relativism, in a radical sense, prevents us from developing criteria or arguments for choosing between various beliefs and, hence, various governance approaches. If various governance practices draw upon very different ideas about what is 'real' and 'rational' (i.e. the underlying philosophical plurality), then where is the common ground that serves as a starting point for developing arguments for choosing between them? Although I share the argument that post-positivism does hint towards relativism, I also argue that such a radical perspective is unnecessary and that post-positivisms hint towards relativism is no appeal to an 'anything goes' perspective to governance.

3.2 A contingency perspective on knowledge

I support the post-positivist acceptance that full objectivity is impossible and that any claim to what is real and rational is, at least to a degree, influenced by the social, cultural, spatial and temporal contexts in which the claim is made. Consequently, I operate from what can be considered as a perspectivist epistemology. As I noted earlier, perspectivism implies an acceptance of the situated nature of how reality and rationality are understood and interpreted (e.g. Deleuze & Gauttari 1994, Hillier 2007, Saint-Andre 2002 and Westin 2014). Perspectivism, therefore, shares with relativism an acceptance that what is considered real or rational is relative to who observes, interprets and understands. It does, however, not equate with a radical form of relativism and its associated 'anything goes' perspective to governance. So while I reject radical realism, I also reject radical relativism. Instead, I suggest that both radical realism and relativism should be considered as ideal type positions that are hardly supported in scientific debates (see also Cromby & Nightingale 1999, Edwards, Ashmore & Potter 1995).

Instead of considering ideal types, I suggest that the differences between positions based on realism and relativism should be appreciated as being much more nuanced. Post-positivism might be sceptical to a realist ontology and foundational epistemology, but it's also weary of a radical form of relativism and the associated 'anything goes' perspective on constructing knowledge and theories. Instead, post-positivism can be associated with the claim that

> The belief that scientific knowledge does not merely replicate nature *in no way* commits the epistemic relativist to the view that therefore all forms of knowledge will be equally successful in solving a practical problem, equally adequate in explaining a puzzling phenomenon or, in general, equally acceptable to all participants. Nor does it follow that we cannot discriminate between different forms of knowledge with a view to their relevance or adequacy in regard to a specific goal.
>
> (Knorr-Cetina & Mulkay 1983; p. 6, italics in original)

Post-positivist perspectives, however, do differ in the kind of responses they give with regard to avoiding radical relativism. While some find inspiration closer to a realist position, others stay closer to a relativist perspective. Therefore, although it is something of a simplification, we can witness a duality within post-positivist strategies between those who stay close to a *realist* perspective and those who stay close to a *relativist* perspective. Next to avoiding an association of these perspectives with radical realism or relativism, I also aim to argue that alternative perspectives that draw on this duality might well be appreciated in a complementary fashion, rather than being exclusive in this section.

My position follows a perspectivist epistemology by first accepting that alternative perspectives exist upon what can be considered real and rational. It second shares with a perspectivist epistemology that these perspectives cannot be judged as being 'true' or 'false' based on absolute and objective criteria. Third, however, I suggest that perspectivism also offers hints towards seeing alternative perspectives in a complementary fashion. Among the arguments following from perspectivism is that we should accept that all alternative perspectives might have something to tell us and can contribute to our understanding of reality and rationality; i.e. all have some potential value. While thus embracing the possibility of

82 *Navigating the plural*

seeing alternative positions in a complementary fashion, perspectivism finally continues by also suggesting not all interpretations and perspectives should be seen as being equal. Rather, as Westin (2014; p. 25) explains, "perspectivism does not imply that all interpretations are equally just – a criticism often levelled against (the in itself diverse and often undefined idea of) relativism" (see also Holmqvist 2009). Instead, she argues, "perspectivism does not imply that the one who claims a multitude of perspectives cannot or should not choose among them" (2014; p. 24, see also Vattimo 2002). Rather, perspectivism suggests that there is a possibility to combine perspectives, differentiate between them and come to judgements. But while thus hinting that judgements can be made between various claims based on such combining and differentiating, this hint remains fairly general. Exactly how such judgements are made and supported, therefore, remains fairly open, both in Nietzsche's original works and subsequent debates. Consequently, perspectivism offers hints, but little direct clarity upon how to develop criteria or arguments for choosing between various beliefs and, hence, also between various governance approaches. It is in drawing from three dominant post-positivist perspectives that I aim to take a next step in finding such criteria and arguments, while highlighting that such criteria and arguments should not be interpreted as absolute or objective.

In this section I will discuss these three dominant post-positivist perspectives as, despite their different origins, they all help express the complementarity between various post-positivist positions. These perspectives are (1) the theories on communicative rationality that originate from the works of Jürgen Habermas, (2) the pragmatic tradition in philosophy and planning theory and (3) a critical-realist perspective. Combined, they help me to establish a 'meta-perspective' on the philosophical plurality that helps us make argued choices between different claims to knowledge and rationality. This is not meant as an attempt at structuring the philosophical landscape but, rather, as an attempt at finding plausible arguments that help us cope with the philosophical plurality we face and, hence, with the plurality in our governance landscape.

Habermas and communicative rationality

One way to avoid the 'relativist trap' of succumbing to an 'anything goes' epistemology can be found in the work of Habermas (e.g. 1981). Habermas' main concern is to rescue the project of Enlightenment from full domination of what he calls instrumental rationality. Instrumental – or technical – rationality is concerned with selecting the most effective and efficient means for reaching a predefined end. Technical/instrumental rationality assumes that there are 'objective standards that are equally applicable to all individuals' (Allmendinger 2001; p. 186, see also Dryzek 1990). Habermas does not deny that instrumental rationality has its merits, so long as people are indeed fairly certain about which standards we should adhere to (i.e. we agree on them). Nevertheless, he also indicates that there are many societal domains and issues in which relying on such objective standards is not a realistic option, as the various groups and people who are involved subscribe

Navigating the plural 83

to different perspectives on what is real and rational. While rejecting the viability of objective standards under such conditions, Habermas does argue that standards can be established on the basis of communicative action. For Habermas, communicative action is a means for people to argue for the validity or invalidity of certain claims. Hence, although people might start with different claims as to what is real or rational, Habermas states that they are able to discuss them, evaluate them or create new constructs to replace the old. During communicative action, what is real and rational can still be established on the basis of the rules and concepts on which people agree.

On the one hand, Habermas considers knowledge to be discursively created and, hence, to be situated within certain societal contexts (discourses). This implies that knowledge is relative to the social and cultural context in which it is produced. On the other hand, Habermas opens the possibility of overcoming discursive differences, as communicative action allows people to (re)construct their ideas. Consequently, Habermas points to the theoretic possibility that ongoing discursive practices produce statements that are universally accepted, i.e. relativity can be counteracted. Habermas acknowledges that this theoretic possibility presents practical challenges. For example, he recognizes the potential suppressive effect of dominant discourses that legitimize certain claims regarding what is real and rational. It is a recognition also shared by (post)structuralism that the 'structure' or 'dominant discourse' of a social context can delimit or dominate 'agency'. Despite this recognition, however, it is exactly this structuring role of power and dominant discourses that is a key challenge to Habermas' theory of communicative action.

Taken to its extremes, structuralism supposes that deep underlying structures determine individual perceptions, interpretations and beliefs. It is a tendency that can be drawn from, for example, the works of Freud and Marx, which are among the predecessors of structuralism. Such an extreme and more deterministic position is not necessary, nor usually propagated by structuralists. As Bevir (2004) makes clear,

> We can accept that people always are situated against the background of a social tradition, and still conceive of them as agents who can act in novel ways and for reasons of their own so as to transform both themselves and this background.
>
> (p. 612, see also Brown 1994)

Supporting such 'situated agency' is thus

> to say only that their [a person's] intentionality is the source of their conduct – they are capable of using and modifying language, discourse or traditions for reasons of their own; it is not to say that their intentionality is not itself influenced by their social context.
>
> (Bevir 2004; p. 612)

84 *Navigating the plural*

Consequently, rationality and reality can be regarded as depending not only on the intersubjective structures in which people are situated, but also on personalized and subjective experiences, interpretations and opinions (e.g. Fainstein 2000; p. 457).

The consequence of allowing for situated agency is, following a *post*structuralist approach, that through continuous linguistic exchange and social interaction, descriptions of rationality and reality and even the concept of one's self are all continuously produced and in flux. Thus, as Burr (2003) explains, for poststructuralists "the meanings carried by language are never fixed, always open to question, always contestable, always temporary" (p. 15, see also Brown 1994). Poststructuralists do not believe in the possibility of developing universal conceptions of reality and rationality, even if discursively created. Rather, poststructuralists "declare war on all categories of transcendence, certainty and foundationalism" (Giroux 1988; p. 60). Clearly, poststructuralists are less optimistic about communicative action than Habermas. They see "language as a site of struggle, conflict and potential personal and social change" (Burr 2003; p. 30). Communicative action, poststructuralists claim, can indeed produce situated knowledge. Poststructuralists refuse, however, to assume that anything is necessarily more rational or real than something else (compare Billig & Simons 1994). Although Dixon and Jones (1998) show that post-structuralism still allows for defendable choices between claims to rationality and reality, it does hold a tendency towards a fairly strong from of relativism, one of the aspects for which it is most criticized (e.g. Bauman 1992).

For Habermas, situated agency has a greater potential. He sees situated agency as allowing us to reflect on each other's concepts and constructions through discursive action. In doing so, language is not only a site of struggle, but also a means for establishing intersubjective truths. Habermas argues for establishing the kind of conditions that can facilitate a free development of communication and discourses without the distorting effect of existing inequalities and power relations. He aims to open up discourses for reflection and, hence, change. Under the right conditions – which Habermas refers to as an 'ideal speech situation' – people are considered able to liaise and debate freely and unconstrained by, for example, inequalities, different communicative skills or serious ideological differences. It is here that Habermas often encounters criticism, as it is often held to be highly utopian that ideal speech conditions can exist. The Habermasian solution is nevertheless important, since it hints at people's capacity for reaching agreement on what we claim to be real and rational, even if it is only partial. It shows that reflection exercised through communicative action can prevent a reliance on an 'anything goes' relativity.

Despite the value of Habermas' ideas, I argue here that there is more we can do to avoid an 'anything goes' relativity. To this end, I highlight the practical consequences of relying on various claims as to what is real and rational or of relying on diverse theoretical positions, which we can at least partly experience. I argue that, when distinguishing between claims to what is real and rational, we do so not only on the basis of various socio-cultural and philosophical positions, traditions and

Navigating the plural   85

beliefs, but also on the basis of the practical consequences we have experienced or anticipate while relying on these claims. In other words, we are able to focus not only on intersubjective-oriented action when deciding between claims as to what is real and rational, but also on object-oriented action. To see how and why both positions matter, I begin with the pragmatic tradition in philosophy.

Pragmatism and a priori statements

Pragmatism builds on the works of philosophers and scholars such as John Dewey, William James, Charles Pierce and, more recently, Richard Rorty. Central to pragmatism is the idea that the value of a statement or belief should be based on the practical consequences of accepting the belief or statement. As Hoch notes, this means that "pragmatists no longer believe in discovering the world as it truly is" (2006; p. 393). Rather, pragmatism aims to "offer beliefs to help improve how we respond to the troublesome situations in the world we inhabit" (Hoch 2006; p. 393). Pragmatism thus focusses on developing knowledge and arguments for guiding actions in relation to the observed consequences of those actions.

Pragmatists were among the first to highlight the constructed nature of knowledge, i.e. early pragmatists already "advocated a focus on the way meanings and conceptions of truth and belief were created in the social contexts of human existence, 'socially constructed' we would say today" (Healey 2008; p. 278, compare Wolfe 1994). Pragmatism thus assumes a relativist epistemology, which implies that it considers our claims to truth and rationality to be (language) constructs. Much in the same way as (post)structural theories and the ideas of Habermas suggest, "pragmatists believe that we make knowledge rather than discover it" (Hoch 2006; p. 392). Consequently, pragmatism acknowledges that multiple and possibly even incommensurable claims as to what is real and rational can exist (e.g. Bernstein 1983, Healey 2008).

Similar to other post-positivist positions, pragmatists relate the emergence of a plurality of claims to the "social contexts in which we live our lives and develop our understandings, make meanings, and develop values" (Healey 2008; p. 278). Pragmatism thus points towards "making and redeeming claims within specific social contexts" (Healey 2008; p. 283, see also Bernstein 1983). Since social contexts differ, pragmatists argue, there will also be "a diversity of ways of thinking and knowing" (Healey 2008; p. 279). Accepting 'a plurality of often conflicting arguments and claims' does not imply that pragmatism assumes that claims cannot be judged and distinguished from each other. Instead, pragmatism calls for us to ask "if all forms of truth are ultimately fictions or myths, what might be more adequate myths for our polity and how might text work contribute to their creation?" (Brown 1994; p. 17). And in answering the question, the key criterion is not which of these 'myths' is actually 'right' or 'true', but rather, which 'makes sense' or 'works'. As Verma notes "what is important to the pragmatic idea of meaning is the way purpose and consequence are linked in a purposive process, rather than through some a priori hypothesis about 'objective' causal properties" (1998; p. 73, see also Healey 2008).

86 *Navigating the plural*

To develop knowledge that improves our responses to situations, Rorty states that

> people live in continual 'conversations' with each other about what we agree and disagree about. In doing so, we situate specific issues and experiences within wider frames of understanding, probing these, in turn, in a continual moving back and forth between attempting to grasp 'wholes' and considering 'parts' until a 'best guess' is arrived at that the 'community of inquiry' in question comes to feel at ease with.
>
> (Healey 2008; p. 283, referring to Rorty 1980; pp. 317–319)

Rorty thus highlights the language games in which we can distinguish between the various ideas or beliefs we hold, or where we can (re)construct them into new ideas or beliefs (see also Healey 2008, Hoch 2006). This is close to the ideas of Habermas, as it supposes we can use critical reflection in intersubjective communication to verify which claims make more or less sense to us. Others (e.g. Coaffee & Headlam 2007, Kloppenberg 1998) follow classical pragmatic writings such as the works of Dewey and Pierce, and focus on the experienced consequences of our actions in addition to language games, bringing us to an object-oriented focus.

As Wolfe explains, "you may believe whatever you like, but that belief itself – and here is the pragmatist imperative – will have consequences because it is subject to 'pressures from the outside'" (1994; p. 104). Although we might believe that smog has no bearing on human health, an incident such as the Great London Smog (1952) and the sickness and deaths occurring during this incident provides a strong argument for altering that belief. Therefore, pragmatism is not just interested in discursive action but also explicitly looks at the practical consequences and our experience of them. As a consequence, pragmatism looks at "establishing truth . . . through practical application to establish what works out most effectively" (Heywood 2002; p. 88). To summarize, pragmatism tries to develop and establish knowledge both on the basis of whether it makes sense to us in debate *and* on the basis of whether the experienced consequences of relying on our established knowledge are consistent with our expectations. Or as Hoch summarizes in relation to planning: "adopting a pragmatic orientation shifts debate . . . to a focus on empirical and interpretive claims about the effect of particular urban changes and planning activity" (2007; p. 280).

Pragmatism accepts that various beliefs and theories can apply at the same time. In addition, it accepts that each of these can be a means for acting in the world. Therefore, as Healey notes, "in public policy terms, the pragmatists argue, the critical question to address is, what difference does it make if we frame our puzzles and problems this way or that?" (2008; p. 287). For Harrison (2002), the solution that pragmatism proposes is to shift attention to

> understanding planning as an arena of social endeavour in which different forms of reasoning and action combine and interact. The appropriate mix is

Navigating the plural 87

contingent on the circumstances and would require the type of practical and experientially based judgement that has been referred to by Dewey.

(p. 165)

Or, different theories and beliefs can help us act in the world. Therefore, according to the pragmatic imperative, we should address theory by looking at what 'works' or 'makes sense' while "paying critical attention to the situated particularities of practices" (Healey 2008; p. 287).

In coping with the application of planning approaches in a specific social and situational context, Allmendinger (2002a) recognizes pragmatism as a tradition in planning thought that "emphasises direct action at specific problems – what works best in a given situation or circumstance" (2002a; p. 114). He thus sees a direct link between pragmatism as an approach in planning and dealing with 'specificity', which requires some kind of practical 'tailor-made' approach. Pragmatism, however, is more than a claim to constantly adjust approaches based on the circumstances faced. As Dewey puts it, a key concern of pragmatism is with the "particular consequences of ideas for future practical experience" (1931; p. 27, see also Harrison 2002). As a result of the flow of experience and learning in planning practice, many knowledge claims have been largely accepted for their usefulness in practice. Or, as Healey notes, pragmatism posits that "through our experiences, both intellectual and sensory, we come to affirm or doubt what we believe to be true" (2008; p. 287). Through theories, inquiries and practical experiences, pragmatists would argue, we have affirmed knowledge claims about the relations between, for example, smog and human health. These claims prove 'useful' in predicting health effects and, hence, we refer to them to guide our planning and policy actions.

When it comes to various approaches to governance, pragmatism argues that practice has also provided us with insights into the likely 'effect of particular urban changes and planning activity' under different circumstances (see also Kloppenberg 1998). This expresses the idea that previous learning and experience have produced knowledge of the contingency between various known planning situations and relying on various theories and planning approaches by recording the likely consequences of matching approaches with certain circumstances. Our a priori statements thus comprise suggestions regarding what to expect (consequences) when applying a certain theory or approach in certain (typical) planning situations.

By rendering it possible to develop a priori statements that guide future action, pragmatism provides an answer to the 'relativist trap'. Pragmatisms helps us to see that this not only relies on intersubjective action to 'make sense' of what we believe but also that, to support our claims, we can draw arguments from practical experiences that something 'works' (i.e. object oriented). While I accept this pragmatic notion, I also note that relying only on the experience that something 'works' or 'makes sense' to ground our a priori statements leaves open the question of where these experiences originate from. I argue that the pragmatic imperative that we should connect the development of knowledge ('best guesses') to the

88 *Navigating the plural*

consequences of our beliefs and actions can be taken a step further so as to give answers to this question. To this end, I will draw upon a critical-realist perspective and dig a little deeper in how object-oriented action can help us find arguments that support claims as to what is 'real' and 'rational'.

Critical realism: towards knowledge

As Allmendinger (2002a) concludes, pragmatists suggest that "we decide what to believe not because it corresponds to the reality of the world, but because an idea or belief makes sense to us and helps us act" (2002a; p. 116). This does not explain *why* a belief might make sense and help us act. I argue, however, that pragmatism holds the seeds of a more thorough understanding of how we come to knowledge and our claims regarding what is true, real and rational. This is already implicit in the statement made by Rorty that "we cannot, no matter how hard we try, continue to hold a belief which we have tried, and conspicuously failed, to weave together with our other beliefs into a justificatory web" (1999; p. 37). In other words, 'pressures from the outside' (Wolfe 1994) confine and delimit what we can plausibly believe. Therefore: "thinking involves relating our choices and our actions to their consequences, which requires reflecting not merely on our words but on the experienced effects of our practical activity" (Kloppenberg 1998; p. 97). I suggest that these 'experienced effects' are not isolated from the social and physical world in which we act. In other words, I still assume a reality that exists independently of our ideas about it, which influences our ideas through the experiences we have of acting in this reality, i.e. what is called a critical-realist perspective (e.g. Archer et al. 1998, Groff 2004 and Sayer 2000).

Although it is something of a simplification, critical realists adopt a realist ontology in the sense that they assume a reality that exists independently of our ideas about it. Critical realism thus assumes that there is a real world with its own mechanisms and structures, be it of a physical nature (e.g. toxic substances, small particles, noise) or of a social nature (e.g. bureaucracies, jurisdictions, discourses). Through our experiences we also gain knowledge about this real material world, resulting in descriptions of the structures and mechanisms and the various interrelated factors and actors of this real material world. That is, while experiencing the Great London Smog, we gain awareness of the breathtaking effect of smog, its effects on public health and the way in which politics and media frame the event. Critical realism is concerned with arriving at descriptions of this material world (i.e. object-oriented action), but accepts that our descriptions are not the same as the material world that they aim to describe, i.e. "the world should not be conflated with our experience of it" (Sayer 2000; p. 3). Therefore, object-oriented action is about how we confront our claims (i.e. descriptions) as to what is real, with our experiences and perceptions of a supposed real material world, but not about developing objective knowledge. Rather, we are faced with the limitations of what we observe, i.e. we only perceive a part of the structures and mechanisms that are at work, while our perceptions are also influenced by the social, cultural, spatial and temporal contexts of the perceiving subject (i.e. a relativist epistemology). The consequence is that our descriptions of the material world are not just approximations that can be fallible, but are also relative to the describing subject and the intersubjective context influencing how a subject comes to his or her descriptions. Hence, critical realists suggest that accepting a relativist epistemology need not result in rejecting that knowledge of the material world can be developed based on object-oriented action: "critical realists argue . . . that notwithstanding the daunting complexity of the [material] world and the fallible and situated character of knowledge, it is possible to develop reliable knowledge and for there to be progress in understanding" (Sayer 2000; p. 30).

From the perspective of the critical realists, we develop and use our knowledge within a context of mechanisms and structures of the material world that constrain what we can plausibly assume to be real. What we know about the material world is not isolated from its mechanisms and structures and our experience of them. Our constructions can turn out to be inconsistent with our experiences and, hence, will be altered to become more consistent with them. For critical realists, this means that knowledge can be altered based on what Sayer (2000; p. 2.) states the "experiences of getting things wrong, of having our expectations confounded, and of crashing into things". In other words, the confrontation of our constructions with experiences of an apparent physical or social reality will alter our constructions. It is here that we can find the origins of the 'pressures from the outside' to which Wolfe (1994) referred.

Despite the importance of experience and, hence, empirical observation, critical realists differ from the empiricism so typical of positivism, where what we experience is considered the same as what is real. For critical realists, the challenge for science is to describe and understand the structures and mechanisms of the world. As experience can only show an image of reality, many structures and mechanisms that could have contributed to the occurrence of an experienced event have not been observed (experienced) or understood. However, they can – at least in part – be logically derived from the occurrence of experiences. Therefore, critical realists are interested not only in practical experience but also in human reasoning to understand the structures and mechanisms, their interactions and their impact on the occurrence of the events and realities we experience. By way of illustration, although we cannot see the gravitational pull of the Sun on the Earth, we can develop a theory of gravitational pull that is consistent with the observation that the Earth does not move away (quickly) from the Sun. In other words, science is not just about experiencing the empirical and accumulating experiences, but also about logically explaining how the empirical could have occurred, i.e. describing the underlying structures and mechanisms.

Also in contrast with a positivist tradition, critical realists are more nuanced in their reliance on object-oriented action for resolving different claims as to what is real and rational. Within the positivist tradition, empirical observations are meant to resolve possible different descriptions of the material world. In other words, object-oriented action is assumed to have the potential to reach definite closure in comparing these differences. Critical realists state that we will always be constrained in doing so, due to the limitations of what we can observe, while the

90 *Navigating the plural*

interpretations of our observations are also mediated by the social, cultural, spatial and temporal context of those observing and interpreting. Hence, our descriptions of the material world are not just accepted on the basis of correspondence with our experiences. They are also about correspondence with our interpretations and beliefs. Hence, accepting such descriptions involves interpretations and discussing these in an intersubjective domain where the social, cultural, spatial and temporal context mediates which descriptions are more widely accepted as meaningful.

Bringing it together

The post-positivist perspectives discussed here all recognize the socially defined nature of knowledge. Hence, they share a relativistic epistemology where our knowledge is at least partly relative to the intersubjective context in which it is constructed and used. This not only explains why we face a plurality of claims regarding what is real and rational, but also explains that we cannot expect definite closure upon these claims. A relativist epistemology thus implies that objective and universal knowledge and standards are impossible. Nevertheless, neither of the perspectives we just discussed succumbs to radical relativism and the associated 'anything goes' perspective on constructing knowledge and theories. Instead, they show pathways that suggest that the difference between realism and relativism might better be interpreted as more nuanced (see also Edwards, Ashmore & Potter 1995). Most notably, it seems the differences reside in especially the approach towards trying to distinguish between alternative descriptions between what is considered real and rational. That is, the degree in which they rely on an object-oriented or an intersubjective-oriented approach.

While a relativist epistemology implies that objective and universal knowledge is impossible, it does not deny the relevance of an object-oriented focus. As critical realists most vividly explain, establishing knowledge about the world in which we act is still possible based on our experiences in this world and on explaining how these experiences could have occurred. Therefore, their main aim is to arrive at descriptions of the structures and mechanisms of the world that best correspond to our experiences and observations. It is a position fairly close to a realist perspective, but one that accepts the relativist argument that these descriptions remain fallible human constructs. Consequently, multiple and competing descriptions and claims towards what is considered real and rational are likely to remain. Object-oriented action can help us distinguish and choose between these descriptions and has the potential to further develop and adjust them. Nevertheless, as our experiences remain partial and subject to interpretation, object-oriented action is not considered to provide closure upon which of these descriptions is ultimately real or rational. Consequently, it is also a position that accepts the relativist imperative that intersubjective-oriented action remains relevant, to at least some degree, for agreeing upon which descriptions make most sense to those involved.

Those who take a stronger relativist position tend to focus more explicitly on the process of intersubjective-oriented action. Rather than taking a radical relativist perspective, however, they also accept that we are constrained in what we can plausibly believe by our experiences of the practical consequences of our actions. Nevertheless, they emphasize how descriptions of what is real and rational are influenced by whether someone considers these descriptions to be meaningful given the dominant values and perspectives held within his or her socio-cultural context (e.g. structuralism and post-structuralism). Hence, those drawing upon a stronger relativist position shift focus away from taking an object-oriented approach and relying on our experiences. Instead, they suggest we shift focus to the interaction between various individuals in an intersubjective context in which we can confront different descriptions of phenomena and the meanings various people attribute to them. We would therefore need to turn to discursive action in which we can establish (i.e. agree) which descriptions are considered to make the most sense (i.e. communicative rationality) to those involved. Ideally, as Habermas would have it, such discursive action could even produce full agreement on what makes most sense given the circumstances faced. In practice, however, as pragmatists would argue, we end up with a variety of different social definitions and beliefs, based on whether they make sense within the situational and social context in which they are developed and used.

The variations in philosophical positions have also manifested themselves in debates on governance. At the risk of again simplifying matters, many newer theoretical approaches to governance draw upon positions that focus on the process of intersubjective action; i.e. they stay close to a relativist position. In that context, planning action is mostly about discussing which descriptions of the phenomena faced, outcomes predicted and courses of action make the most sense to those involved. Some largely follow the argumentative model of governance and rely on collaboration between various stakeholders (e.g. Fisher & Forester 1993, Healey 1997 and Innes 1996). Others also draw upon the argumentative model but are also more explicit in accepting the relevance of the competitive model of governance. Hence, they also point to the importance of (power) struggles and competition that are needed to arrive at plans and policies (e.g. Booher & Innes 2001, Flyvbjerg 1998 and Susskind & Cruikshank 1987). Either way, each of these theoretical positions accepts that what is considered real and rational is mostly a human construct that is decided on the basis of struggle and debate by those involved. It has resulted in theoretical positions regarding planning and governance that often stay close to the notion of communicative rationality.

Other theoretical proposals are more modest in their acceptance of a relativist perspective and communicative rationality (see also Allmendinger 2002b, Fainstein 2000 and Harvey 1997). These proposals consider that, despite the socially mediated nature of knowledge, planning action begins with developing knowledge about the structures and mechanisms of the world in which we act. Oriented towards an object-oriented approach, these proposals stay close(r) to a (critical) realist position. They are approaches that try to model and understand planning issues and translate such models and understanding into physical and institutional designs assumed equipped for altering these issues. As such, these approaches also typically stay close(r) to a technical/instrumental rational perspective.

92 *Navigating the plural*

Beyond a duality

Inspired by the simplifications just made, the duality between realist and relativist positions can be seen to contribute to the theoretical plurality we face in our governance landscape. Ideally, we would be provided with criteria or arguments to choose between the different governance approaches linked to these positions. Finding such arguments and criteria is not self-evident, however, as this same duality has also manifested itself in conflicting philosophical assumptions regarding how we might distinguish between or combine these theoretical approaches. Consequently, we seem to be left without a common ground that serves as a starting point for choosing between different approaches to planning and decision making; including debates on decentralization in environmental policy.

Despite the limitations of finding a common ground, however, I suggest it is possible to develop well-founded arguments for choosing between various approaches to governance. It is a suggestion following the pragmatic stance that the dual philosophical positions can be considered as complementary. This is based on the argument that it is difficult to ignore the fact that people both draw from their experiences (object-oriented focus) and from intersubjective debates (sense making) when they distinguish between and attribute meaning to the objects, cases and situations they face. Hence, I suggest it valuable to assess how choices between various approaches to governance are based on two complementary arguments, namely one following an assumed real world of structures and mechanisms that constrains what we can plausibly believe, and, alternatively, one following the intersubjective debates helping us establish descriptions and understandings that are meaningful to us. It is also in drawing upon both object-oriented and intersubjective-oriented action that I continue with my proposal for a post-contingency approach.

3.3 A post-contingency approach

My proposal for a post-contingency approach can be traced to the emergence of contingency theories during the 1960s and 1970s in organizational sciences and public administration. A contingent perspective considers a dependency between two conditions or two sets of conditions, where given the presence of one condition or set of conditions another condition or set of conditions would be the logical consequence. Within organizational sciences and public administration, this logic was translated into the argument that a governance approach or an institutional organization should be adapted to the environment in which it intends to operate, i.e. a good 'fit' is a necessity (e.g. Burns & Stalker 1961, Bryson & Delbecq 1979, Chandler 1962, Donaldson 2001, Fiedler 1994 Lawrence & Lörsch 1967). The resulting contingency theories proved important, also within planning and administrative sciences.

Most notably, contingency theories rejected the 'one size fits all approach' towards governance that was in practice equated with the coordinative model. Rather, contingency theories produced the important message that the performance of different organizational structures and strategies depends on the contextual circumstances encountered. This message has shown to be an important advance in organizational and policy sciences, also gaining empirical support. Furthermore, it seems to offer us an entry point for trying to develop arguments for choosing between alternative approaches to governance. It is an entry point, however, that is not without problems. Most notably, contingency theory became the subject of important critiques for being largely embedded in a logical positivist tradition. It is in responding to these critiques that a post-contingency approach can be set apart from what I here refer to as 'classical approaches' to contingency. Hence, before I explain how a post-contingency perspective can help us develop arguments for navigating the plural governance landscape, I will first discuss why and how a post-contingency perspective is different from and can add value to classic approaches to contingency theory. It is a discussion that will first bring me to discussing these classic approaches to contingency (section 3.4) and their main critiques (section 3.5).

From contingency to post-contingency

What I here call the 'classic approaches' to contingency of the 1960s and 1970s became popular by expressing a clear and valued message: select organizational structures and strategies that match with the circumstances encountered. The process of coming to such a selection, however, proved largely coherent with the ruling doctrines of the day: an object orientation and an instrumental/technical rational approach. An object orientation comes forward in these 'classic approaches' by trying to develop uniform knowledge and standards upon the performance of different structures and strategies under different circumstances. This knowledge would subsequently inform the choice for a structure or strategy, where given the circumstances encountered and objectives set, this choice is a mere logical consequence of having such knowledge; i.e. where a deterministic logic consistent with a technical/instrumental rationale comes forward. With the rise of post-positivist notions in planning and administrative sciences, it was hardly a surprise that critiques soon emerged upon these 'classic approaches' to contingency theory.

Dominant among the critiques was the notion that contingency theory largely ignores the societal context in which knowledge is interpreted and contingency choices are made. It led to an acceptance within contingency approaches that people face limitations in observing and interpreting contextual circumstances and in observing and interpreting which organizational structures and strategies perform best given these circumstances. Although this nuances the deterministic logic of classic approaches, it hardly challenges its object orientation and rationale. Furthermore, these nuances also do not explicitly acknowledge that what is considered as performing 'best' is also a possible matter of interpretation and debate. Post-contingency implies we should further redefine and interpret the idea of contingency choices. Post-contingency does not deny the value of taking an object-oriented focus, but understands that its limitations force us to look beyond only taking such an object-oriented focus. Hence, it moves beyond the object-oriented

94 *Navigating the plural*

and deterministic logic of these classic approaches by adding an intersubjective focus to the process of making contingency choices and adjustments.

Post-contingency takes inspiration from the idea that contingency theories offer rationales for categorizing our (spatial) environment or situations and contingently relate (planning) approaches, actions and consequences to these situations. Post-contingency does so by expressing that this process of categorizing and relating is based on joining an object-oriented and intersubjective-oriented approach. Its focus on an object-oriented approach draws from one of the most dominant contingency arguments developed in the past; i.e. the argument that different conditions of complexity would logically coincide with different planning approaches and organizational formats (section 3.6). This argument can, for example, be found in the works of scholars such as De Roo (2003), Lawrence and Lörsch (1967), Miller (1993), Mintzberg (1983) or Gresov and Drazin (1997). It is in using the degree of complexity as an argument for making planning choices that post-contingency draws from an object-oriented approach, albeit by accepting the limitations of such an object-oriented approach and while acknowledging that what is considered as more or less complex is also not self-evident.

A first limitation lies in the acceptance that the perceived degree of complexity does not *determine* choices regarding planning action. Post-contingency does accept that knowledge can be gained about circumstances and about the performance of different organizational structures and strategies under various circumstances. It is such knowledge that also allows us to gain a better understanding of the character of the issues and circumstances encountered and of the possible or likely consequences of relying on alternative planning approaches under such circumstances. More specifically, through taking an object-oriented approach we are able to gain knowledge of the degree of complexity encountered and express expectations upon the consequences of relying on different governance approaches when applied under these perceived conditions of complexity. In other words: the idea of different degrees of complexity helps us categorize the world in which we aim to intervene and helps us to formulate expected consequences of relying on different approaches or taking different actions. Despite such knowledge, however, post-contingency suggests that valuing and judging these consequences remains a socially mediated matter of choice: knowledge does not determine action. Consequently, I argue, we are also subject to intersubjective-oriented action for making sense of the alternative values and preferences decision makers adhere to.

A second limitation lies in the acceptance that an object-oriented approach does not result in uncontested and 'objective' knowledge. Knowledge is subject to doubts, uncertainties and will be interpreted differently in different socio-cultural, temporal and spatial contexts. Especially under conditions of conflict, doubt and uncertainty, generated data or knowledge might even be to such a degree contested that it has little constructive value in coming to mutually agreed interpretations of the conditions we face. In addition, practice also forces us to accept that many planning issues are defined also on the basis of dominant political or institutional beliefs and interpretations and not just on knowledge collected. Hence, establishing the degree of complexity of a situation is itself also a matter of perception, interpretation and debate. As a result, the process of relating circumstances, approaches and expected consequences to each other might be *informed* by taking an object-oriented approach, but is also an *argumentative process* that allows decision makers not only to define a situation differently, but also to come to different expectations of their proposed actions.

It taking contingency theory beyond its deterministic and object-oriented tendencies, post-contingency thus proposes that defining issues in terms of their degree of complexity and subsequently choosing how to respond to these definitions are, by and large, socially mediated matters of choice (section 3.7). It is a proposal that does not imply that these choices are 'free', or to better phrase it, without argument. One key argument lies in simply accepting that an object-oriented approach can still produce valuable knowledge that helps us support our decisions; i.e. it does constrain what we can plausibly believe and expect. Furthermore, existing political or ideological arguments might prevail, whilst existing power relations also constrain the freedom of making choices. Within this context, a post-contingency approach argues, we are nevertheless also able to find additional arguments for making choices without relying on an object-oriented approach. That is, arguments can also be found without trying to understand and interpret the situation we face, but from the consequences that we aim to achieve. The argument follows the idea that different approaches are known to have alternative benefits (consequences) that we might value regardless of the exact situation we face. To illustrate: if it is uniformity of outcomes we are after, it might not be smart to rely on decentralized governance approaches. Alternatively, if it is tailor-made solutions we are after, it might well be smart to use such decentralized approaches. Instead of now starting from an object-oriented approach by working with 'what we do think we know' about the situation, we start by judging alternative consequences that we a priori desire or refute. That is, we start within the intersubjective domain with judging alternative consequences, contingently relate this to alternative approaches so as to make our choices. It is in section 3.8 that I will further discuss how we can develop arguments based on contingently relating consequences to approaches. It is a discussion that will particularly focus on the idea that the 'function' (desired outcomes) and 'structure' (organization or process) of a governance approach are contingently related (section 3.8).

Debates on the contingency between function and structure emerged in the wake of contingency theories, most notably during the 1980s and 1990s (e.g. Gresov & Drazin 1997, Miller 1986). The main argument is that some combinations of function and structure are considered to perform better than others. It implies that choices regarding what decision makers intend to achieve (their functional focus) can be related to choices regarding how they organize their approach (the structural configuration). To use the same illustration: I know certain functional outputs (uniform) relate to certain structural formats (centralized) and that these are opposed to alternative functional outputs (tailor made) and structural formats (decentralized). Therefore, when I judge certain functional or structural outcomes or, to use the common jargon of this chapter, 'consequences' as desirable or undesirable, I can indeed consider some planning approaches to be better

96   *Navigating the plural*

equipped to rely on than others. It implies that the contingency between function and structure helps me to translate judgements about desirable and undesirable consequences into limiting the range of eligible planning approaches.

Finally, in section 3.9, I will conclude that a post-contingency approach becomes especially valuable once we combine both the object-oriented focus and the intersubjective-oriented focus. One set of arguments can then be drawn from taking an object-oriented approach, where we use knowledge of how situations influence the consequences we can expect when relying on different approaches. The other set of arguments follows from which consequences we prioritize and how these influence our a priori preference to rely on alternative approaches. In combining both sets of arguments, we essentially come to choices on which approach to pursue, based on contingently relating the situations we face and consequences we desire to alternative approaches and actions we could pursue. But before I continue with elaborating on this conclusion, it seems more suitable to start at the beginning. It is therefore the emergence of contingency theory itself that will be the start of the next part of this chapter.

3.4 Classical contingency theory

Contingency theory became a relevant line of thought in organizational sciences and public administration during the 1960s and 1970s. These early approaches draw upon the ideas that were commonly accepted during these years. Contingency theory is therefore often associated with a reductionist and deterministic approach, accompanied by a reliance on objective knowledge and an instrumental or technical rationale. In this section I begin by discussing these early approaches, which I refer to here as classical approaches to contingency. In the following section (3.5) I will then discuss the criticisms of such approaches following the post-positivist acceptance of the socially constructed nature of knowledge and rationality.

Central beliefs

Among the dominant arguments within contingency theory is that the right way to organize management and governance should be contingent on the circumstances encountered (e.g. Bryson & Delbecq 1979, Fiedler 1994 and Lawrence & Lörsch 1967). It is a rejection of the 'one size fits all' notion that dominated the rational-instrumental perspective on policymaking and implementation. A 'one size fits all' notion implies that there is in principle a single best way to conduct governance and organize institutions such as governments dealing with environmental issues. In practice this 'single best way' was equated with the coordinative model of governance and the associated instrumental or technical-rational approach. During most of the 20th century, this approach was dominant in planning, public administration and organizational science, inspired by the writings of, for example, Fredrick Taylor (1911), Lyndall Urwick (1929, 1953) and Max Weber (1922). A contingency approach denies that there is a 'single best way' of organizing.

Instead, it argues that a governance approach or an institutional organization should be adapted to the environment in which it intends to operate, i.e. a degree of 'fit' is required (e.g. Bryson & Delbecq 1979, Donaldson 1996, Fiedler 1994, Lawrence & Lörsch 1967). Hence, "choices regarding planning phases and tactics are seen as dependent on planning goals and contextual variables" (Bryson & Delbecq 1979; p. 167, see also Tarter & Hoy 1998).

Early writings

Lawrence and Lörsch (1967) are often seen as the founding fathers of contingency theory (but see also Burns & Stalker 1961, Fiedler 1967, Woodward 1958). It was in their work that 'contingency theory' was first introduced as a concept. In their research, Lawrence and Lörsch conclude that the design of an organization should be adapted to the uncertainties of the environment in which it operates (see also Dubbeldam & Goedmakers 2003, Geurtsen 1996). They identify two main criteria that inform the design of an organization: the level of differentiation in an organization and the level of coordination or integration needed. Based on the contextual conditions, varying levels of differentiation and integration can then be chosen. Their main conclusion was that in a complex, uncertain and dynamic environment, the levels of both differentiation and integration need to be high (see also Dubbeldam & Goedmakers 2003). In other words, organizational design depends on environmental conditions faced.

During the early years of what came to be known as contingency theory, various schools of inquiry emerged. Some, for example, took a clear reductionist approach, focussing on isolating individual environmental variables and considering their relations with isolated organizational or decision-making variables in a pairwise fashion (see Pozzebon 2000; p. 3). Others considered 'fit' from a more holistic perspective, which "takes into account co-alignments as simultaneous and holistic patterns of interlinkages between large numbers of variables" (Pozzebon 2000; p. 4, see also Venkatraman & Prescott 1990).

Schools also differed in their object of study. Writers such as Burns and Stalker (1961), and Woodward (1958) focussed on the contingency of organizational structure upon the circumstances faced. Hence, today it is often referred to as either a contingency theory of organizations, or structural contingency theory (e.g. Pfeffer 1982). Others focussed on the influence of varying circumstances on the role of leadership (e.g. Fiedler 1967) or the style of decision making (e.g. Vroom & Yetton 1973). For them, it is not just the 'fit' between the organization and the physical environment that is to be addressed, but also the constraints faced by managers or decision makers, i.e. a 'fit' with both environmental constraints *and* institutional constraints such as the willingness and ability of staff, the information processing capacity of units, levels of autonomy, laws, etc. (see also Gresov & Drazin 1997, Lawrence & Lörsch 1967). Finally, others focussed on the contingency between an organizational function and the circumstances (see Miles & Snow 1978, Mintzberg 1973, Porter 1980 see also Gresov & Drazin 1997). For them, it was not the structure of an organization that needed adaptation

98   *Navigating the plural*

to environmental circumstances but, rather, the function (i.e. the strategy) of an organization or organizational unit. As we will see in section 3.7, classical approaches to contingency largely overlook the contingency between function and structure that might further enhance our capacity to find arguments for choosing between various approaches to governance.

Innovations: towards a nuanced perspective

Eventually, after the 1960s and early 1970s, contingency theory established itself in organizational sciences and subsequently also in public administration and planning. Despite the differences in the focus of contingency researchers, all were united in working on the idea that the function and/or structure of organizations and governance approaches are contingent on constraints faced. In doing so, classical contingency theory also stayed close to the instrumental or technical-rational perspective on policymaking and implementation. Over the years, however, criticism has fuelled important innovations and led contingency theory to evolve into a more elaborate and nuanced set of theories. In the following section I will first explain how this has pushed contingency theory outside of the realm of 'objectivity', where confidence in instrumental/technical rationality and reductionist analysis were common.

3.5 Beyond classical approaches to contingency

Classical contingency approaches came into fruition during a period when there was still a lot of confidence in acquiring objective knowledge of the environments and organizations faced, whilst also relying on technical/instrumental rationality and the coordinative model of governance. As such, early contingency approaches often relied on the idea that 'fit' occurs based on the rational choices made by members of organizations, related to the objectified and 'certain' knowledge they hold about sets of contingency relations between contextual and organizational variables. As we have seen, widespread confidence in objective knowledge and instrumental/technical rationality has gradually faded in the latter part of the 20th century. The classical approaches to contingency and their associated objectivity and deterministic explanations subsequently became increasingly criticized. Instead of undermining contingency theory, this criticism can also be seen to have pushed contingency theory into new directions. After addressing the main criticism to classical approaches to contingency in this section, these new directions are discussed in both this and the following sections (3.6 to 3.8).

Criticism

Prominent among the early criticism of contingency approaches were those that focussed on its assumed static and reductionist character. Classical approaches were considered static, as they focussed on finding a 'fit' between organizational structure and strategy, with the organization's environment at only one specific

Navigating the plural 99

point in time. Criticism focussed on the notion that contingency constraints are in practice constantly changing (e.g. Donaldson 1996, Miller 1981 and Mobach et al. 1998). The second critique focusses mostly on those following a reductionist perspective. To critics, this is considered to be based on assumptions about the process of finding the right 'fit' that are too simplistic (e.g. Miller 1981, Van de Ven & Drazin 1985 and Venkatraman & Prescot 1990). Rather, it is suggested, "organizations are more than a set of linear relationships. Researchers fail to look at rich and complex adaptive models and to discriminate among the different models that can arise in different contexts" (Pozzebon 2000; p. 5).

In the meantime, contingency studies were also criticized for their philosophical basis. In conceptualizing contingency, classical contingency theories relied on the idea that analysing both environmental circumstances and existing organizational strategies and structures would produce objective knowledge about the existing contingency relations. Once identified, these contingent relations would be translated into decisions by assumed rational decision makers. Being framed within a technical/instrumental rational paradigm, contingent relations would simply determine the choices made; i.e. only one choice was rational, and hence, contingent relations would be viewed from a deterministic perspective. However, this ignores the role of people and how they act when they face and cope with these contingencies. Responding to environmental contingences, the argument goes, is influenced by the values, perceptions, interpretations and attitudes held by those stakeholders and people who are involved (see also Pozzebon 2000, Weill & Olson 1989). Consequently, critics suggest, assuming that 'the' contingency relations can be objectified is too simplistic. Instead, the relation between the environment and the organization is also being 'created' by members of an organization while they try to make sense of what they experience and what these experiences mean to them and their organizations. Here, they draw upon the idea that organizations, their environment and the contingency relations between them are essentially social constructs (e.g. March & Olsen 1976). This has led to important criticism on the process of contingency change, which by many was described as too simplistic in deterministic terms (e.g. Miller 1981, Mobach et al. 1998 and Parsons 1995).

Innovations in contingency theory

Contingency theory has not been immune to its critics. Instead, it now offers a more nuanced perspective on the relation between environmental circumstances and organizational variables and how organizations and policies respond. To start with the main nuances, Donaldson (1996, 2001) first explains that a 'good fit' need not be a static condition at all, but rather can be approached as a cyclical and dynamic concept (see also Mobach et al. 1998). Strategies and organizational structures can be altered along the way to find a better 'fit' with the organization's environment. This also includes a more holistic assumption about the process of finding a 'fit', where multiple related variables are supposed. In that case, 'fit' is being pursued not based on simple static and bivariate links between two

100 *Navigating the plural*

variables, but rather as a process inspired by monitoring and evaluation contributing to organizational learning (e.g. McElroy 2000). As Mobach et al. (1998) further point out, such a process of change is also not just based on deliberate (rational) actions. Contingency also follows changes in people's behaviour while they learn from experiences. In the tradition of Weick (1993), change then follows a multitude of decisions made by various actors within a frame of the options and constraints these actors perceive. Choices are made explicitly *and* implicitly, sometimes resulting in changes to a small-scale strategy or unit, while other times they affect the organization as a whole. Combined, they produce changes in organizations meant to increase their potential for dealing with the environmental challenges they face, a process Weick describes as 'bricolage'.

Second, shifting attention to the role of people in bringing about contingency responses to environmental constraints has also led to diminished reliance on objective knowledge. This starts with acknowledging that it is "the members of the organizations – interpreting what they understand as the environment, interpreting meanings and common definitions – who do the regulating and adapting" (Pugh & Hickson 2000; p. 123 on Silverman, see also Mobach et al. 1998). This acknowledgement follows the post-positivist criticism of the possibility of objective knowledge and is supported by the idea that contingency adjustments are based on the way in which people perceive and interpret the environment and how their organizations 'fit' in with this environment. This offers a more nuanced perspective on how contingency adjustments take place.

Innovations: accepting the role of the social context

By shifting focus to the role of people in making contingency adjustments, a nuanced perspective has drawn contingency theory away from reliance on objective knowledge. However, by highlighting the role of people in making contingency adjustments, I argue that a second key alteration to classical approaches to contingency should be pursued. Contingency theory is about improving the performance of organizations and governance approaches based on improving their match with the environmental circumstances. A nuanced perspective explains how this match should not be described in terms of objective knowledge, but seen as a matter of perception and interpretation, where we are limited in what we know about the actual degree to which a match occurs. Nevertheless, such a nuanced perspective still assumes that, given what we claim to know about this match, we should proceed with adjusting organizational structures and strategies to the circumstances faced in order to improve organizational performance and thus better fulfil our objectives. It therefore remains to follow an object-oriented approach, but does not explicitly address the possibility that values and judgements differ due to the different ambitions and objectives that people have.

As I explained in section 3.3, planning and policymaking often take place in a societal context where different groups have different perspectives, interests and ambitions. Although classical approaches largely fail to take account of a societal context where objectives can be a matter of debate, they can easily be expanded

Navigating the plural 101

in response to this criticism. A first step is to accept that the societal context can be seen as *part of the contextual environment* to which organizational structures and strategies can be adjusted contingently. That is, the contextual environment includes the circumstances regarding the material object of intervention, such as the issue faced, its causes and effects, how these relate, etc., but also the societal context where people might attribute different meanings to this material object and its contextual environment, such as different preferences, interests and ambitions of stakeholders. In the following section (3.6) I will expand contingency theory in response to this acceptance by defining the contextual environment in terms of its degree of complexity. In doing so I begin with discussing how this further nuances following an object-oriented approach. After that, I continue by moving beyond an object-oriented approach by bringing contingency into an intersubjective realm. It is there that I will explain how both defining issues in terms of their degrees of complexity and the ways in which people respond to these degrees of complexity are also socially mediated matters of choice.

3.6 Contingency as a matter of degree

In this section I follow the argument that differences in contextual circumstances to which organizational structures and strategies can be contingently adjusted can be categorized in terms of their *degree of complexity*. The argument will be that this degree of complexity helps inform choices between various approaches to governance. In doing so, I still stay close to many studies that have come to a similar conclusion, both within contingency research (e.g. Geurtsen 1996, Gresov & Drazin 1997, Lawrence & Lörsch 1967, Miller 1993 and Mintzberg 1983) and in planning and policy sciences (e.g. Christensen 1985, De Roo 2003 and Van der Graaf & Hoppe 1996). However, I also differ from these studies on two counts.

Complexity – part I

First, many earlier contingency studies (e.g. Lawrence & Lörsch 1967, Mintzberg 1983) chose to define complexity without paying attention to the different actors involved. They merely equated complexity with uncertainty regarding the factors or causes that contribute to the phenomena we mean to address, how these factors relate and change over time (their stability) and the possible intervening circumstances that might influence these cause and effect relations. Here I also relate complexity to the actors involved and the different perspectives, interpretations and behaviours they have. Therefore, I take into account that complexity is also influenced by how circumstances are interpreted, valued and translated into opinions and preferences. To us then, an increased degree of complexity involves, among other things, diminishing certainties, increased contextual instability, diminishing direct causal relations and increasingly diverse and possibly conflicting interpretations and judgements of the issues, their contextual circumstances and the preferred approaches to them. The result is that it becomes increasingly difficult to 'objectify' the issue and its circumstances and that instead, there is an

102　*Navigating the plural*

increased need to intersubjectively establish how the issue and its circumstances are seen and which approaches might make sense to those involved.

Second, with my chosen description of complexity, I also argue that improving knowledge of the issues and circumstances faced is not sufficient to resolve the uncertainties faced. This is different from some contingent perspectives on planning and policymaking, which do show a tendency to believe that uncertainties can be resolved by improving our knowledge of issues and circumstances. This tendency can, for example, be found in the work of Christensen (1985) and Van der Graaf and Hoppe (1996). These authors accept that improving knowledge does not mean that all uncertainties will be removed, as people might still disagree on the goals to pursue. Nevertheless, uncertainties regarding the circumstances faced and the right means or technology for approaching should be solved based on 'experimentation' and 'learning' (Christensen 1985) or 'scientific inquiry' (Van der Graaf & Hoppe 1996). I here do accept that such an approach can help to *reduce* uncertainty. Nevertheless, I also argue that the crucial point about complex systems is that *resolving* uncertainty is impossible. Resolving uncertainty based on improving our knowledge and technology is merely possible when it comes to *complicated* systems; i.e. a complicated system is uncertain because we still lack sufficient knowledge on its components and the relations between them. *Complex systems* are something different from complicated systems and point to a type of uncertainty that is fundamentally impossible to overcome (e.g. Kauffman 1995, Prigogine 1997). Over the last three to four decades various scholars in multiple scientific disciplines have also begun to understand and conceptualize complexity as something fundamentally different from complicatedness. It has spawned a rich collection of research, theories and debates that are often summarized as 'complexity theory' (e.g. Heylighen 2008, Holland 1999, Lewin 1992, Prigogine & Stengers 1984, Waldrop 1994 and Wolfram 2002). Complexity theory is focused on the non-linear and adaptive behaviours of social and physical phenomena. It stands for a world view where reality is seen as constantly changing in often unpredictable ways due to its non-linear character. It has in recent years also found its way in debates on planning and governance (e.g. Batty 2005, Byrne 2003, De Roo & Silva 2010, De Roo et al. 2012, Portugali 2000 and Zuidema & De Roo 2004). In defining complexity in this book, I also draw from these perspectives. It is, however, again important to specify how I do so.

Complexity – part II

Complexity, as it is used here, stays close to what is also known as *static complexity*. Static complexity is used in contrast to dynamic complexity within various scientific disciplines that address complexity theory. Among the examples are information theory (Watt & Welch 1983), mathematics and computational science (e.g. Allender et al. 1999), manufacturing technology (e.g. Frizelle & Woodcock 1995), engineering (e.g. Dodder & Dare 2000), systems theory (e.g. Casti 1979,

Sterman 2000) and policy sciences (e.g. Wijen 2002). Static complexity, as Dodder and Dare explain, relates to

> the structural aspects of a system's complexity. This includes notions of hierarchy, connectivity, detail, intricacy, variety, and levels/strength of interactions; most easily visualized as a network with complex patterns of links and nodes. Static complexity is to a certain extent context dependent, since the structural complexity would appear much differently on the micro versus macro-level scale, and would change as one redefines the scope and boundaries of the system.
>
> (2000; p. 8, see also Lucas 1999; p. 2 and Deshmukh et al. 1998; p. 645)

It is called static for relying on the assumption that it makes sense to freeze a situation in time. It is a choice allowing for a simplification of reality when analysing systems and phenomena we aim to deal with, obviously including environmental issues as an example. The idea of freezing a situation in time does not imply that a system or its contextual conditions do not change; static complexity does not deny that time is crucial in understanding the behaviour of complex systems, neither does it deny that that this behaviour might at least to a degree be fundamentally unpredictable. Static complexity merely assumes that if we are to respond to such systems, we need to at least come to some kind of a description and interpretation of the system that helps us act in the here and now.

In contrast to static complexity, *dynamic complexity* explicitly incorporates the element of non-linear change. Dodder and Dare (2000; p. 8) explain how a dynamic perspective is closely related to the notion of complex adaptive systems and hence, to the adaptation of systems to changing circumstances (see also Holland 1999). Hence, dynamic complexity can be seen as "the unpredictability in the behaviour of the system over a time period" (Deshmukh, Talavage & Barash 1998; p. 645). Fundamental to dynamic complexity, therefore, is the concept of change. Over the last decades planning scholars have also started to conceptualize and describe planning issues and their surrounding circumstances from such a dynamic complexity perspective (e.g. Batty 2005, Bertolini 2010, De Roo & Silva 2010, 2012 and Portugali 2000). These approaches went beyond what had already been done by existing contingency researchers that discussed how planning approaches and organizational structures and strategies dynamically adapt to changing circumstances (e.g. Donaldson 1996, 2001 and Mobach et al. 1998). Such adaptations, essentially, are based on making adjustments which are informed by new information about the contextual circumstances. Hence, these researchers start from a static perspective, albeit by accepting its limitations, and hence aim to continuously revisit choices made based on new understandings of the situation and complexity faced. This is already an important first step forward in assessing how governance can adapt to changing circumstances.

A focus on dynamic complexity goes beyond just discussing how we might adapt organizational structures and strategies to new knowledge, ideas and

104   *Navigating the plural*

changing circumstances. Instead, a dynamic complexity perspective also looks at how organizations, plans or strategies can themselves be made adaptable and resilient, so as to co-evolve with their changing characteristics and contexts (e.g. Dietz et al. 2003, Stacey 1996, but see also Batty 2005, Bertolini 2010, De Roo & Silva 2010, De Roo et al. 2012 Portugali 2000, and Senge 1990). In other words, it explicitly aims to rethink the organization of governance, institutions and policy measures for coping with dynamically changing phenomena such as cities, neigh-bourhoods, regions or social systems that are understood to behave as complex systems (i.e. not in equilibrium, self-organizing, adaptive and evolving).

Dynamic complexity poses some vast challenges to conventional ideas and theories on planning and policy sciences. Not only does dynamic complexity explain and show that change in the social and physical world which planning and policy sciences deal with is often far from linear and can be highly unpre-dictable, offering serious and sudden surprises. It also challenges the traditional focus of planning and policy sciences on predicting, controlling and directing developments and change. Instead, dynamic complexity challenges the possibil-ity of control and urges us to reframe our understanding of 'influence' in relation to concepts such as co-evolution, resilience and adaptability. These new ways of looking at our world and at planning and policies are still fairly revolutionary and are to a large degree also beyond the scope of my study here. I here aim to navigate the plural governance landscape and formulate criteria or arguments for choosing between various approaches by emphasizing the complementarity between an object-oriented and intersubjective-oriented approach. This comple-mentarity is not explicitly addressed by (dynamic) complexity theorists (but see Portugali 2006), while the idea of finding arguments for navigating the plural governance landscape is neither a focus for complexity theorists. I instead draw from those perspectives that, at least partly, have tried to do so, which are the contingency proposals that draw largely from a static perspective. It is a choice that follows my belief that the claim made within contingency theory that choices regarding organizational structures and strategies should depend on the circum-stances encountered, is a good starting point for finding the criteria or arguments I am after. If we are to make choice dependent on the circumstances, I propose, we indeed need to at least come to a description of these circumstances to make a choice in the here and now. This is why a static perspective makes sense as a starting point. Importantly, within such a description, acceptance of non-linear and unpredictable change should also be a part. Hence, a choice for an approach to pursue might well be a choice for more adaptive and time-sensitive plan-ning approaches fitting with a more dynamic perspective. Nevertheless, it was a choice made based on a static perspective. In Chapter 7 I will come back on the choice for a static perspective. There I hope to explain that taking a static perspective can indeed be a starting point to also respond to dynamic complex-ity (compare De Roo & Rauws 2011) and also address possibilities for future research into linking dynamic complexity perspectives to the post-contingency argument presented in this book. For now, I continue by addressing how (static) complexity can be considered as a criterion for choosing between various gov-ernance approaches.

Complexity as criterion

I here argue that we can relate differing degrees of complexity to a choice between various governance approaches. On the one hand, I thus argue that issues and circumstances can be interpreted to have different degrees of complexity (see also Größler et al. 2006, Jorand et al. 2009, La Porte 1975, Stacey 1996, Steward 2001, Watt & Welch 1983 and Zolo 1992). On the other hand, as I will explain, I argue that we can use the difference between technical/instrumental rationality and communicative rationality to categorize these governance approaches (see also Allmendinger 2002a, De Roo 2003, Dryzek 1987 and Healey 1997). Technical/instrumental rationality and communicative rationality are seen here as underpinning the two contrasting and ideal-type planning approaches that help us develop a spectrum along which to position alternative approaches to governance considered; i.e. they help us categorize the governance landscape.

A technical/instrumental rational approach is here associated with a reliance on an object-oriented focus and the coordinative model of governance. Therefore, I consider it associated with aspects such as hierarchy, centralization, formalization, standardisation, specialization, routine and performance (see also Vroom 1981, Pugh 1973). Alternatively, a communicative rational approach celebrates the importance of an intersubjective-oriented approach. It is an approach I also connect to aspects such as negotiation, societal networks, increased informality and, more generally, a more flexible time- and place-specific search for solution strategies that make sense to those involved. It is an approach that stays close to the argumentative model of governance (section 2.1), but is not fully disconnected from the competitive model either. In integrating the need to negotiate, bargain and come to compromises, it draws partly upon, or at least can be associated with, elements of the competitive model. Nevertheless, it is a reliance on Habermas' intention that arguments can result in finding and pursuing commonly agreed outcomes that prevails over the merits of relying on market mechanisms to create such outcomes. It is with these descriptions of the ideal types of a technical/instrumental and communicative rationale that I proceed to categorize the governance landscape.

I subsequently argue here that increased conditions of complexity provide arguments to shift from a technical/instrumental rational approach and the related coordinative model of governance towards a more communicative rational approach, including more flexible governance practices. In the following section (3.7) I will also show how such a shift has consequences. So while a shift towards communicative rationality might allow us with means for better adapting to conditions of increased complexity; it also has consequences we might not appreciate. These consequences relate to aspects as varied as the efficiency of policies, economies of scale or legitimacy. Adopting a post-contingency perspective means to help us identify these consequences and raise the question as to which consequences we consider most important.

Relying on technical/instrumental rationality

Technical/instrumental rationality is concerned with selecting the most effective and efficient means for reaching a predefined end. Governance, then, is about

106   *Navigating the plural*

the rational design of policies and instruments in order to best reach the objective, i.e. actions are object oriented. Technical/instrumental rationality thus starts with assuming that planning action is all about fulfilling one or a series of clearly defined ends. To do so, technical/instrumental rationality is subsequently focussed on mapping and understanding the structures and mechanisms of the world and using this understanding to express the kind of actions that will deliver on the predefined goals most effectively and efficiently. In doing so, technical/instrumental rationality assumes the existence of uniform criteria that help decide what is rational to do given a certain situation. In practice, as is well understood, a technical/instrumental rational approach is closely related to the coordinative model of governance. The coordinative model relies on the legitimacy of representative democracy, where it is the state that, based on being well informed by specialists, decides upon the goals that are considered to benefit the public good. Subsequently, the model relies on policy implementation of clearly defined goals based on rationally ordered organizations. Efficiency is promoted by specialization to accommodate skilled administrators to work, often routinely, on clearly defined tasks and goals. Furthermore, clear lines of control, often in a hierarchical structure, would not only increase the effectiveness and reliability of organizational output, but also would result in predictable output.

Under conditions of low complexity, there are little problems to rely on a technical/instrumental rational approach (compare Christensen 1985, De Roo 2003 and Gresov & Drazin 1997). After all, conditions of low complexity imply that there is much agreement on the set of objectives that are to be pursued; we are well aware of the causes and effects that need to be addressed to meet these objectives; and we thus share a similar perspective upon what is rational to do given the issue we face. That is, we have a unified perspective on what the issue is we should be dealing with and what to achieve. Hence, we can focus on achieving our objectives as effectively and efficiently as possible. This brings us to a technical/instrumental rational approach, where "rationality is being equated with an instrumental or goal rationality where the basic assumption is that the subjectively created connection between means and ends should correspond with the objective relation between means and ends" (Berting 1996; p. 21, translation CZ). That is, we assume we both know how to deliver on our objectives and agree on the objectives to be achieved. Given these assumptions, there is also little reason for relying on a communicative rational approach.

As we have seen in Chapter 2, societal changes have gradually undermined our confidence in the coordinative model. These societal changes can be connected to the acceptance that many problems are simply more complex than assumed earlier. This complexity can, in a more practical sense, first be associated with the interrelatedness of policy agendas and issues. This interrelatedness implies that a focus on fulfilling one set of isolated goals becomes problematic, as fulfilling such an isolated set of goals has implications for reaching alternative and related goals. Second, conditions of complexity also imply that we are uncertain about the cause and effect relations that are involved in contributing to the existence of the issues we face. Partly, this uncertainty might be caused by our limited overview of the situation (i.e. the system is complicated). Partly, this uncertainty might also be caused by non-linear processes and relations that can exist between causes and effects or the instability of the context in which issues are embedded (i.e. the system is complex). Both these uncertainties imply that we face serious limitations concerning the predictability of our actions. We simply are uncertain of the exact impact of our actions, both regarding whether we will achieve set goals and whether these actions have implications for meeting alternative goals. Finally, governance often involves people and groups with different values, preferences and objectives. People will not only have different opinions, interests and preferences, they might also have widely diverging perceptions and interpretations of the issues faced. This is especially relevant if we are already uncertain about causes and effects and how one issue relates to another issue. The result is not only that we will face conflicting perspectives. Also, it is likely that people will try to advance or defend their own perspective. The dispersal of power in society subsequently allows for many of them to employ resources and strategies for protecting their own interests and influence decision making and policy implementation. The result is a dynamic interaction between the various influential actors and groups that is also difficult if not impossible to predict adding to the complexity faced. In the meantime, a technical/instrumental rational approach typically "fails to reflect the dynamic interplay of interests groups, public attitudes and corporate business influence, all of which may shape the policy agendas and practices" (Healey 2007; p. 35).

Combined, interrelatedness, non-linearity and dynamically interacting actors all contribute to an issues complexity. Rather than having a unified perspective on what the issue is we should be dealing with and what to achieve, we now face 'fuzzy' entities that are strongly embedded in their changing and unstable contexts. This undermines the capacity to rely on a technical/instrumental rational approach. Rather, when we try to describe, understand and deal with these complex issues, we are urged to respond to many (contextual) factors, interrelations and mutually dependent actors that interact in unpredictable ways, constantly altering how issues are interpreted and valued. It is also therefore that an increased degree of complexity can be seen as an argument for shifting away from a technical/rational approach. Instead, it calls for an approach that does respond to various stakeholder interests and attempts to balance and combine these so as to develop a shared perspective on which action is considered appropriate, i.e. we shift focus towards a communicative rational approach.

Relying on communicative rationality

Communicative rationality shifts perspective to the *meaning given* to an action, i.e. it is not just about reaching an end as effectively and efficiently possible, but also about what action is considered appropriate by those involved. Hence, communicative rationality also shifts perspective away from a sole focus of planning action based on the object of intervention (i.e. object-oriented action) towards the intersubjective process of making decisions (i.e. intersubjective-oriented action).

108 *Navigating the plural*

Communicative rational perspectives are intended to open up organizations and bureaucracy to social and physical contextual influences on planning issues (e.g. De Roo 2003, Forester 1989 and Voogd 1995). Hence, "the communicative/ interpretative turn in planning theory and policy analysis emphasises the social dynamics of . . . encounters between different forms of knowledge and different frames of reference" (Healey 2007; p. 33). In adopting a communicative rational approach, knowledge will still be produced, while rationality also remains to serve as a benchmark for making decisions. However, a shift in focus takes place from relying on object-oriented action for establishing what is real and rational, to intersubjective action to do so. We interpret, connect and value our knowledge in a process of 'making sense together'. In doing so, communicative rationality first intends to reduce the uncertainties involved in not knowing what to address and aim for, by agreeing on what we think we *do* know and what we *do* agree about. This also helps to set an agenda for planning action, based on a shared perspective that this action is appropriate.

A second key argument for shifting focus to communicative rationality under conditions of increased complexity is, as Healey (2007) states, that such conditions increase the need for connecting various interests, ideas and objectives for producing strategies that carry meaning for those involved, i.e. people argue and assume these strategies 'make sense'. In response, a communicative rational perspective allows for the kind of competition and bargaining that can be associated with both the competitive and the argumentative model of governance. Various actors can deliberate over which objectives to pursue and which courses of action to take. Different priorities can be reconciled by either bargaining for compromises between them, or by developing ideas and strategies that can achieve multiple objectives at the same time. It is also this approach that, as was discussed in Chapter 2, can be related to area-based approaches where we aim to respond to the interrelatedness of policy issues, social fragmentation and power dispersal. As a consequence, knowledge, goals and rationality are not given a priori, but are actively constructed during the decision-making process itself: we try to overcome differences in interpretation and balance or combine the various interrelated and competing objectives. Doing so, consequently, also requires that various stakeholders, their interests and resources should be part of the process establishing which combinations are considered appropriate.

Third and finally, a communicative rational approach can be an important first step to make plans and policies more adaptable to changing circumstances. After all, actively constructing how issues and circumstances are defined and approached often has somewhat of an open end. This implies that choices are not only made in the here and now so as to be readily implemented, but partly remain open to reflection, critique and redefinition. It is during the process of policy implementation that debates remain to take place between alternative stakeholders, while consequences of planning actions often directly influence socio-political debates that can partly re-open closed debates. Consequently, adopting a communicative rational approach often implies that stakeholders keep questioning choices and actions made, based on their ongoing interpretation and valuing of changing circumstances and appreciated consequences. Planning processes then become an 'ongoing process' (De Roo 2003) where changing opinions, interpretations and circumstances are actively translated into new definitions and approaches. In other words, a communicative approach can help open up planning and decision-making processes to reflection. Hence, although not necessarily sufficient, a communicative rational approach might be a helpful approach for plans and strategies to adapt to changing characteristics and contexts (see also Dietz et al. 2003 and Stacey 1996, but see also Senge 1990).

Responding to circumstances: a matter of degree

Summarizing, increased levels of complexity provide arguments for shifting focus from a technical/instrumental rational approach towards a communicative rational approach. It is a conclusion, however, that needs further nuancing. Most notably, technical/instrumental rationality and communicative rationality are now regarded as mere philosophies for ideal-type planning approaches (see also De Roo 2007, compare Van der Valk 1999). They fit in with situations where we either face a great deal of certainty and control or – in contrast – complex conditions "where uncertainty prevails, particularly in relation to actors' perceptions, motivations and behaviours" (De Roo 2007; p. 116). Most planning issues are, unsurprisingly, not characterized by either certainty and control or highly complex or uncertain circumstances. Instead, most planning issues have some characteristics that we understand quite well and are fairly certain about, whilst they are also surrounded by at least a degree of controversy concerning either the nature of the problem, the desired means for solving them, or the actual objectives to aim for. Hence, as De Roo also continues, "nearly all issues in planning practice include technical or object oriented aspects as well as communicative or institutional aspects" (2007; p. 117). Therefore: "by far the majority of issues cannot be dealt with by either a purely technical or a communicative approach" (ibid, p. 117, compare Kreukels 1980). In response, the shift from technical/instrumental rationality to communicative rationality is best seen as a contingency shift, where it is about the degree to which we rely on the object-oriented approach associated with technical/instrumental rationality or the intersubjective-oriented approach associated with communicative rationality.

To continue with our first key nuance in using a spectrum between a technical/instrumental and communicative rational approach, we can try to better grasp the grey area in between both ideal type approaches. By way of illustrating, we draw upon two planning approaches that can be positioned somewhere between technical/instrumental rationality and communicative rationality. The first is *scenario planning* (e.g. De Roo & Porter 2007), which is an example of an approach that accepts that we might face serious margins of error in describing the mechanisms and structures we deal with. Scenario planning responds by showing how large these margins might be, given different assumptions about the structures and mechanisms faced, and hence what kind of consequences our interventions in these structures and mechanisms might produce. It intends to reduce uncertainty

110 *Navigating the plural*

by gaining knowledge about the consequences of misinterpretation. By also comparing these consequences to the kind of values and objectives that are prioritized, a better image can be developed upon which to base decisions. Clearly, it is an approach that stays close to an object-oriented approach where it is information about the world around us that means to reduce uncertainty. Also, it stays close to a technical/instrumental rational approach were we still aim to choose the solution that seems most probable to help us achieve our objectives as effective and efficient as possible. Nevertheless, it does shift away from it to some degree, as it does address the possible margins of error in our knowledge. What it does not do, though, is consider that there is a need to enter into the kind of discursive action where we can really compete over our interpretations, ideas and objectives.

The second example is more focussed on coping with stakeholder influences and is the *actor-consulting model* (De Roo & Porter 2007). Instead of fully opening up the decision-making process to actor influences, stakeholders are merely consulted. It is an approach typically applicable under conditions where there are already some common agreements and perspectives, but additional choices need to be made, i.e. what can and will stakeholders contribute in translating commonly shared ideas into action? It thus aims at reducing uncertainties by understanding various stakeholder positions and how the interests they have relate to each other. Based on this understanding, the aim is then to develop policies that sufficiently integrate these interests and can benefit from the employment of resources and actions by key stakeholders. It is an approach which accepts degrees of complexity, but assumes that retaining a degree of control over the process and its outcomes is still possible by ensuring that the interests of key stakeholders are taken into account. Therefore, instead of really shifting to a communicative rationale, it is an approach that opens up approaches based on a technical/instrumental rationale to competing interests, interpretations and opinions.

These two examples illustrate that we are indeed capable of shifting focus between approaches that rely on technical/instrumental rationality and those that rely on communicative rationality. The result is a spectrum along which we can shift between approaches using and combining ingredients of both ideal type approaches, including on aspects such as the earlier mentioned notions of hierarchy, centralization, formalization, standardization, time and place sensitivity, etc. In this, the perceived *degree of complexity* encountered provides arguments for inspiring such a shift (see also Christensen 1985, De Roo 2003, Geurtsen 1996, Lawrence & Lörsch 1967, Mintzberg 1983, Stacey 1996, Van der Graaf & Hoppe 1996; compare Emery & Trist 1967). Distinguishing situations in terms of their degree of complexity, therefore, can help us to recognize a contingency relating various circumstances to various governance approaches. The idea is then that "effective planning begins by confronting the problem at hand and assessing conditions of uncertainty, rather than misapplying theories and methods without regard to the particular problem characteristics" (Christensen 1985; p. 63). While supporting this idea, post-contingency does suggest we need to take the discussion yet another step further. That is, it argues that how we respond to increased conditions of complexity is, by and large, also a matter of choice.

3.7 Contingency as a matter of choice

In the previous section (3.6) we have seen how an object-oriented focus emerges in a post-contingency approach by realizing how differences in the circumstances perceived, defined in terms of their degree of complexity, provide us with an argument for choosing between various approaches to governance. An object-oriented approach helps us see how the degree of complexity influences the consequences of relying on different governance approaches. Gaining an improved understanding of the degree of complexity concerned can therefore help inform choices between governance approaches.

A post-contingency approach takes contingency theory past a singular reliance on an object-oriented approach. Instead, it brings it into an intersubjective realm where both defining issues in terms of their degree of complexity and how they are subsequently approached are also, by and large, socially mediated choices. There are two main arguments I use to support this claim. First, I accept that the degree of complexity does not *determine* the choice between governance approaches. Classical approaches to contingency focussed on finding a match between different circumstances and various organizational strategies and structures. They assumed that the objectives to achieve were already known. This meant that, given the circumstances, contingency studies would be able to show which governance approach should be chosen. It is a deterministic perspective, where the circumstances directly inform the choice of a governance approach (see also Figure 3.1).

As was explained in section 3.3, I accept the post-positivist assertion that there are no fully objective standards to inform our choices. Rather, we argue that people can have different and possibly conflicting values and standards. So while

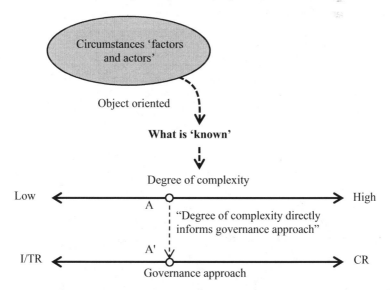

Figure 3.1 A contingency approach: complexity as a criterion

112   *Navigating the plural*

an object-oriented approach helps us gain an understanding of the likely consequences we can expect when relying on different governance approaches, how decision makers want to respond to these consequences remains an intersubjectively mediated choice. In other words, the perceived or deduced degree of complexity is an argument for choosing between various approaches to governance; how decision makers want to respond to this argument remains a *matter of choice* (Figure 3.2; Line Y). It is a choice that, obviously, is also constrained by the existing discourses, power balances and political contexts. Nevertheless, what I aim to express here is that complexity is merely an argument to make such choices, an argument that can be interpreted and valued differently by alternative stakeholders and decision makers.

Second, I also explained that the role of the intersubjective emerges when accepting that defining issues in terms of their degree of complexity is influenced by the different frames of reference from which people perceive and interpret what they experience, i.e. it is intersubjectively mediated. It follows from accepting that relying on an object-oriented focus should not be conflated with developing fully objective knowledge regarding the degree of complexity faced. Defining issues in terms of their degree of complexity is therefore always influenced by how stakeholders and decision makers perceive and interpret the circumstances faced. We might be able to gain an improved understanding of the circumstances or degree of complexity we face. We also know, however, that we are constrained in what we can observe and can come to different interpretations and perspectives concerning the circumstances we face. Hence, an object-oriented approach will not produce definite answers regarding these circumstances and might even fail to help us move beyond highly uncertain and disputed knowledge. As was also noted in section 3.3, knowledge is subject to doubts, uncertainties and will be interpreted differently in different socio-cultural, temporal and spatial contexts. Also, it is especially under conditions of conflict, doubt and uncertainty that generated data or knowledge might even be to such a degree contested that it has little constructive value in coming to mutually agreed interpretations of the conditions we face. It is, to start with, therefore merely the *perceived* degree of complexity that provides us with an argument when taking an object-oriented approach (Figure 3.2; Line X). It is this perceived degree of complexity that might provide us with arguments to choose, but such choices take place in a societal realm where planning issues and objectives are defined also on the basis of dominant political or institutional beliefs and interpretations that influence how we interpret and value the knowledge collected. It also shows us that relying on an object-oriented approach has clear limitations in providing us with arguments for choosing between various governance approaches.

A post-contingency approach now continues by arguing that even if we accept the limitations of collecting knowledge of the circumstances encountered and the consequences we predict of the alternative choices we make, we are still able to make well-argued choices. It does so by suggesting that instead of starting with trying to understand and interpret the situation we face, we

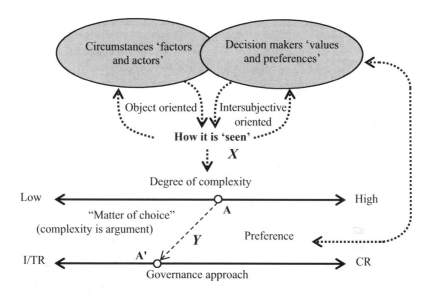

Figure 3.2 The intersubjective in a post-contingency approach

X: The degree of complexity is not objectively established, but is also influenced by the perception and interpretations of those involved. Y: The choice in favour of a governance approach (A') is informed, *but not dictated*, by the perceived degree of complexity (A); it is also informed by dominant values and preferences when judging issues.

can also start by highlighting the consequences that we aim to achieve. This suggestion, as was also mentioned in section 3.3, follows the idea that different approaches are known to have alternative benefits (consequences) that we might value regardless of the exact situation we face. Instead of now starting from an object-oriented approach by working with 'what we do think we know' about the situation, we start by judging alternative consequences that we a priori desire or refute. That is, we start within the intersubjective domain with judging alternative consequences, contingently relating this to alternative approaches so as to make our choices. To do so, I take our debate further in explaining that when we choose between various approaches to governance, we do so in the midst of two related choices.

These two related choices relate to on the one hand the functional ambition of our approach (i.e. what do we mean to achieve) and, on the other hand, the structural configuration or our approach (i.e. the allocation of power, influence and responsibilities between various actors). As I will explain in the following two sections (3.8 and 3.9), these two choices are related, as some configurations of function and structure are considered to perform better than others. I will subsequently explain that this first allows us to see that there will be consequences in terms of how the chosen approaches perform, which depends on the

114   *Navigating the plural*

degree to which function and structure match. This limits the range of possible choices regarding our approaches to governance, as decision makers are advised to choose only configurations between function and structure that match. I will subsequently explain how both a technical/instrumental rational approach and a communicative rational approach can be considered to rely on matching configurations of function and structure. Second, we can see that when we shift from a relative reliance on a technical/instrumental rational approach to a communicative rational approach this has consequences for both the functional ambitions we can expect to be achieved and which structural configurations can be considered effective and efficient in doing so. It allows us to see that the choice between various approaches to governance has consequences without relying on an object-oriented focus. As such, we are again enabled to develop an argument that can help us make choices between alternative approaches to governance without it being based on knowledge of the situations and circumstances we face.

3.8 Function and structure

The relation between the function and structure of an approach has for a long time been largely overlooked in most social sciences and early contingency research. Authors often considered a structure selected or designed for a specific function, and hence to be equated with this function (e.g. Miller 1986). This consideration supposes a deterministic relation between function and structure. Gradually, criticism has emerged that suggests that such a deterministic relation is not correct. In practice, a specific structure can perform various different functions, while various different structures can also perform a specific function. Various researchers have subsequently examined the relationship between function and structure. They concluded that, although a deterministic relation between the two should be rejected, there is still a contingency relation between the structure and function of government approaches. So rather than rejecting the close connection between function and structure, this research highlighted the importance of understanding how both are related. The result is that we have begun to understand that certain configurations of function and structure can be expected to perform better than others. In this and the following section I will proceed by explaining this contingency relation and show how it helps us to formulate arguments that enable us to navigate the pluriform governance landscape.

Function and structure

To begin with, the *function* of a system is related to what the system means to achieve with regard to other systems or the larger environment (Gresov & Drazin 1997; p. 406, compare Morgeson & Hofmann 1999 and Scott 1981). It has traditionally been equated with the effects or output that the system is intended to produce and was therefore defined in relation to a technical/instrumental rational perspective. Some authors use the concept of function not in contrast to structure, but instead refer to a difference between *strategy* and structure (e.g. Amburgey and Dacin 1994, Miller 1986, 1987 and Payne 2001). Strategy is a somewhat broader concept than function, as it addresses both the outputs to be produced by a system and the idea or plan behind it. Still, as is also illustrated by its use in contrast to structure, the aforementioned authors all link strategy to the desired output or effects to be produced by a system. Strategy, used by these authors, is thus close to what I here describe as a governance approach's function. Payne, for example, relates strategy explicitly to the "intentions and actions" (2001; p. 57) of an organization for achieving an objective, while Miller sees it as a reference to the "achieving of certain objectives" (Miller 1986; p. 56). I thus consider the use of strategy by these authors as referring to the function of a system.

Box 3.1   Interpreting function

The first EU daughter directive on sulphur dioxide, nitrogen dioxide and oxides of nitrogen, particulate matter and lead in ambient air (1999/30/EC) sets minimum standards regarding the ambient air quality. In practice, we have learned (e.g. Backes et al. 2005) that doing so can result in excessive consequences, such as closing highways, factories and energy plants or setting a stop on urban and economic developments. It implies that these policies will in practice mean more than just meeting limit values. If we interpret the directive as taking a technical/instrumental rational perspective, these consequences are not considered meaningful as its function is solely about ensuring that these minimum levels are all met. From a communicative rational perspective, this is considered way too simplistic. Rather, the meaning we attribute to the directive should take account of these consequences and how they are interpreted and judged in relation to meeting air quality ambitions. Therefore, the directive will no longer be considered to function as a means for meeting pre-selected objectives, but to be meaningful merely in relating environmental targets to their consequences in relation to other priorities and interests.

The choice to connect function to the effects or output a system intends to produce tends towards a reliance on the technical/instrumental rational perspective and therefore needs nuancing. In the wake of the rise of the post-positivist perspective, also in light of policy as an organizational science, we have seen that relying on a technical/rational perspective has gradually been nuanced and complemented by perspectives that find that organization, strategies and functions should be considered as socially mediated constructs (e.g. March & Olsen 1976, Mobach et al. 1998, Pugh & Hickson 2000 and Yanow 2000). In such perspectives, the function

116   *Navigating the plural*

of a system is not equated with the output it intends to produce. Rather, I suggest, the function of a system should be considered more widely as a socially mediated perspective on what the system means to those involved, i.e. a function is the meaning attributed to the system (Box 3.1).

Next, the *structure* of a system relates to the organization of the system. In social systems it can be expressed as "a pattern of relationships between individuals that transfers and modifies information and physical objects", i.e. they are "social arrangements or patterns" (Gresov & Drazin 1997; p. 406, see also McMillan 2002). In adopting a technical/rational perspective, a structure refers to the organization of the various elements of a system, which emerges in characteristics such as the levels of hierarchy, specialization and links between various divisions and sub-divisions. Such a perspective is thus focussed on the allocation of tasks between elements such as units, divisions and people, while meant for enhancing the effectively and efficiently performing function of the system. It is also this philosophy that has manifested itself in the coordinative model of governance and its hierarchical control and functional specialization.

The rise of post-positivist perspectives in policy and organizational sciences has again resulted in a far wider perspective on the structure of social systems in relation to policy development and delivery. Fuelled by social fragmentation and power dispersal, policy development and delivery increasingly is understood to take place in a context of various societal or market parties claiming their place in the governance process and, as expressed with power dispersal, also having the resources to exercise influence. In response, the structure of governance approaches is hardly about the design of blueprints of tasks, responsibilities, units and divisions. Instead, it is about how policymaking and delivery can be organized so as to make sense of the various ambitions and tasks that should be pursued and performed in relation to the issue faced and how and by whom these can subsequently be delivered. It implies that we then operate from a communicative rational perspective where the structure of a governance approach revolves around activities such as debating, bargaining and negotiating, the structure relates to a division of tasks, responsibilities and investments between actors that is accepted and considered to make sense by those involved.

Post-contingency now follows the idea that once we make a contingency shift from a technical/instrumental to a communicative rational approach, we also make a contingency shift in perspective on both the functional and structural focus related to our governance approach. In the following section I will also address this contingency relation between a choice between various approaches to governance and the functional and structural focus which is associated with such a choice. We will then also see how this allows us to understand the typical consequences of such a shift, which will thus provide decision makers with arguments for choosing between various approaches to governance. In this section I will first continue the discussion on why a choice between various approaches to governance is related to both choices regarding function and structure. I will do so by explaining that function and structure can also be seen to relate in a contingency fashion.

Relating function and structure

In academic debates on function (or strategy) and structure there is widespread acceptance that these are closely related. Gresov and Drazin (1997; p. 406) address this relation in stating that "a structure can fulfil a function in relation to some other structure or to the whole system". Function thus gives meaning to a structure. A commercial department *is* a commercial department as it *intends* to perform commercial activities. Furthermore, without a function, a structure is empty, i.e. it has no meaning. In line with how function and structure are used in adopting a technical/instrumental rational perspective, the relation between them for a long time was even considered in a deterministic fashion. Or, in other words, many scholars in organizational sciences assumed that any given function would best be performed by a single structure, while any single structure was meant to perform a distinct function (e.g. Merton 1968, Miller 1986). Here, I agree with those criticizing this assumption and consider this deterministic perspective to be too simplistic (see Gresov & Drazin 1997, Merton 1968, Miller 1986, 1994, Mobach et al. 1998, Morgeson & Hofmann 1999 and Payne 2001). I accept that a single structure can perform more than one function, and that a single function can also be performed by different structures (see also Box 3.2). As is illustrated by Amburgey and Dacin (1994), this is a more widely accepted idea, which in the 1980s resulted in growing attention for the way in which function and structure are related.

Box 3.2   An example of structure and function

A municipal environmental department has environmental quality aims. This is its main function, which is related to the larger function of municipal planning: to deliver and guarantee a good quality of life for its citizens. However, a focus on environmental quality can be translated in different structures. The department can be organized based on making different people responsible for different themes, such as air quality, noise nuisance, soil pollution, etc. The department can also be organized by making different people responsible for different areas, such as one person for this neighbourhood, another for the following, etc. Yet another option is basing the responsibilities of people on specific priority issues such as traffic, businesses and industries, households, etc. Clearly, each of these structural configurations can be chosen to perform the same function (i.e. structure is not determined by the function). However, it is also likely that some configurations will perform better than others. It is hardly a surprise that most municipalities focus on specialization based on different themes as there are economies of scale involved in acquiring expertise in these different themes (i.e. function and structure do relate).

118 *Navigating the plural*

Miller (1986, 1996) is one of the primary scholars to argue why overlooking the relation between structure and function is unfortunate. To him, "previous typologies and taxonomies found in general strategy and organizational literature have failed to account for both aspects of strategy [function] and structure adequately" (1996; p. 30, compare Payne 2001). On the one hand, many contingency thinkers focussed on finding correlations between certain environmental circumstances and organizational structures (e.g. Burns & Stalker 1961, Mintzberg 1979 and Woodward 1965). However, if a structure can perform different functions, whether or not you find such correlations does not necessarily say much. On the other hand, other contingency researchers focussed mostly on relations between environmental circumstances and specific strategies to adopt or functions to perform (e.g. Miles & Snow 1978, Porter 1980). However, if there are different structures that can perform a similar function, this again provides limited clarity regarding the kinds of structure that would match these functions. In both cases, the relation between function and structure is simply assumed to be there, and is consequently largely overlooked (e.g. Gresov & Drazin 1997, Mobach et al. 1998, Payne 2001 and Van de Ven & Drazin 1985). To Miller, "a central gap in the literature to date is that the rich content of strategies [function] has never been related to structure" (1996; p. 234).

In recent decades the relation between function and structure has been paid more attention in academic works. It has provided support for the claim that function and structure are related in a contingent fashion. Prominent is the work of Gresov and Drazin, who take the perspective that "functional requirements do not determine a particular social structure, but rather permit a range of structures that will fulfil the functions required" (1997; p. 407, see also Merton 1968, Mobach et al. 1998). It is a suggestion that resonates with the works of Miller (1986, 1996), Amburgey and Dacin (1994) and Payne (2001). As Miller puts it, "given a particular strategy [function] there are only a limited number of suitable structures and vice versa" (1996; p. 234). This argument is thus "built around the basic premise that certain patterns of strategy [function], structure and/or process will tend to certain levels of performance" (Payne 2001; p. 2, see also Amburgey & Dacin 1994, Miller 1996). In other words, some combinations of function and structure do match (i.e. they perform well), while others do not match (i.e. lead to lower performance). The consequence is that once we shift focus regarding our functional focus, we should also contingently shift our focus in terms of our structural focus and vice versa.

Over the past decades research has also given us more empirical support for the claim that there indeed is a clear contingency between function and structure. As Miller (1996) illustrates, for example, research has revealed that certain combinations of structural and functional (strategic) elements are more likely to be found successful than others. This is also empirically illustrated in his work with Friessen (Miller and Friesen 1984, compare Meyer et al. 1993, Miller 1987), suggesting that some combinations survive and can be seen to 'perform' better than others. It suggests that a contingency relation exists between function and structure, contributing to the existence of 'good' and 'bad' configurations.

More support can be found in the works of researchers that have also explicitly addressed the contingency between function and structure. Much of this spawned from the work of Chandler (1962), who was one of the first to conceptualize it. Chandler found that companies (organizations) face a variety of strategic purposes, related to the phases in a company's life cycle. In this, he found evidence that the structure of an organization follows the strategy pursued, i.e. the functional objectives of the organization. Other examples of studies that support this conclusion include those by Woodward (1958), Mintzberg (1990) and Rumelt (1974). To Miller (1986; p. 236), these studies show that organizations are "interrelated in complex and integral ways", which to him implies that certain strategies 'find' structures that suit them, while existing structures also influence the kinds of strategy adopted. A routine function, for example, produces more stable and formal structures, whilst stable and formal structures often reduce the possibility for innovative functions (compare Payne 2001). More recently, Amburgey and Dacin (1994) also provided evidence for a relation between function (strategy) and structure. They conclude that their results offer "substantial support for the common conception of a contingency relationship between strategy and structure" (p. 1446). While doing so, they also suggest that structure does not only follow function (strategy), but that function (strategy) also follows structure. This is very similar to what Mintzberg concludes when he states that

> Structure follows strategy as the left foot follows the right in walking. In effect, strategy and structure both support the organization. None takes precedence; each always precedes the other, and follows it, except when they move together, as the organization jumps to a new position.
>
> (1990; p. 183)

Some conclusions

Instead of adopting a deterministic perspective on the relation between function and structure, recent scholars have conceptualized this relation as a more intricate and reciprocal one where function and structure interact in a dynamic way (e.g. Amburgey & Dacin 1994, Miller 1996, Payne 2001, Ramanujam & Varadarajan 1989). However, these studies have most of all enhanced our understanding of how function and structure relate. Most notably, the conclusion remains that function and structure relate in a contingency fashion. On the one hand, this implies that the choice in favour of a certain structure is contingent on the kind of function to be performed. On the other hand, the choice in favour of a certain structure also limits the kinds of function that can be properly performed, i.e. the functions to be performed are also contingent on the structure in place. Some combinations of function and structure match, while others don't.

The contingency between function and structure functions as advice when we choose between various approaches to governance. After all, while following Payne (2001), there is "little doubt as to the importance of matching both strategy and structure to environmental conditions, *and matching them to each other*"

120 *Navigating the plural*

(p. 171, italics CZ). So when we choose between various approaches to governance, we are advised to search for a match between function and structure, or, as Miller (1986) puts it, *successful configurations*. The following point to address is which configurations of function (strategy) and structure match and which do not, i.e. which configurations are more and which are less successful given a certain standard of success and given a certain context. While doing so, we have already touched upon the possibility to use the difference between a relative reliance on a technical/instrumental rational approach or on a communicative rational approach as a way of categorizing various approaches. After all, as noted earlier, when shifting from a relative reliance on a technical/instrumental rational to a communicative rational approach, we also make a contingency shift with regard to both the kind of functional ambitions that we pursue and the kind of structural configurations that we rely on. The following section (3.9) will show how this categorization can indeed be used in relating function and structure.

3.9 Limiting choices: towards common configurations

Finding successful configurations is not very easy, as there is a wide variety of functional and structural options to choose from. For example, when it comes to structural options, Vroom (1981) notes options related to, for example, hierarchy, centralization, formalization, standardisation, specialization, routine and performance (compare Pugh 1973). Faced with so many options, a high number of possible combinations of structural and functional characteristics emerges (see also Miller 1986; p. 235). Hence, it becomes hard to predict which configurations could emerge, as so many variables can influence the process. As Miller (1986) explains, it is simplistic to try to understand these complex configurations just by doing bivariate studies, i.e. to focus only on relations between two variables (compare Berting 1996). After all, as Miller (1986) continues, "we would be in a position of having to formulate a myriad of bivariate or circumscribed multivariate hypotheses. These would be extremely numerous and perhaps conceptually intractable. Any coherent theme might be obscured by the mist of atomistic speculation", while "an even more serious shortcoming of this approach is that reality usually cannot be expressed in terms of linear bivariate or even multivariate relationships" (p. 235).

In response, as Miller (1986) states, we can also follow an alternative approach to find successful configurations of structure and function. If we accept that there is a contingency between function and structure, we can also assume that we will see a convergence towards combinations of structure and function that perform well at the expense of those performing less well. Such convergence can be seen in the development of theoretically defined ideal type approaches to governance, which have proven valuable in practice. In addressing these ideal types – or what Miller (1986) calls stereotypical or common approaches – we can thus gain information about how function and structure relate. Thus, we can use these common configurations as the starting point for understanding *why* these configurations emerged, i.e. why they perform well. This is also the route taken here. In doing

Navigating the plural 121

so, we have seen that the spectrum between technical/instrumental rationality and communicative rationality has provided us with a means for finding the ideal type of approach. Furthermore, in the previous section (3.8) we have seen how they both relate to a different focus on both function and structure. This is also the spectrum that I therefore will use to distinguish between various configurations of function and structure, based on arguments concerning why some are expected to perform better than others.

Configurations uncovered

Typologies of planning and policy approaches are rather common in the academic literature (e.g. Christensen 1985, De Roo 2003, Douglas & Wildavsky 1983, Friedman 1973, Mintzberg 1983, Teisman 1998, Thompson & Tuden 1959 and Van der Graaf & Hoppe 1996). Although many of the configurations assume a relation between function and structure, few of them make this explicit. A useful exception is the model developed by De Roo (2003, 2004). Here, an explicit distinction is made between structure and function and I therefore use it as a tool for illustrating and explaining how function and structure can be related. In doing so, many of the concepts used in this model will be related to the findings of other authors. Consequently, this will help us to develop a well-supported navigation tool for operationalizing the contingency between structure and function.

Function and structure are expressed in the model of De Roo (Figure 3.3) as the 'scope of goals' (i.e. the function or strategy), which is focussed on the question *what* is to be achieved, and the 'pattern of relationships' (i.e. structure), which is

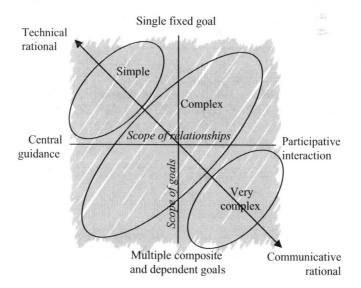

Figure 3.3 Framework for planning-oriented action, with complexity as a criterion (source: De Roo 2003)

122 *Navigating the plural*

focussed on *who* is involved in decision making and implementation. The scope of goals is depicted as a spectrum between 'single fixed goals' on the one hand and 'multiple composite goals' on the other. The idea is that the function of an organization or approach can either be fully focussed on maximizing a single objective, or can be to facilitate combining or balancing a theoretically infinite number of competing or congruent objectives. The pattern of relationships is expressed as a spectrum between a fully centralized structure (i.e. 'central guidance') with one actor exerting control and decentralized structure (i.e. 'participative interaction') with a theoretically infinite number of actors involved. It is a choice that also relates to previous studies on organizations' functions and structures. Gresov and Drazin (1997) also focus on differentiating between single and integrated goals or, to use the words of De Roo, 'multiple composite goals'. Furthermore, although many other characteristics can be associated with structure (e.g. Vroom 1981), the degree of centralization is often considered the primary structural characteristic as it is considered to correlate with other structural characteristics (see also Amburgey and Dacin 1994, Friedman 1973 and Miller 1986, 1996). Centralized structures are typically associated with formalized, hierarchic and bureaucratic structures. Instead, decentralized structures are often linked to informal, horizontal and organic structures (e.g. Geurtsen 1996).

To De Roo (2004; p. 25) "every position within the model stands for a policy choice (how), an associated objective (what) and a constellation of associated actors (who)". Theoretically, all positions in the scheme of Figure 3.3 can be chosen as a planning or policy approach. Hence, all possible combinations of function and structure are in principle viable options. However, De Roo continues by explaining that not all of these combinations can be expected to be equally successful. Instead, he uses the spectrum of rationality – again between technical/instrumental rationality and communicative rationality – as a third spectrum for connecting function and structure. Similar to what was discussed in section 3.5, the perceived degree of complexity is used by De Roo as the informant for shifting focus from a technical/instrumental rational approach towards a communicative rational approach. While doing so, this shift also has direct repercussions for both the functional and the structural focus. In other words, the spectrum between technical/instrumental rationality and communicative rationality connects functional and structural choices. Hence, it becomes a means to establishing which configurations of function and structure match.

Functional choices

De Roo distinguishes between 'single fixed goals' on the one hand and 'multiple composite goals' on the other hand as the extreme choices on the functional spectrum, which here refers to 'the scope of goals'. To De Roo, the degree of complexity encountered can be used as an argument for decision makers in choosing between a focus on either a single fixed goal or on multiple composite goals. In doing so, a focus on a single fixed goal is associated with a limited degree of complexity. This is similar to what, for example, Gresov and Drazin (1997;

Navigating the plural   123

pp. 408–409) explain: "under conditions of low conflict, the functions that the organization must perform can be characterized by 'simplicity' (Miller 1993), as having a 'dominant-imperative' (Gresov 1989; pp. 438–439) or as having a 'dominant logic' (Bettis & Prahalad 1995)" (citations by Gresov and Drazin). Thus, conditions of low complexity provide an argument for fully focussing on achieving the central objective, following on from the demands faced. The functional choice expressed here, as De Roo (2003) explains, is to try to achieve a 'single fixed goal', and in doing so to try to maximize the outcome by disregarding potential alternative demands as these are assumed to be congruent or irrelevant (see also Christensen 1985). It is typically a focus which fits in with a technical/ instrumental rational style of planning and policymaking.

A focus on fulfilling a single a priori selected objective as effective and efficiently as possible is not always evident. After all, we can face conditions where there are multiple and conflicting objectives that are interrelated, i.e. we are faced with conditions of increased complexity. Then, as Gresov and Drazin (1997; p. 409) explain: "achieving one function must come at the expense of not achieving another", while "the organization does not have the luxury of considering one function as dominant and ignoring the other". Essentially then, a choice is required based on a trade-off or, as Van der Graaf and Hoppe (1996; p. 48) suggest, the exercise of power in favour of one objective. To De Roo (2003), there is also the alternative where incongruence between the functional demands can potentially be solved by attempting to develop 'multiple composite goals'. This is an "approach involving the linking of issues, the bundling of solution strategies, and multiple goal realization by combining different issues within a single solution strategy" (2003; p. 106). A similar distinction between trade-offs and combining multiple objectives is also made by Christensen (1985; p. 65), who calls these two options 'bargaining' over competing objectives and the "accommodation of multiple principles". In both cases, as Christensen also notes (1985, compare De Roo 2003, Van der Graaf & Hoppe 1996), decision making does not rely on certainty, control and technical/instrumental rationality. Instead, there is a reliance on communicative rationality for arriving at consensus, compromises or other communicatively produced shared meanings. The functional focus thus shifts from a focus on fulfilling predefined ends and outputs, to attributing meaning to the choices made based on which objectives and actions are considered to make most sense to those involved. In addition, instead of focussing on maximizing the outcome, the focus is on optimizing the process in which these meanings, objectives and strategies must be produced. Ignoring crucial stakeholders and arguments can easily produce resistance to the outcomes, frustrate successful implementation or action and produce undesirable outputs.

In between these extremes, intermediary positions can of course be found. We can prioritize a single objective, but allow for a degree of bargaining, accommodating alternative objectives that can come at the cost of fulfilling this prioritized objective. In other words, there is a 'matter of degree' in which the focus is on a single objective to be maximized or multiple objectives to be dealt with. This can be informed by the degree of complexity encountered, which is, for example,

124   *Navigating the plural*

manifested in the degrees of conflict and clarity, the degrees of incongruence and compatibility between objectives and the number of varying opinions concerned. Indeed, we see how complexity can be considered as an argument for shifting from a technical/instrumental rational approach and its focus on a single fixed objective, towards accommodating multiple objectives within a communicative rational approach.

Structural choices

When it comes to *structure*, or what De Roo (2003) calls 'the pattern of relationships', the two extreme choices are a strongly centralized or a strongly decentralized structure. Full 'centralization' is closely related to the 'governance through coordination' model. It is the kind of organizational structure that Teisman (1998) refers to as unicentrism, which he also associates with hierarchy, formalized control and centralization. It is also a model that can be associated with what Mintzberg (1983) calls a centralized bureaucratic one, Hanf and Scharpf (1978) call top-down steering, and which Smith (1996) refers to as 'autocratic', according to her a top-down approach where managers decide a plan and others follow. De Roo (2003) summarizes this position as 'central guidance', which he associates with a limited number of responsible participants in the decision-making process who focus on achieving their prioritized objective(s). Important benefits of a centralized approach relate to the effectiveness associated with hierarchical control and the efficiency of relying on routine bureaucratic procedures as associated with the governance-through-coordination model and a technical/instrumental rational approach.

At the other extreme we find full decentralization, i.e. a pattern of relationships that is based on the interaction of a theoretically infinite number of players (De Roo 2003). It is a structure associated in the literature with decentralized governance (e.g. Börzel 1998), pluricentrism (Teisman 1998), what might even be called 'participative' (Smith 1996), implying a 'bottom-up' approach (Hanf & Scharpf 1978). Similarly, while referring to Chapter 2, it is also an extreme that is associated with participative and interactive planning approaches (e.g. Healey 1997, Innes 1996, Nijkamp, Begeer & Berting 1996 and Sager 1994).

Under conditions of limited complexity there is wide societal support for the kind of objectives that need to be met, while there are also no conflicting alternative objectives that are prioritized. Given the benefits associated with a centralized approach, such an approach is typically considered an appropriate way of organizing governance under conditions of limited complexity (e.g. Mintzberg 1983), and hence with limited uncertainties (see also Christensen 1985, Teisman 1998 and Thompson & Tuden 1959). It is, as Christensen (1985) explains, a situation where "public action may be prescribed through standard, routine procedures set into a replicable program" (1985; p. 64). However, under more complex situations such an approach is often not considered appropriate. Under such conditions, multiple intervening variables exist, such as competing interests and power struggles (e.g. De Roo 2003, Mintzberg 1983), and situations are

Navigating the plural 125

characterized by unique circumstances and more dynamics (e.g. Keuning 1978, Lawrence & Lörsch 1967); hence, we are surrounded by increased uncertainties (e.g. Christensen 1985, Geurtsen 1996, Teisman 1998, see also Van der Graaf & Hoppe 1996). Including more parties is now relevant – if not crucial – for both 'making sense' of the different perspective and preferences held, and for responding to the specific local circumstances. We therefore shift focus towards a communicative rational approach where people can connect various interests, ideas and objectives for producing strategies that carry meaning within the specific context in which these strategies are applied. The underlying idea is that these policies can identify and respond to (local) relations between different policy issues and objectives, and take the related interests and perspectives of the actors involved into account. Identifying and responding to such relations and their specific manifestation in a specific (local) context is problematic for a single actor, and certainly the central state (i.e. it will be outside of their span of control). Instead, the proximity of decentralized units allows them to bring collaborating and competing stakeholders together in a network of bargaining and collaboration, and hence for adapting to context-specific and often local power dispersal and social fragmentation (compare Hooghe & Marks 2001). Therefore, shifting to a communicative rational approach is not just a shift away from a focus on maximizing the delivery of a predefined goal to multiple composite goals. It is also a shift away from a structural focus on centrally guided bureaucracies towards a focus on optimizing the process that can benefit from the self-governing ability of various local stakeholders and actors in a decentralized structure.

In between these two extremes, there are again multiple positions to be considered. Degrees of multiple dependencies, centralization and the number of actors involved are among the causes of this. After all, full centralization and decentralization are merely theoretical positions (e.g. Flynn 2000, Prud'homme 1994, Rhodes 1981, compare Barret & Fudge 1981, Elmore 1979, Sabatier 1986). Again we find that it is a 'matter of degree' in which structures are dominated by central guidance or decentralized interaction and participation. And again, the degree of complexity can be considered as an argument for shifting from the central guidance associated with a technical/instrumental rational approach, to a decentralized structure associated with a communicative rational approach.

Bringing it together

We can now see how the technical/instrumental rational approach and the communicative rational approach are associated with opposing extreme positions along both the spectra of functional and structural choices. A technical/instrumental rational approach is associated with both a focus on a single fixed goal and a centralized structure. A communicative rational approach, contrastingly, is associated with a focus on multiple composite goals and a decentralized governance structure. Consequently, if we shift the focus from a technical/instrumental rational approach to a communicative rational approach, we also contingently alter both our functional ambitions from 'single fixed foals' to 'multiple composite goals'

126 *Navigating the plural*

and our structural configurations from 'central guidance' to 'participative interaction'. If we fail to do so, we will encounter combinations of function and structure that show mismatches (see also Box 3.3). Alternatively, in the words of Payne (2001), they are combinations with a lower degree of 'internal harmony'.

The configurations that show the kind of internal harmony just described are found close to the diagonal between technical and communicative rationality as depicted in Figure 3.3. These are configurations that can be considered to match well or, as De Roo (2004) puts it, 'optimal' in terms of their effectiveness (performing the function) and efficiency (structuring smartly). This is not to say that each function has a detailed structure to go along with it. It merely indicates that only a range of structural characteristics can be expected to work well (match) given a certain set of functions to be performed. The result is that any choice regarding the functional focus to pursue has consequences for the structural focus and vice versa. In other words, the contingency between function and structure allows us to formulate arguments for relating choices regarding the function of an approach with choices regarding its structure.

Box 3.3   Matching and non-matching configurations

We can illustrate choosing function-structure configurations that are far removed from the diagonal of Figure 3.3 indeed have drawbacks as compared to those close to the diagonal. On the one hand, we encounter such a situation if we focus on participative interaction while holding on to a single fixed target. Participative interaction requires various stakeholders and individual citizens to defend the objectives they consider most important. However, there is no room for altering the objective already set. Consequently, participative interaction is not just rather pointless, but it might also increase frustration and hence resistance to policies, while it claims resources for organizing debates that make no contribution to solving the issue. On the other hand, we also encounter drawbacks if we focus on central guidance while accommodating multiple composite goals in a tailor-made approach. As I argued, proximity of local authorities to the local circumstances, where the groups and people affected by the issue and its possible solution strategies are part of the process, puts them at an advantage over central agencies. Certainly, central agencies will have difficulties overseeing all the detailed circumstances involved and balancing all competing objectives as it is simply outside of their control.

First, we are advised that a decision to prioritize fulfilling a single fixed goal as effectively as possible should be accompanied by a centralized structure. Alternatively, a decision to prioritize combining or balancing various objectives in multiple composite goals should be accompanied by a decentralized structure and the related structural focus on optimizing the process of policymaking and delivery. Second, we are advised that a decision to prioritize the efficiency of routine procedures within centrally guided bureaucracies should be accompanied by a focus on single fixed goals. Alternatively, a decision to optimize the process of policymaking and delivery, allowing various stakeholders to 'make sense' of the specific problem conditions faced and the kind of actions they consider appropriate, should be accompanied by a focus on multiple composite goals.

These contingency arguments do not just help decision makers to find matching configurations of function and structure. They also help decision makers see that choosing a governance approach will have consequences in terms of both the kind of functional ambitions and structural configurations they either rely on or not. Decision makers know in advance that the degree to which they prioritize the benefits of fulfilling predefined objectives as effectively as possible is negatively associated with the capacity to involve various groups in society. Similarly, they know that the degree to which they prioritize the benefits of combining various conflicting interests into strategies that 'make sense' to those involved will reduce the capacity to benefit from the efficiency associated with centralized bureaucracies. The result of these arguments is that decision makers are suggested to choose between these different sets of benefits. In other words, decision makers have access to a priori knowledge of the kind of consequences they can expect when they choose a specific governance approach. This knowledge can help them to choose between different approaches, based on which of these consequences they prefer. Importantly, this knowledge is not based on an object-oriented focus and assessment of the contextual circumstances faced. Instead, it is based on accepting that institutional variables related to the functional and structural focus are best combined into matching configurations.

3.10 Navigating the plural

This chapter was about finding arguments that can help us choose between various approaches to governance and hence for navigating the plural governance landscape. As we noted, finding such arguments is constrained by the philosophical and socio-cultural plurality that underlies the plurality we face in our governance landscape. This manifests itself in accepting that claims to what is real and rational are at least partly influenced by the intersubjective context in which the claim is made. As we all have our own context, we will also value claims to reality and rationality in light of this context. As such, we are faced with a tendency to relativism that, in a radical sense, prevents us from developing generically accepted criteria or arguments for choosing between various claims and hence various governance approaches. After all, if various governance practices draw upon highly disparate ideas about what is 'real' and 'rational', then where is the common ground serving as a starting point for developing arguments for choosing between them that can be generically accepted?

While I accept that objective knowledge and absolute claims to rationality are impossible, I also argue that a radical relativist perspective need not be taken. Despite a rejection of the possibility of a definite closure between various

128   *Navigating the plural*

plural perspectives, I support most post-positivist perspectives in arguing that there are still viable arguments that can help us choose between various governance approaches. To do so, I have embraced the post-positivist conclusion that both the object-oriented focus associated with a realist perspective and the intersubjective-oriented focus associated with a relativist perspective can help us. Consequently, my proposal for a post-contingency approach draws from both an object-oriented and intersubjective-oriented focus.

A post-contingency approach: conclusions

An object-oriented focus emerges in a post-contingency approach by accepting that the performance of different organizational structures and strategies is contingent on the contextual circumstances encountered. These contextual circumstances are defined in terms of the perceived degree of complexity encountered. I on the one hand associate complexity with the uncertainty we face regarding the phenomena we address, the cause and effects relations involved, and the stability of the contextual influences on these phenomena. On the other hand, I also associate complexity with the uncertainty regarding the societal context where people might attribute different meanings to the issues faced, manifesting itself in diverging preferences, interests and ambitions. I subsequently concluded that the perceived degree of complexity provides us with an argument for choosing between various approaches to governance. To explain this argument, I have differentiated between various approaches to governance by the difference between technical/ instrumental rational or communicative rational approaches. Defining issues in terms of their degree of complexity has subsequently allowed us to gain insight in the consequences we can expect if we shift from a technical/instrumental rational to a communicative rational approach under different circumstances.

The object-oriented focus of a post-contingency approach takes inspiration from the classical approach to contingency in accepting that the performance of different organizational structures and strategies is contingent on the contextual circumstances encountered. However, my proposal for a post-contingency approach does differ from classical approaches to contingency. These classical approaches suggest that we can adjust organizational structures and strategies for reaching a set of predefined objectives, given the possibility to acquire objective knowledge about the circumstances concerned. In accepting predefined objectives to be achieved as effectively and efficiently as possible, they furthermore operate from a technical/instrumental rational perspective. In doing so, these approaches on the one hand rely on the assumption that planning and policymaking take place in a context of known and widely accepted predefined objectives. On the other hand, these approaches also rely on the assumption that knowledge acquired about the issues and circumstances faced will be sufficiently clear and convincing to be undisputed by various groups and people. In holding on to the two aforementioned assumptions, however, classical approaches to contingency tend to overlook that planning and policymaking take place in a societal realm where interpretations and values often matter more than facts. Therefore, these approaches overlook the

Navigating the plural 129

fact that issues and circumstances are often perceived and interpreted differently, and that people tend to have different preferences and interests. It is in accepting the importance of this social context in which issues and objectives are interpreted and carry different meanings that a post-contingency approach adds value to classical approaches to contingency.

A post-contingency approach first accepts that different and conflicting perspectives and preferences in the social context should be considered part of the contextual circumstances to which we can adapt organizational structures and strategies. It does so by defining contextual circumstances by their degree of complexity, which also includes possible differences regarding the interpretation of the issue faced and the objectives to pursue. Second, post-contingency also addresses differences in the societal context by bringing contingency into an intersubjective realm. It does so by shifting perspective from contingency as a response to a *matter of degree* of complexity to contingency as a *matter of choice*.

In shifting perspective to contingency as a matter of choice, the societal context is no longer confined to a contextual condition that influences the degree of complexity involved. Instead, I argued that defining issues in terms of their degree of complexity and subsequently choosing the approach to be used is done by people who themselves also hang on to their own values, interpretations and preferences. Obviously, relying on an object-oriented approach still allows us to gain an understanding of the issues and circumstances we face. It thus helps us to also gain an understanding of the consequences we can expect if we shift from a relative reliance on a technical/instrumental rational approach to a communicative rational approach. However, this does not imply that we will either be certain about this degree of complexity, or that we will agree on how we can best respond to this perceived degree of complexity. I therefore in the first place accept that *defining issues in terms of their degree of complexity is socially mediated*, as decision makers perceive and interpret the issues and circumstances using their own frames of reference and dominant socio-cultural values. Second, how decision makers subsequently *respond* to the perceived degree of complexity is also socially mediated, as people will value and judge the circumstances differently, i.e. what they think should be done will be different. Hence, I argue that how issues are seen and how they are subsequently approached is, by and large, a socially mediated choice.

In bringing contingency into an intersubjective realm, we also are brought back into a relativist realm. With my proposal for a post-contingency approach I have, however, explained that this does not deny us the possibility to formulate contingency rules that help us to investigate and predict the consequences of choosing between various approaches to governance. On the one hand, we can still rely on an object-oriented focus for doing so. Albeit limited, this still helps us gain understanding of the consequences of shifting from a relative reliance on a technical/instrumental rational to a communicative rational approach given the perceived degree of complexity encountered. On the other hand, the contingency between function and structure helps us see these consequences even without having a priori knowledge of the situations encountered or, more specifically, the degrees of complexity encountered. Decision makers know in advance that the degree to

130 *Navigating the plural*

which they prioritize fulfilling predefined objectives will reduce the capacity to involve various societal groups in a decentralized setting. Similarly, they know that the degree to which they prioritize the benefits of combining various conflicting interests in strategies that 'make sense' to those involved will reduce the capacity to benefit from the efficiency associated with centralized bureaucracies. The result is that decision makers will have to choose between these different sets of benefits. So even if we accept that this is indeed a choice, a post-contingency approach makes them aware of the consequences, and thus informs making such a choice.

Applying a post-contingency approach

We can now also combine the object-oriented and intersubjective-oriented focus and see how a post-contingency approach helps us navigate the plural governance landscape. Based on an object-oriented focus, decision makers will gain information about the issues they face and the circumstances surrounding this issue. This information will help them to estimate the degree of complexity they face, where increased conditions of complexity provide an argument for shifting from a technical/instrumental rational approach to a communicative rational approach.

How we respond to differences in the perceived degrees of complexity is, however, by and large a matter of choice. Hence, responding to a perception of increased conditions of complexity by shifting the focus to a communicative rational approach is not the only option. Given the contingency between function and structure, we know that if we prioritize a certain set of objectives that must be met, we are advised not to allow for the kind of participative interaction associated with communicative rational approaches. After all, objectives will become debatable in a decentralized setting. This in no way guarantees that all objectives will be addressed or to what degree possible objectives will be met. Rather, we end up with multiple composite goals where, for example, the knowledge of local circumstances, the relative power base of the stakeholders involved and their argumentative power will decide the actual course of action. Similarly, we know that if we prioritize the benefits associated with functional specialization of centralized bureaucracies, we are advised to accept that pursuing multiple composite objectives becomes difficult. Doing so will be easily outside the span of control of central agencies, and hence we are advised to decentralize so as to enable local stakeholders to develop strategies based on the local circumstances in an area-based approach.

The consequence is that we are advised to pursue a shift towards a communicative rational approach if the benefits of allowing stakeholders to be involved and of developing strategies that are tailored to the local circumstances outweigh the benefits of the effectiveness of fulfilling predefined objectives and the efficiency of central guidance in bureaucracies and vice versa. A post-contingency approach thus argues that a perceived increase in the degree of complexity should indeed not be considered to dictate a shift from a relative reliance on a technical/instrumental rational approach towards a communicative rational approach. Instead, it argues that the perceived degree of complexity provides a reason for such a shift, but whether this reason is pressing enough should *also* be interpreted within a context of alternative arguments related to the benefits of a technical/instrumental rational approach, such as the effectiveness and efficiency of delivering policy outcomes and the desire for protecting 'weak interests'. Vice versa, benefits of a communicative rational approach for allowing multiple stakeholders and local citizens to be involved might also be considered as motives for shifting towards a communicative rational approach under conditions of limited complexity. We however now also know that this can have consequences such as a reduced capacity to pursue a priori objectives and to benefit from the routine implementation of common policy formats. Either way, the question is simply which of the arguments in favour or against various approaches to governance are accepted and prioritized. A post-contingency approach does not provide us with an answer to this question but provides us with clarity as to which arguments are concerned for navigating the plural governance landscape.

Towards navigating environmental policy

In the following chapter I will continue our discussion by employing a post-contingency approach within the realm of environmental policy. As I argued in Chapter 2, environmental policy is an example of a policy field where we should be prudent when stepping out of the confines of the coordinative model and its associated central guidance. As discussed, the protection of humans and ecosystems from adverse environmental consequences provides a key argument for a swift achievement of policy outcomes that also seem to have a high degree of certainty. They are motives for holding on to 'single fixed goals', at least to some degree, and thus, in line with the contingency of function and structure, motives for being prudent when pursuing decentralization in environmental policy. I also discussed that the 'weak profile' of environmental interests and the social dilemma that is involved in dealing with many environment issues provide further stimulus for support by central policy objectives or procedural formats. Finally, the equality of delivering similar levels of protection in various jurisdictions is yet another argument for supporting these central policies. Again, decentralization in environmental policy is confronted with important counter-arguments. These arguments matter even when conditions of complexity seem to urge that decentralization accommodate communicative rational approaches.

A key conclusion is that if we are pursuing governance renewal in environmental policy, we should be especially prudent in responding to increased conditions of complexity with more communicative rational, flexible and adaptive or area-based approaches. I acknowledge that it might be hard to hold on to a technical/instrumental rational approach in the face of the complexity caused by social fragmentation and power dispersal. Power dispersal means that governments have lost their relative monopoly in structuring the governance landscape, i.e. governance renewal is also an autonomous process. This also implies that the arguments a post-contingency approach can produce do not imply that decision

132 *Navigating the plural*

makers are 'free' in making their choices based on the arguments. Our societal reality simply forces us to accept that dominant political and ideological arguments exist, that previous choices, contracts or investments also constrain current choices and, also, that power has a key influence on such a freedom. I argue, however, that governance renewal usually does involve deliberate choices made by governments, even if they are constrained by the pressure due to processes such as social fragmentation and power dispersal. I propose arguments that can help to make these choices, in particular when it comes to renewing and restructuring the governance landscape. In other words, I aim to provide arguments that help us to establish when, to what degree and within which framework of 'checks and balances' decentralization in environmental policy can be pursued.

4 Making decentralization work

In recent decades, we have witnessed an increased role of the local level in environmental policy in the Western world. It is both a response to the decreased support for the coordinative model of governance and its related central guidance, and to the emergence of new governance ambitions such as the pursuit of proactive and integrated approaches to the environment. Consequently, decentralization has become one of the strategies for renewing environmental policies in many western European states (e.g. Bergström & Dobers 2000, de Roo 2004, Gibbs & Jonas 1999, Hovik & Reitan 2004, Lemos & Agrawal 2006, Van Tatenhove et al. 2000 and Wätli 2004).

Decentralization is often surrounded by much optimism regarding its capacity to respond to the limitations of the coordinative model. This optimism can easily cause decentralization to take place without a keen understanding of its advantages and disadvantages. Prud'homme (1994), for example, recognizes that decentralization is often pursued merely for being a 'fashionable idea', dominated by political motives (see also Fleurke & Hulst 2006). However, as we discussed in Chapter 2, decentralization is also surrounded by counterarguments whilst not being exempt from important risks. In addition, many authors conclude that there is still a limited understanding of why and how decentralization measures should be taken (e.g. De Vries 2000, Fleurke et al. 1997, Gershberg 1998 and Prud'homme 1994). Therefore, as Prud'homme (1994) argues, we should be well aware of the 'dangers of decentralization'.

With a post-contingency approach I have developed arguments that help us to choose between various governance approaches and, hence, also to choose between levels of centralization or decentralization. In this chapter I will explicitly connect a post-contingency approach to debates on decentralization operations in environmental policy. In doing so, I will also return to the argument made in Chapter 2 that discussions on decentralization can be positioned in a multilevel perspective on governance. Decentralization, then, is viewed as a relative shift of power and authority within a context of remaining or newly created central policies. In this, I follow up on the suggestion made by De Vries (2000) that debates on decentralization should be positioned within a search for "an optimal institutional arrangement [that] fits the specific situation in a specific area in a specific country given the specific problems at stake" (p. 220). In this chapter I thus draw

134   *Making decentralization work*

upon a post-contingency approach to find arguments for both choosing between various degrees of (de)centralization and for establishing how these can be linked to discussions on how central policies and decentralized governance can interact (compare Ashford 2002, Fleurke et al. 1997, Gershberg 1998 and Prud'homme 1994).

Section 4.1 will pick up on the first suggestion following a post-contingency approach, i.e. that the perceived conditions of complexity as defined in Chapter 3, provide an argument for more centralized or decentralized governance formats. Here I will discuss how relying on centrally issued common policy formats is constrained by their ability to respond to local circumstances. The second suggestion made in following a post-contingency approach is that we should refrain from adopting a deterministic perspective on connecting the degree of complexity with the degree of decentralization. Rather, responding to complexity is a matter of choice that can be inspired by alternative arguments. In sections 4.2 and 4.3 I discuss such alternative arguments. I do so by explaining how decentralization has consequences. The outcomes of governance will become increasingly dependent on local performance and, hence, be influenced by the available local willingness and ability to perform decentralized tasks and responsibilities. As I explained earlier, this reduces the capacity of governance to focus on predefined single fixed goals, while it also increases risks associated with undermining environmental performance given the 'weak profile' of the environment in relation to alternative local priorities and stakeholder interests. Decentralization thus takes place in a realm of more arguments than just the desire to respond to conditions of complexity.

Section 4.2 begins by exploring arguments relating to the functional ambition to *guarantee* minimal levels of performance in local environmental governance. Following the contingency between function and structure, decentralization will involve a shift from focusing on single fixed goals to multiple composite goals. Such multiple composite goals are designed to integrate or balance various local interests and perspectives. There might, however, be single fixed goals that continue to be prioritized and are not considered suitable for balancing with other goals. I will explain how the ambition to maintain minimal levels of protection for ecosystems and people against environmental stress can constitute such a reason. It provides an argument for retaining central governance formats and the associated coordinative model, even if we face conditions of complexity.

Section 4.3 explores arguments that relate to possible limits to the *willingness* and *ability* of local authorities and stakeholders to perform tasks that are decentralized. Decentralization is supported by the assumption that local authorities are at an advantage in comparison to the central state when it comes to developing and delivering more proactive, integrated and tailor-made approaches. Hence, it is also assumed beneficial under increased conditions of complexity that require such approaches. This assumption leaves open the question of whether and to what extent local authorities and stakeholders are willing and able to do so. In section 4.3 I will, however, discuss some important constraints on local willingness and ability. These constraints subsequently provide us with important arguments

Making decentralization work 135

for retaining at least a degree of central control. I will explain that the coordinative model can play a key role in setting the conditions required for decentralization and the associated use of proactive, area-based and integrated approaches in order to realize their envisioned outcomes. Hence, if we are to pursue governance renewal and its associated desire to rely on more dynamic approaches in a local setting, we should still address the question of how the coordinative model can help us in providing a 'robust and stable' foundation upon which to build these new and dynamic approaches. The constraints discussed relate to the economies of scale involved in policy development and delivery, the possibility of issues with external effects and the 'weak profile' of the environment.

Following on from sections 4.1 to 4.3, I will conclude that central governance formats and the associated coordinative model of governance still have some important benefits. These benefits are not only associated with conditions of limited complexity. Central policies can also be relied upon to respond to the constraints of local willingness and ability and to guarantee minimum levels of local performance. Consequently, central policies can be an important instrument for setting the conditions required for decentralization to result in its envisioned outcomes. While doing so, however, we must realize that central policies put pressure on local resources and constrain local freedom to act. In addition, we have seen how reliance on central governance formats and the coordinative model of governance involve the risk of promoting fragmented policy agendas, complicated regulations and excessive bureaucracy (see also section 2.3). This can undermine local authorities' abilities to deal with decentralized tasks and responsibilities. In section 4.4 I therefore discuss possibilities to prevent such problems.

In section 4.5 I will repeat and combine the arguments presented in this chapter. The key conclusion I will work towards in this chapter will also be formulated, namely that central guidance and the coordinative model of governance can be important in providing a 'robust foundation' of policies for employing and supporting decentralization so as to improve the governance capacity for developing and delivering proactive, integrated and tailor-made approaches.

4.1 Decentralization in response to conditions of complexity

Decentralization, seen here as the devolution of power and authority away from the centre towards the locality, is one of the means for reducing the reliance on the coordinative model and its associated technical/instrumental rational approach. Instead, by bringing decision making closer to local stakeholders and circumstances, it can facilitate a shift towards more flexible, adaptive and communicative rational approaches. Therefore, decentralization can be seen as a means for adapting to conditions of increased complexity.

Increased social complexity

The popularity of decentralization in recent decades can be associated with the increased acceptance that the coordinative model on which governance

136   *Making decentralization work*

approaches in the early post–Second World War period largely relied is often incompatible with the societal complexity we face (e.g. Hooghe & Marks 2001, Kooiman 1993, Nelissen 2002, Pierre & Guy Peters 2000 and Van Tatenhove et al. 2000). In Chapter 3 I associated complexity with uncertainty we face regarding both the cause and effect relations of the phenomena we mean to address (the knowledge we have of the issue and circumstances faced), and uncertainty regarding the perceptions, interpretations and ambitions various actors might hang on to (the knowledge we have of the different meanings that surround the issue). Some important societal changes have contributed to the acceptance that such uncertainties are widespread and have even increased in recent decades.

On the one hand, social fragmentation has fuelled an increase of different interpretations regarding the challenges that policies mean to address and the various ambitions that groups and individuals hold on to. In the meantime, various societal and market parties are also claiming their place in the process of governance and, as expressed with power dispersal, also have the resources to exercise influence. Social fragmentation and power dispersal combined therefore challenge a monistic approach to governance whereby the central state controls policy development and delivery. In the face of the resulting socio-cultural plurality, support for the supremacy of central state control is undermined, while the central state is also increasingly confined in its capacity to dominate societal or market parties. On the other hand, we also face an increased acceptance of the interrelation between problems, their causes and effects. Consequently, there are usually multiple and potentially conflicting objectives to which policies should respond, while multiple parties with their own interests and resources are also involved. This is further stimulated by the increasing focus on more proactive and integrated approaches, such as sustainable development, that are intended to pursue cross-sectoral policy approaches. Relying on the coordinative model is not only undermined in such cases due to the mutual dependence between public and private parties. It has also been shown to have practical drawbacks, manifesting itself in, for example, fragmented policies, implementation deficits, a reactive focus on solving issues, and symptoms of management overload. The development of cross-sectoral policies is therefore problematic within the coordinative model of governance. This is especially true in also connecting such cross-sectoral approaches to unique local circumstances, as doing so is often outside of the 'span of control' of central governments. Consequently, the coordinative model is also increasingly considered incompatible with the need and desire to respond to interrelated policy issues.

In bringing decision making closer to the local level, decentralization facilitates local parties and people to influence policy development and delivery. It allows political constituents to exert influence over the issues they face in their own areas based on representative democracy. It also increases the possibility for direct democratic involvement that is hard to organize at higher levels of authority. While decentralization increases the possibility for local groups to exert influence over governance, it also increases the capacity of governance to respond to local circumstances. Indeed, a key purpose of decentralization is to help "local government to act pragmatically and develop locally contingent solutions to

Making decentralization work 137

problems rather than feeling compelled to fit with guidance" (Coaffee & Headlam 2007; p. 1595). The proximity of local units to local circumstances and related stakeholder interests can help local parties to replace the pursuit of fragmented policy ambitions in the face of interrelated policy issues with connecting issues and ambitions and various stakeholder interests related to these issues and policy ambitions. It allows local authorities to develop policy approaches in a dynamic local setting that are tailored to the local circumstances.

Inspired by these benefits, decentralization can, among other things, be seen as a response to increased social complexity. As there is a wide acceptance of the growing complexity of our late 20th- and early 21st-century societies (Chapter 2), it is hardly surprisingly that decentralization is among the popular means of governance renewal. However, I argue that we should not conflate an increased social complexity with the idea that 'all is complex'. Nor should we deny the benefits associated with the coordinative model and its associated central guidance. In response, I argue first that conditions of complexity can be considered as a criterion in choosing between various degrees of centralization and decentralization.

Complexity as criterion

I associated complexity with interrelated policy issues and diverging political, societal and stakeholder perspectives on how to deal with these issues. Strategies for solving such issues should identify and respond to relations between different policy issues and objectives, and take account of the related interests and perspectives of the actors involved, not rarely manifested in unique local configurations. However, not all or even most issues are characterized by conditions of complexity. Instead, many issues still have a rather common manifestation in various localities and are surrounded by a high degree of consensus as to how they ought to be addressed.

Conditions of limited complexity are here associated with a relatively high degree of certainty regarding the issues faced, their causes and effects and converging ideas regarding the objectives to pursue and how best to pursue them. In that case, relying on a technical/instrumental approach and the related coordinative model of governance has some important benefits. After all, a technical/instrumental rational approach is about fulfilling predefined objectives as effectively and efficiently as possible. Efficiency should then be enhanced due to a routine implementation of policies based on functional specialization in bureaucracies and its related economies of scale (see also section 2.2). Effectiveness is to be enhanced through strong lines of hierarchical control in the coordinative model, which means to enhance the capacity of delivering policy outcomes in various jurisdictions both with a high degree of certainty and equally. Under conditions of limited complexity it is also possible to focus on fulfilling predefined objectives without the risk that doing so will conflict with other local issues or objectives and diverging stakeholder interests. After all, such conflicts are either limited under conditions of low complexity, or there is consensus on which objectives to prioritize. Generic policies that are issued by central governments are now likely to be

138   *Making decentralization work*

both effective and efficient. In conclusion, under conditions of limited complexity, the coordinative model of governance, and its related technical/instrumental rational approach and central guidance, is a well-substantiated choice.

While centralization can be related to conditions of limited complexity, decentralization is a response to increased conditions of complexity. In Chapter 2 I discussed decentralization as a response to dealing with issues that are strongly embedded in their local context. These are examples of issues that are often more complex. Such complexity manifests itself in a strong relationship between different issues (interrelatedness), between different priorities and associated stakeholder interests (power dispersal) and in a divergence of local preferences (social fragmentation). Central policy objectives are now more difficult to deliver. Fulfilling them will have consequences for fulfilling other policy objectives that are also considered a priority. Furthermore, there is uncertainty over how causes and effects relate. This not only limits our ability to predict the consequences of our actions, but also results in different interpretations and expectations of what causes should be addressed and what consequences of our actions to expect. While this already provides a strong ground for societal debate, various societal groups and stakeholders also tend to have different opinions about balancing and prioritizing alternative objectives. Within such a societal debate under more complex circumstances, many stakeholders tend to have access to resources such as land, funding, knowledge, political pressure or the media. They also have the capacity to block or frustrate policy delivery. A communicative rational approach can be pursued in response to this, although power relations might constrain this (see Chapter 7).

A communicative rational approach (section 3.5) relies on more dynamic policy approaches. This involves an interactive process in which different actors can express and negotiate their differences and, on that basis, come to a joint perspective on which actions are considered to 'make sense'. The underlying idea is that the policies developed are *area-based* and, hence, tailored to the local circumstances. Therefore, these policies mean to identify and respond to local relations between different policy issues and objectives, and take account of the related interests and perspectives of the actors involved. Identifying and responding to such relations and their specific manifestation in a local context is problematic for central states (i.e. it will be outside their span of control). Decentralization is supposed to "bring decision making closer to those affected by governance, thereby promoting higher participation and accountability; and finally, it can help decision makers take advantage of more precise time- and place-specific knowledge about natural resources" (Lemos & Agrawal 2006; p. 303, compare section 2.5). The proximity of decentralized units is argued to be beneficial in allowing local authorities to bring together collaborating and competing stakeholders in a network of bargaining and collaboration and, hence, adapt to local circumstances, dynamics and the existing power dispersal and social fragmentation (compare Hooghe & Marks 2001). In other words, decentralization is intended to facilitate a more flexible, adaptive and communicative rational approach. Furthermore, decentralized units are also considered to have more potential to respond to the interrelatedness and the related balancing of various interests and objectives. Therefore, they are also considered a good place for developing more dynamic area-based approaches and for developing and delivering proactive and integrated approaches to the environment (see also section 2.6).

As explained in Chapter 3, a technical/instrumental and communicative rational approach can be seen as opposing ideal type approaches. The same goes for a theoretical distinction between full centralization and decentralization. In practice, there will instead be a more nuanced distinction where central guidance will influence local governance and possible interactive approaches with differing degrees. The complexity of the circumstances encountered can be seen as an argument for increased levels of decentralization. The idea is then to establish a match between the structure of the decision-making process and the external environmental conditions expressed in the degree of complexity. We can also visualize this relationship, as shown here in Figure 4.1. In Chapter 3 I related increased levels of complexity to both a shift from technical/instrumental to communicative rationality and shifts considering the function and structure of various approaches to governance. In this context, decentralization essentially involves a shift from central guidance towards participative interaction, i.e. it is a change in the structure of governance. However, in accepting the contingency between structure and function, decentralization is also about shifting from single fixed goals towards multiple composite goals. Decentralization is thus both a structural change (transforming the intersubjective focus towards increased participation and debate) and a functional change (decision makers focus on combining and balancing objectives).

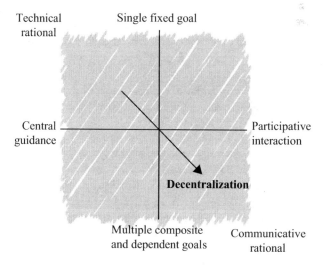

Figure 4.1 Decentralization in the framework for planning-oriented action (see also De Roo 2004)

140   *Making decentralization work*

More to be said

Variations in the perceived conditions of complexity provide decision makers with a first argument for choosing between different degrees of centralization or decentralization. The message: we should retain centralized governance formats and the associated technical/rational approach if we are faced with conditions of limited complexity. Decentralization is instead seen as a means to adapt to conditions of increased complexity, where governance approaches can be developed that are tailored to the unique circumstances surrounding these more complex issues. In adopting a post-contingency approach I have also argued that doing so is by and large *a matter of choice*. Although conditions of increased complexity might provide an argument in support of decentralization, decentralization also has consequences that might result in arguments against it. It is in the next sections (4.2 to 4.4) that I will explain that such arguments indeed exist and also are important for informing choices regarding degrees of centralization and decentralization and also, how we might combine central and decentralized governance formats.

4.2 A focus on protection

Decentralization involves a shift away from a focus on fulfilling single fixed goals towards developing multiple composite goals in an area-based setting. This is also among the motives supporting decentralization. Decentralization allows local authorities, supported by local democratic decision making, to balance the various costs and benefits involved in pursuing environmental ambitions as compared to other local priorities and interests. Single fixed goals, previously of an obligatory nature and supported by hierarchical control, therefore become debatable in the local realm. Decentralization results in a relative shift away from such hierarchical control towards an increased dependence on local authorities and stakeholders to set their own targets and ambitions. Thus, while decentralization *increases* the potential for developing integrated policy approaches and multiple composite goals in a decentralized setting, it *decreases* the capacity of governance to meet single fixed goals. It shows us that decentralization also involves a functional choice regarding the degree to which governments intend to ensure that single fixed goals are prioritized in a local realm.

One of the dominant functions of environmental policies is the protection of humans and ecosystems from certain degrees of environmental stress. Informed by scientific data, we also have access to the kind of harmful effects that can be expected if humans and ecosystems are subject to certain degrees of environmental stress. Based on this data, choices are made regarding the levels of harm that are considered tolerable and to what degree we might tolerate differences between various groups and localities regarding the degrees of harm experienced. Responding to this question is by and large an ethical choice. My intention here is not to take a position in these ethical debates. Instead, my intention is to express that *if indeed* a choice is made to install minimal levels of protection that apply equally in multiple jurisdictions, the contingency between function and structure tells us that central guidance is advisable to control local performance.

Making decentralization work 141

Guaranteeing: a focus on equal levels of protection

Many governments have chosen to install a set of environmental limit values that are designed to be implemented generically in all jurisdictions (e.g. Andersen & Liefferink 1997, Van Tatenhove et al. 2000, Weale 1992 and Weidner & Jänicke 2002). This is also true for the EU and the Dutch national government, whose approaches to stimulate proactive and integrated approaches to the environment I will address in the following two chapters. Many of these environmental limit values were developed in the early days of environmental policy, specifically the 1970s and 1980s (see also Andersen & Liefferink 1997). The aim of decentralization is to increase the power and responsibility of local governments to decide on their own environmental targets and objectives, based on assessing local circumstances and priorities. As a consequence, it often involves a reduced impact of (inter)national standards. Rather, the outcomes of environmental policy become increasingly dependent on local intentions and performance. In doing so, decentralization has at least two key consequences related to pursuing minimum levels of protection.

The first consequence is increased uncertainty regarding the outcomes of environmental policy. After all, deciding on the environmental quality ambitions which will be pursued increasingly becomes a local responsibility. The outcomes of such decision-making processes will depend on aspects such as local policy priorities, the political colour of governments and the relative power base of the stakeholders involved. Central governments must accept that they have little or no guarantees as to the kind of ambition levels that are pursued locally. In addition, they have to accept a second consequence. As different localities will face different problems and priorities and have different social or political preferences, decentralization also increases the diversity of environmental quality ambitions that will be pursued in different localities.

Both these consequences might result in a conflict between decentralization and the desire of central governments to retain minimum levels of protection that apply equally to all localities. Decentralization involves the risk that some localities will fall below these minimum levels of performance if they are no longer supported by central guidance. If this is considered unacceptable by central governments, they should retain policies in order to guarantee that this will not occur. It is essentially a decision to focus on fulfilling a set of single fixed goals, such as environmental standards, that obliges localities to maintain minimum levels of protection. To this end, we know on the one hand that it is efficient to rely on central guidance to translate the single fixed goals into common policy formats that can be routinely implemented. On the other hand, we also know that the hierarchical control exercised in a centralized setting can promote effective implementation of such goals in all localities. Here I confirm the message of Chapter 3, namely that a decision to focus on single fixed goals is associated with centralized governance formats.

We can also visualize the consequences of installing minimum levels of protection for local performance, as in Figure 4.2. The left-hand side of Figure 4.2 shows the kind of normal distribution of the performance of various localities

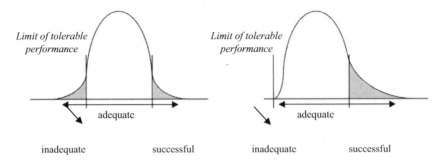

Figure 4.2 Possible spread of performance of local governance (see also De Roo 2004; p. 69)

that could be expected without any central guidance. Variations in local problems, priorities and social or political preferences will contribute to this type of normal distribution. We can also see the possible consequence that the performance of some localities might not meet the standard considered adequate by central governments (the limit of tolerable performance). Central guidance can respond by implementing policies that urge municipalities to meet minimum performance levels. The consequence is shown on the right of Figure 4.2. Although there is still a wide distribution of local performance, performance levels are no longer below the limit of tolerable performance.

The desire to maintain minimum levels of protection of humans and ecosystems against environmental stress provides an argument for retaining central governance formats, even if conditions of complexity increase. This does not deny the potential for a degree of decentralization, for example, in response to increased conditions of complexity. Proactive and integrated approaches can still be developed in dynamic local settings, while local authorities can also be given greater responsibilities to cope with unique local circumstances. However, minimum levels of protection do set important boundaries on local policy development and delivery. They thus impose limits on the degree of decentralization that is possible, since choices made in a local realm with regard to balancing and combining various objectives and interests should at least result in the fulfilment of minimum requirements. It is a provision that has important consequences.

Fulfilling minimum requirements will usually require investments, such as reducing the stress produced by local factories or traffic, the implementation of mitigating measures such as sound buffers or banning harmful industries or activities from sensitive areas. Such investments will be fairly modest and will hardly be disputed under conditions of limited complexity, as few interests and objectives will conflict. This is different under conditions of increased complexity. Due to the interrelatedness of policy issues and objectives, fulfilling limit values might have serious or even extreme costs. We gave some examples in

Making decentralization work   143

section 2.3, such as the need to demolish large city areas, or closing down most local industry (Borst et al. 1995, see also Chapter 7). There is little question that this could be accompanied by strong public resistance and political discussions in the local realm. The result is that we must ultimately choose between accepting the consequences of fulfilling minimum levels of protection (e.g. social, financial, economic) or accepting that these cannot always be maintained.

If central governments do not accept non-compliance with minimum levels, the choice is clear: continue to guarantee that minimum levels of protection are met and accept the consequences of this choice. The alternative is that central governments accept degrees of non-compliance with minimum levels, based on local balancing of priorities. Such a choice need not be equated with handing over full responsibility to the local level. After all, central governments can retain a degree of control over local performance. This can even be a fairly high degree of control. To illustrate this, national and international standards can be introduced that allow for a degree of deviation in the face of complex and unique local circumstances (compare De Roo 2004). Central governments can also decide on the conditions that need to be met locally to allow for such deviations (e.g. Creedy & Zuidema 2007). In addition, they can also prescribe procedural requirements that local authorities need to satisfy in the face of such deviations. These can include compensation measures, an obligation to continue improving conditions over time, using best available techniques, etc. It is an example of a modest shift away from a focus on single fixed goals and central guidance towards allowing conflicting objectives and interests to be balanced in a local setting.

Decentralization in the face of protection

In considering complexity as a criterion for decentralization, the role of the centre would be confined to dealing with issues of limited complexity. Decentralization is, then, related to conditions of increased complexity so as to allow local authorities to respond to unique local circumstances. I have argued that responding to conditions of increased complexity should, however, be considered a matter of choice. In this section I have also explored a first argument that might be used in making such a choice. The desire to retain minimum levels of protection that must be met in a local realm can be an argument for retaining central governance, even if we face conditions of increased complexity. It implies that decentralization has limits if central governments decide to retain minimum levels of protection that must be met in a local realm. Although decentralization is still possible, it should take place in a context of generic policies and hierarchical control in order to provide the desired guarantees or pressures on local performance. The consequence is that localities will be faced with generic policy ambitions that constrain their capacity to respond to unique and complex conditions. However, the desire to maintain minimum levels of protection for humans and ecosystems against environmental stress can be an important argument in favour of accepting this consequence.

144 *Making decentralization work*

4.3 Constraints on local willingness and ability

Guaranteeing minimum levels of protection is a first argument that can be used in defence of central governance formats even if we face conditions of increased complexity. In this section I will discuss yet another set of arguments that can be used to this end. They follow on from an acceptance of the fact that local authorities might be constrained in their *willingness* and *ability* to perform decentralized tasks. This can also be the case when it comes to environmental policy. We have argued that decentralization is supported by the assumption that local authorities have benefits over the central state in terms of developing and delivering proactive, integrated and tailor-made approaches. The proximity of local authorities to local circumstances and related stakeholder interests fuels the assumption that such benefits exist. In the meantime, we can also encounter important limits when it comes to the potential of decentralization if local authorities and stakeholders are constrained in their willingness and ability to develop and deliver proactive, integrated and tailor-made approaches. In this section I explain that there are indeed some key constraints upon the willingness and ability of local authorities to perform decentralized tasks and responsibilities as compared to central governments.

The literature on decentralization suggests that not all government functions are liable to be decentralized, even if conditions of complexity might stimulate decentralization (see also De Vries 2000, Gershberg 1998, Marks & Hooghe 2001 and Prud'homme 1994). Rather, this literature suggests that there are important constraints upon what can be assumed realistic and desirable when discussing decentralization. In response, decentralization can be pursued to invite and enable localities to develop their own course of action, but *within* the frames of references and stimuli provided by central policy imperatives. It is an attempt to establish a hybrid mixture of interacting central governance formats and decentralized governance. It is, in other words, an explicit attempt to position debates on decentralization in a search for what De Vries (2000) called 'optimal institutional arrangements'.

In reviewing the literature on decentralization, two main variables are discussed that affect local willingness and ability to perform government functions. These are the economies of scale involved in dealing with government functions and the external effects relating to the issues these functions are designed to address. First, economies of scale refer to the extent to which we deal with functions requiring repetitive actions and hence routine action. Second, they refer to government functions that are knowledge-intensive and hence require agencies to attract expert knowledge. In both cases, the literature asserts that increased economies of scale provide arguments against decentralization. The same goes for external effects. External effects refer to geographical spill-overs of issues within one jurisdiction into other jurisdictions. In this section I will address both of these variables. In addition, I pay special attention to what I have called the 'weak profile' of the environment (Chapter 2). As I will explain, this 'weak profile' is also an important condition affecting local willingness and ability to pursue local environmental policies.

Making decentralization work 145

Economies of scale

The coordinative model is built on the assumption that there are economies of scale involved in organizing policy development and delivery at higher levels of authority. As such, economies of scale typically provide arguments against decentralization, where "factors influencing such economies of scale are the repetitiveness, and the knowledge required" (De Vries 2000; p. 2010). To begin with, *central guidance can benefit from the repetitiveness of implementing, at a central level, common policy formats that can be applied at all lower levels of authority.* These formats include environmental standards, licensing and permit systems and common solution strategies – for example, the size of buffers needed between the source and recipient of environmental stress (e.g. a factory and nearby houses). If common formats are implemented at a high level of authority, it is no longer necessary to implement them at each lower level of jurisdiction. Local or regional governments do not need to 'reinvent the wheel' each time.

Under conditions of limited complexity, the benefits of relying on common policy formats and routine implementation confirm our conclusion that there are good reasons for focusing on common policy formats and central guidance. Under conditions of increased complexity, these benefits are accompanied by some important disadvantages. After all, routine implementation is constrained by possible local consequences of fulfilling common policy objectives in the face of unique local circumstances. There might, for example, not be enough space to apply the buffer zone just mentioned. Central governments will face serious constraints in terms of responding efficiently to such unique circumstances. This again confirms our conclusion that it can be more effective and efficient to rely on local governance in order to respond to these circumstances. This implies that local authorities and stakeholders will develop their own strategies. They should have sufficient resources for this purpose, i.e. time, expertise and staff. This brings us to a second aspect of to economies of scale: *central governments' greater capacity to attract knowledgeable or experienced staff as compared to smaller (local) units.*

As Prud'homme argues,

> Central government bureaucracies are likely to attract more qualified people, not so much because they offer higher salaries, but because they offer better careers, with a greater diversity of tasks, more possibilities of promotion, less political intervention, and a longer view of issues. Furthermore, central government bureaucracies invest more in technology, research, development, promotion, and innovation.
>
> (1994; p. 9)

Prud'homme (1994) therefore considers that, before a task is decentralized, the technical and managerial expertise required to perform that task should always be assessed. In this context, I argue that *technical* and *managerial* expertise should, however, not be considered similar.

When it comes to technical expertise, central governments can benefit from attracting specialists who work on distinct tasks following a functional

146   *Making decentralization work*

specialization in distinct government agencies. In environmental policy, functional specialization can indeed be beneficial, as it often involves fairly technical tasks (see also section 2.6). Many environmental tasks involve technical calculations (e.g. the effects of new developments on existing noise levels), computer models (e.g. air pollution, noise), risk assessments (e.g. chemical hazards, accidents, soil remediation) and are based on scientific data (such as dose-response relationships). Smaller units in particular – such as many municipal departments – cannot be expected to employ experts on all these areas. However, even large municipalities face problems in attracting or accommodating individual experts in different fields. Research in, for example, the EU (ICLEI 2004), Sweden (Burström and Korhonen 2001) and the Netherlands (Flameling 2010) also illustrates these problems. Functional specialization, so common in central government, is simply difficult to arrange at a local level. This confirms that decentralization in environmental policy runs the risk of handing over tasks to local units that will find it difficult to perform those tasks.

Managerial skills are especially relevant when it comes to proactive and integrated approaches (e.g. Bakker et al. 2005, Mackay 2005). Integrated and proactive policies require a strategic approach and a need to link various interests, priorities and ideas. As Prud'homme (1994) argues, central governments have some benefits over local authorities in terms of attracting skilled staff. In addition, they are usually able to invest more in research and development, innovation and new experimental policy approaches. However, I argue that such economies of scale should not necessarily be interpreted as a claim towards a centralized approach when it comes to developing proactive and integrated approaches. Let us not forget that the proximity of local authorities to local circumstances and stakeholders give them some important benefits in connections between various interests, priorities and ideas. Integrated and proactive policies can thus benefit from a decentralized approach. What I do argue is that decentralization with a view to stimulating local integrated proactive policies should pay attention to the kind of managerial skills needed.

Proactive and integrated approaches require a holistic attitude and skills in bargaining and communicating with key stakeholders and developing partnerships with them (see also section 2.6). These are quite different from the skills needed for implementing central policies and regulations, such as making calculations in the face of existing standards, or issuing permits. In other words, integrated and proactive policies require a different skill set. It remains to be seen whether local authorities are able to cope with them. As, for example, Burström and Korhonen (2001), Bouwman et al. (2005) and Maas (2005) illustrate on the basis of one Swedish and two Dutch case studies, local authorities are not necessarily sufficiently staffed and competent to do so. In addition, attracting new staff to respond to new tasks can also be problematic. After all, local financial resources (see also Fleurke & Hulst 2006) are limited, especially in the context of regular budget cuts that have affected many Western nations in recent decades. As a study by ICLEI (2004) into a large number of EU towns and cities indicates, this is also true when it comes to environmental policy. In my empirical study I will further assess whether local authorities have, or are able to attract, the necessary managerial skills for developing and delivering proactive and integrated approaches. In my study of Dutch practice in particular, we will also see that such skills are far from evident.

Both a lack of technical and managerial expertise can seriously constrain local performance in environmental governance. Therefore, *decentralization should be confined to tasks that require skills and resources which local units should, realistically, be able to acquire*. If this is not the case, decentralization should either not be pursued or be accompanied by central government support in order to enable localities to cope with decentralized tasks. When it comes to the development of proactive, integrated and tailor-made approaches, there are clearly benefits associated with decentralization. However, these benefits should not hide the fact that local authorities can face financial constraints, a lack of staff, limited managerial skills and unfamiliarity with tasks such as proactive and integrated approaches. In response, I argue that it is important to assess, *before* decentralization operations are instigated, whether local authorities can realistically be expected to perform decentralized tasks. If some or most local authorities are constrained in this regard, a degree of central support is a prerequisite. Central governments can, for example, offer expertise, train local officers or other actors, disseminate knowledge through policy recommendations, present strategic ideas and visions, or help localities cope with new situations by offering additional resources such as legal competences, funding and staff. The question remains as to whether such additional support is sufficient to equip local authorities. Especially when it comes to tasks that require high levels of technical expertise, it is well known that there are important benefits in relying on the coordinative model and its related functional specialization. Central support for localities might not be sufficient to compensate for their limited expertise. Even if it does, the question remains as to how much additional support is needed in comparison to relying on central guidance. It might be the case that the possible benefits of relying on local governance do not outweigh the costs associated with the decrease in efficiency that results from providing additional support to the local level, as compared to relying on central guidance.

External effects

External effects are the positive or negative effects of an activity in a domain or jurisdiction that 'spill over' to other domains or jurisdictions (Fleurke & Hulst 2006, compare Lemos & Agrawal 2006, Prud'homme 1994). External effects can constrain the local sphere of influence over dealing with issues, while they can also reduce local willingness to deal with them in the face of the potential social-dilemma character associated with such issues. Consequently, decentralization should go only as far as demanding contributions from local authorities that they can realistically be assumed to have control over and that they can be expected willing to perform. If these conditions are not met, I argue that decentralization should either be avoided or should be accompanied by central government support to increase local willingness and ability.

148　*Making decentralization work*

External effects mean that the causes of environmental issues that local authorities face are fully or partly outside their jurisdictional sphere of influence. There are many examples of environmental issues that are partly outside the sphere of influence of local authorities. The most common examples are issues that manifest themselves on multiple scales. On a regional scale, for example, we find issues such as urban sprawl, congestion and ambient noise. Other issues manifest themselves on an even larger scale, air pollution, acid rain and global warming, for example. External effects thus result in issues that manifest themselves on multiple spatial scales. This multiscalar character results, on the one hand, in a limited local sphere of influence over such issues. Local decisions might have only a marginal impact on solving such issues, as decisions of adjacent municipalities or higher-level authorities are also relevant. To illustrate this, issues such as air pollution and global warming cannot be solved by an individual municipality alone, but call for regional or even national and international cooperation in order to share responsibilities. On the other hand, multiscalar issues also tend to behave as social dilemmas (section 2.6). Therefore, ignoring external effects can cause 'free rider' behaviour and uncontrolled 'spill-overs', and can undermine the production of collective goods (e.g. Fleurke et al. 1997, Prud'homme 1994).

External effects mean that decentralization can result in a reduced willingness and ability to address environmental policies. In response, central governments can decide to rely on centralized governance and coercive tools such as sanctions, permits and protocols to force local authorities to make contributions. This is essentially a decision to retain centralized governance formats where local governance is only involved in implementing central policies. However, external effects do not deny that the proximity of local authorities to local circumstances and stakeholders involved in producing environmental stress puts them in a good place to contribute to reducing this stress. Congestion, air pollution and energy consumption are all examples of issues that are subject to external effects, but where there is also much potential for local action. External effects mean that local contributions should be confined to those which are within the local sphere of influence, while there is also a need for persuasive policies to prevent 'free riders' and stimulate local initiatives (e.g. Bressers & Kuks 2003, Lemos & Agrawal 2006, see section 2.6). Therefore, central governments can decide to decentralize power and authority to the local level, but within a context of more modest central influence over local performance.

Central governments can, for example, issue financial rewards and penalties, possibly on the basis of market-based instruments such as taxes, subsidies or tradable permits. Such financial tools can not only stimulate local willingness to add value to minimal environmental requirements but can also improve the efficiency of policy delivery, particularly since financial rewards act as an incentive to those municipalities that are able make substantial contributions against limited costs. Contracts and agreements can also be made between central governments and local authorities, or between central governments and distinct target groups. The contracts and agreements can summarize the contributions that are expected of different authorities and stakeholders and the possible sanctions if these are not delivered. Finally, central governments can issue policies to set up (regional) cooperative bodies to improve coordination between localities in dealing with issues subject to external effects. These examples all illustrate that the response to issues subject to external effects need not be to refrain from any form of decentralization. However, they do show that, again, a degree of central guidance is advisable if decentralization is to be pursued when dealing with issues subject to external effects.

A weak profile

Traditionally, environmental policies in most of the Western world rely on the coordinative model of governance and the associated central guidance. Supported by scientific research, central governments have set many environmental standards, such as, for example, the maximum amount of emissions or imissions tolerated for the quality of the environment. As a result of decentralization, environmental ambitions increasingly become a matter of debate in the local realm. However, as I also discussed in Chapter 2, the 'weak profile' of environmental issues means that environmental policies are not necessarily generating a sense of urgency in (local) political and societal debates.

Environmental issues can be difficult for politicians, stakeholders and planning professionals to understand, as these issues are often surrounded by rather complicated technical aspects and described in jargon. In addition, environmental stress is not always tangible. It can be hard to express in financial terms (e.g. noise nuisance), is often invisible (e.g. safety risks), is diffuse (e.g. air pollution) or has effects that are highly subjective (odour). This also makes it difficult to compare the benefits of removing environmental stress with losing the benefits of, for example, economic growth, project development or increased accessibility. Finally, environmental ambitions often focus on a long time horizon (e.g. global warming, ozone layer, biodiversity). This adds to the ambiguity of the consequences of environmental stress on people, since long-term effects can be difficult to envisage. Furthermore, short-term costs and benefits are typically given priority over long-term costs and benefits.

Decentralization is a means of shifting power and authority to the local level so as to promote the development and delivery of proactive and integrated approaches to the environment in a dynamic area-based setting. Therefore, environmental ambitions should be balanced or possibly combined with alternative policy ambitions based on the local circumstances. Due to their 'weak profile', it might be difficult for environmental ambitions to stand up to alternative policy ambitions. More tangible and short-term policy outcomes that are also financially attractive are often easier to recognize and defend within public administrations or the political domain. Indeed, as noted by Kamphorst (2006), De Roo (2004), Eckersley (1992) and Oates (2001), for example, local political and governance choices are often biased towards more economically attractive policies. Oates (2001) even suggests that there might be a tendency to breach environmental targets if possible, so as to favour the economic position of localities. The 'weak profile' of the

150 *Making decentralization work*

environment, therefore, tends to constrain political or administrative willingness to pursue proactive and ambitious environmental policies.

Problems associated with the 'weak profile' of the environment are not limited to political or administrative willingness. Local governments often have a limited influence over local stakeholders such as project developers, businesses, housing corporations, etc. Power dispersal has left local authorities without direct influence over many of the resources needed for developing and delivering policies. Within the confines of the coordinative model, local authorities are given legal responsibilities and competences that can help them *force* local actors to comply with minimum requirements. Examples include permits, licences, subsidies and fines. Proactive environmental policies are not necessarily accompanied by such legal responsibilities and competences. If they are not, local authorities should seek the promotion of environmental qualities based on the employment of local resources or collaborate with local stakeholders who, for example, control funding, knowledge, housing stocks, land or other key resources. However, these local actors are not necessarily interested in cooperative efforts to improve environmental quality beyond minimum requirements, especially since there is often a lack of direct gain (including financial gain) or societal controversy regarding such improvements. The 'weak profile' of the environment now interacts with the local balance of power, again with the possible consequence that local proactive and ambitious environmental policies are difficult to deliver.

The 'weak profile' of the environment and the local balance of power can constrain local willingness and ability to pursue environmental policies. As a result, there is a risk that local authorities choose or are forced by local stakeholders to adopt development-oriented paths that degrade rather than improve the environment (Eckersley 1992, compare De Roo 2004, Jordan 1999, Oates 2001). This, again, is an argument in favour of central governments retaining at least a degree of control over local performance in the realm of environmental policy. In the previous section (4.2), I argued that the desire to retain minimum levels of protection for humans and ecosystems against environmental stress can be an argument for implementing generic policies supported by centralized control to guarantee that such minimum levels are met. However, a degree of control by central governments over local performance is also relevant when it comes to environmental policies in which minimum levels are not considered important. After all, if decentralization in environmental policy is to result in more proactive and integrated approaches, the 'weak profile' of the environment can be an important bottleneck.

In addition to coercive tools designed to *guarantee* certain policy outcomes in a local realm, central policies can also be used to *stimulate* local performance in a more modest way. In such cases, central policies are not intended to result in full compliance, where success is defined by "the degree in which the actions of implementing officials and targets groups coincide with the goals embodied in an authoritative decision" (Matland 1995; p. 146). Rather, these policies are designed to inspire, invite, stimulate or guide (Faludi & Mastop 1997, see also Matland 1995). In doing so, they can be used to counteract the possible weak position of environmental interests in a local realm.

Counteracting the weak profile

First, central policies can be installed to function as a *frame of reference* for subsequent decisions and actions. These are policies of a more visionary or strategic nature. In such cases, success should not be measured by the level of compliance but rather by the degree to which the initial central policy objectives are taken into account during subsequent actions and decision making. Faludi and Mastop (1997) call this method of valuing influence 'performance' (see also De Lange 1995, compare Barret & Fudge 1981). Central policies allow for local governance to develop its own course of action *within* the frames of reference and stimuli provided by central policy imperatives. Harvesting local opportunities and adapting general policy ambitions to local interrelated issues and stakeholder interests is intended to optimize the effect of these policies under various local conditions. Central policy imperatives are not 'copied' but are instead *translated* by individuals and organizations into their personal, organizational and situational context (see also deLeon & deLeon 2002, O'Toole 2000 and Yanow 2000). As deLeon and deLeon (2002) suggest, such translations allow implementation to move away from what they call the 'one-sided rationalization' of policies by legislators or managers. Instead, as Matland (1995) and Schofield (2001) indicate, allowing for alterations opens up implementation processes to local and personal expertise, networks and skills.

Second, central government can again turn to financial tools and contracts, covenants or voluntary agreements in order to exercise influence. Subsidies and the allocation of budgets can stimulate local willingness to act, and can counteract possible local power imbalances. Examples include subsidies for buyers of energy-efficient cars, subsidies for sustainable housing programmes or additional budget for environmentally stressed localities. However, financial leverage is not always realistic. Market-based instruments – such as eco-taxes, pollution charges, tradable permits, tradable development rights and government subsidies – can, however, be relevant methods for exerting influence, limiting the additional financial burdens for central governments. This is also true for the use of covenants, contracts and other partnership constructions. These are usually not based on central financing but rather on negotiations regarding the sharing of resources, costs and benefits of policy measures between governments at various levels of authority and/or stakeholders.

Finally, influence can also be based on merely persuasive and argumentative policies. Examples include policy recommendations, example projects, informative campaigns and the use and dissemination of planning concepts (compare Van Duinen 2007). These tools inform local authorities and stakeholders about the kind of possibilities and ambitions that exist. They are thus fairly weak means of exerting influence and tend to be considered as merely symbolic policies (see also Edelman 1977, Matland 1995). However, such "symbolic policies play an important role in confirming new goals, in reaffirming a commitment to old goals, or in emphasizing important values and principles" (Matland 1995; p. 68, see also Olsen 1970). In addition, they can be supported with rewards such as eco labels or certificates to give them more meaning.

152   *Making decentralization work*

Decentralization in a context of central influence

In this section I argued that choices regarding decentralization are contingent upon the degree to which local authorities can be expected to be willing and able to perform decentralized tasks. In addition, I discussed some important constraints on local willingness and ability. Each of these constraints can undermine the potential for decentralization, even if we are faced with conditions of increased complexity. On the one hand, these constraints can thus be arguments against decentralization. Such arguments are especially convincing if we face decentralized tasks that local authorities cannot reasonably be expected to be willing and able to perform. We also saw how certain government functions qualify as such tasks. Examples include tasks involving high levels of technical expertise, and tasks that involve external effects. They are governance functions for which the consequences of decentralization are such that we are advised to be highly prudent with pursuing decentralization.

On the other hand, I argue that constraints on local willingness and ability should not be equated with an argument against pursing decentralization altogether. Instead, I discussed how central policies can be important means of generating local willingness and ability. They can enable local ability by, for example, offering policy advice, training local officers or providing localities with additional resources such as funding, staff and expertise. They can also stimulate local willingness through financial rewards or penalties, contracts and agreements and flexible regulations, or through merely persuasive policies.

The key conclusion to which this brings me is that central policies and the associated reliance on the coordinative model can be crucial means for setting the conditions required for decentralization so that its envisioned outcomes are achieved. Therefore, relying on the coordinative model is not confined to circumstances of limited complexity or situations in which the focus is on fulfilling a series of predefined priorities. The coordinative model of governance is also essential in providing a robust and stable foundation upon which to build if we pursue proactive and integrated approaches within a dynamic area-based setting. It means that decentralization is now indeed positioned in a context of interacting central policies and local governance approaches that should result in the kind of 'optimal institutional arrangements' De Vries (2000) referred to. Contrary to popular beliefs in recent debates in planning theory and practice, the coordinative model can thus be a key part of such arrangements, even if pursuing decentralization (compare Ashford 2002).

4.4 Enabling local governance

In the last three sections (4.1 to 4.3) I discussed arguments that urge us to consider decentralization as a relative shift of power and authority towards the local level within a context of central policies. Practice also shows that decentralization does not take place in a vacuum. Many central policies and regulations remain in place while decentralization is pursued. These can be informed by the same arguments

Making decentralization work 153

discussed in the previous (4.2 and 4.3) sections. They can, for example, be used to deal with conditions of limited complexity or to provide possible guarantees for fulfilling certain minimum targets in a local realm. They can also be used to stimulate and support local governance in dealing with decentralized tasks and responsibilities. But while central policies can be important instruments for enabling, stimulating and guaranteeing levels of local performance, they are also potential constraints on local performance.

Central policies and regulations constrain local freedom to act and claim local resources for performing the tasks and functions demanded by these regulations and policies. Hence, they can also undermine decentralization processes in delivering their envisioned outcomes. Some constraints are, of course, intended. They are policies, regulations and procedures that mean to influence local performance. Some, constraints, however, are not intended. Central governance formats and the coordinative model of governance can, for example, result in fragmented policy agendas, complicated regulations and excessive bureaucracy. The model involves a genuine risk that local authorities will be constrained in their attempts to cope with central policies and regulations. The idea is now to try to remove unnecessary constraints on local freedom to act, and to minimize the local resources needed to implement central policies and regulations. Or, as Fleurke and Hulst (2006; p. 43) explain, decentralization should indeed involve assessing the "volume and type of remaining rules" that might constrain localities in dealing with decentralized tasks and responsibilities (see also Gresov & Drazin 1997). I will now reflect on this type of assessment.

Removing constraints

To begin with, decentralization involves a shift of policy-making powers from the centre to the locality (e.g. Flynn 2000, Marks & Hooghe 2004, Prud'Homme 1994 and Rhodes 1981). As noted earlier (section 4.3), local communities need a sufficient sphere of influence over decentralized tasks and functions in order to perform them. Existing policies and regulations can limit this sphere of influence. By way of illustration, decentralization of noise policy can increase the capacity of local authorities to set their own targets. However, if many installations or roads remain under central government control, local authorities are constrained in their sphere of influence in terms of actually pursuing their own targets. Another example: national housing programmes might require those same local authorities to build a substantial number of homes in cities (i.e. compact city development). This can constrain their attempt to reduce the number of houses subject to noise nuisance, since houses need to be built in existing urban fabrics where noise levels are usually significant.

It is not always easy to remove constraints on the local sphere of influence. This is especially true if constraints follow from government functions other than those that have been decentralized. The decentralization of noise policy in a context of national housing programmes is a good example of this. The problem with constraints relating to alternative policy domains is that they are often controlled

154 *Making decentralization work*

by other political and administrative units. Even if coordination between these policies is possible, the alternative policies might not be suitable for deregulation and decentralization. To illustrate, decentralizing the authority to issue permits and licences to limit the number of trains allowed to pass on a railway line can both enable local authorities to pursue their own targets. However, it can conflict with regional or national interests to increase rail traffic on this particular line. Therefore, there are reasons for retaining regional or national policies to counteract local choices that could conflict with regional or national interests. This is a typical example of a social dilemma for which I already argued that a degree of centralized coordination is necessary.

As it is not always desirable to remove constraints on the local sphere of influence, local authorities will continue to face constraints on local behaviour. Decentralization operations can run into problems if local authorities and stakeholders face too many constraints while coping with their new tasks and responsibilities. The question is thus whether there is sufficient potential for removing such constraints when discussing decentralization measures. *If serious constraints cannot be realistically removed, important doubts arise as to whether decentralization should be pursued at all.*

Robust and dynamic

Central policies and regulations put pressure on local resources such as time, budget and staff that localities need to perform decentralized tasks. In addition to removing unnecessary constraints on local performance, it is also important to minimize the local resources needed to implement central policies and regulations. The willingness and ability of localities to respond to newly decentralized tasks and responsibilities also depends on the existence of a *robust* and well-coordinated set of central policies that can be routinely implemented. Robust, here, does not imply an unchangeable body of policies or regulations. Rather, what I here refer to as robust implies a 'healthy, well-functioning and durable' set of policies and procedures able to withstand changing political fashions or external shocks without falling apart (see also Oxford English Dictionary). While such policies also change, they are less prone to abrupt, rapid and extreme shifts. Hence, they help provide stability in conducting governance by, for example, providing clarity and (legal) security. Also, they can promote efficiency in the face of conducting many relatively straightforward, well-established and accepted tasks and responsibilities. In the meantime, however, we must accept that such robust and well-coordinated policies are not evident in practice.

I have already discussed (in Chapter 2) that the coordinative model of governance and its associated central guidance run the risk of problems such as fragmented policy agendas, excessive bureaucracy, complex regulations, incoherencies, etc. This is also a serious risk when it comes to decentralization operations. Decentralization often takes place against a background of perceived ineffectiveness and inefficiency of the coordinative model of governance (see also Martens 2007, Pierre & Guy Peters 2000). It is often a response to previous efforts

Making decentralization work 155

by central governments to improve the organization of governance by amending their central policies (e.g. de Leeuw 1984). As discussed in Chapter 2, there is a common tendency to first try to make changes *within* the coordinative model of governance, i.e. add regulations, policies and control mechanisms (see also De Roo 2004). These attempts to regain control over societal issues therefore carry the risk that central policies and regulations only become more fragmented, voluminous and complex, up to the point where local authorities even face incoherent or contradictory regulations and policies (compare Baldwin 2005, Mayntz 1980, Miers 2001 and Van Gestel & Hertogh 2006). It can result in what I referred to earlier as conditions of 'management overload' (De Leeuw 1984).

The existence of fragmented, voluminous and complex central policies and regulations has at least two important consequences when it comes to choosing between relying on centralized or decentralized governance. In the first place, it will undermine one of the key benefits commonly associated with central policies and regulations. If organized well, central policies and regulations are supported by the efficiency associated with economies of scale such as the routine implementation of common policy formats and functional specialization. As such, it can limit the local resources needed to cope with issues of a limited complexity and maintain minimum levels of protection for humans and ecosystems against environmental stress. It reduces the need for local bargaining and debates to set policy ambitions or to develop local strategies to deliver on such ambitions. Routine implementation of common policy ambitions and formats will, however, be problematic in the face of fragmented, complex and voluminous regulations. Local resources such as time, funding and staff are needed to navigate the bureaucracy with which localities are faced. Under such conditions, the efficiency of relying on central policy formats in response to environmental issues with a common manifestation is less evident.

Second, fragmented, voluminous and complex central policies and regulations will reduce the local resources available for fulfilling decentralized tasks and responsibilities. Implementing or dealing with the central regulations and policies will become difficult and complicated. Instead of a routine implementation of policies and regulations, local units find themselves in a constant state of crisis management to try to navigate the complexity of policies and regulations. As a result of decentralization, overstressed local authorities and stakeholders see their workloads increase, but are hardly, if at all, able to cope. However, as Geelhoed (1984) notes, practice shows that there is a genuine risk that decentralization is used as a way to divert problems associated with certain policies and governance practices to local arenas (see also De Vries 2000). The risk is that it might well be that decentralization indeed takes place in a context of complex and fragmented policies and regulations.

It is hardly realistic to suppose that local authorities can add value to central policies and requirements if they are faced with extensive efforts to cope with the central policies and requirements coming their way. It then makes little sense to develop a governance framework based on cooperation between central and local authorities to stimulate dynamic local approaches meant to add value to

156 *Making decentralization work*

central policies and requirements. Rather, it would suggest a search for first simplifying and reducing the regulations and fragmentation that local authorities, stakeholders and citizens face, in order to enable them to cope with their new tasks and responsibilities (e.g. De Roo 2004, Geelhoed 1984). Before developing a framework of cooperation between central and decentralized policies to support and stimulate dynamic local approaches, there should be a robust framework of well-coordinated central policies and regulations in place that can be routinely implemented. That developing such a robust framework is not necessarily easy is, I argue, a truism not to be denied and that should thus be taken into account when considering decentralization operations.

4.5 Making decentralization work – a summary

Despite its potential benefits, I have argued in this chapter that the benefits of decentralization should not be interpreted as an argument for letting go of reliance on the coordinative model of governance and its related central guidance. Instead, I argued that the coordinative model still has a pivotal role to play. It has manifested itself in four key arguments that can assist decision making with regard to when, how and why decentralization should be pursued. In this final section I summarize the main arguments.

Argument 1: complexity as criterion

In following a post-contingency approach I have asserted that variations in the perceived conditions of complexity provide decision makers with an initial argument for choosing between different degrees of centralization and decentralization. Relying on central guidance is considered effective and efficient under conditions of limited complexity. Under such conditions, it is possible to benefit from the effectiveness and efficiency of functional specialization and the routine implementation of common policy formats associated with central guidance. Common policy formats supported by central guidance are, however, constrained in their ability to adapt to local circumstances and the pursuit of proactive, integrated and tailor-made approaches. By contrast, the proximity of local authorities to local circumstances gives them advantages over the central state in terms of responding to complexities such as interrelated policy issues, associated stakeholder interests (power dispersal) and diverging preferences (social fragmentation). Therefore, decentralization can be seen as a means of adapting to conditions of increased complexity, where governance approaches can be developed that are tailored to the unique circumstances surrounding the more complex issues. In adopting a post-contingency approach, however, I also noted that decentralization has consequences that can provide arguments in favour of prudence when it comes to decentralization operations. Reflecting on these arguments, we can also see that decentralization under conditions of increased conditions of complexity is indeed a *matter of choice*.

Making decentralization work 157

Argument 2: a focus on protection

Following the contingency between function and structure, decentralization will involve a shift of focus from single fixed goals to multiple composite goals. Although it one of the main motives supporting decentralization, a consequence of a focus on multiple composite goals is that single fixed goals that were previously supported by hierarchical control become debatable in a local realm. Therefore, decentralization not only increases the potential for developing integrated policy approaches in a dynamic decentralized setting, but also decreases the capacity of governance to meet single fixed goals. There can, however, be reasons for retaining single fixed goals that need to be met in the local realm and are therefore not considered suitable for balancing with other goals. As I have argued, the ambition to maintain minimum levels of protection for ecosystems and people against environmental stress can be such a reason.

The contingency between function and structure tells us that, if a choice is made to retain single fixed goals to be met in the local realm, central guidance is needed to control local performance. On the one hand, it is efficient to rely on central guidance to translate single fixed goals into common policy formats that can be routinely implemented. On the other hand, hierarchical control exercised in a centralized setting can also promote the effective implementation of such goals in all localities. Therefore, the choice to retain minimum levels of protection to be met in a local realm is an argument for retaining central governance, even in conditions of increased complexity. This does not preclude the possibility of decentralizing power and authority to the local level, but does imply that decentralization has limits. Local governance will need to take place in a context of generic policies and hierarchical control for providing the desired guarantees. Consequently, localities will be faced with generic policy ambitions that constrain their capacity to respond to unique and complex conditions. Although this might in itself be considered an undesirable consequence, the desire to maintain minimum levels of protection for humans and ecosystems against environmental stress can be an important argument for accepting it nevertheless.

Argument 3: responding to constraints on local willingness and ability

A second key consequence of decentralization is that the outcomes of decentralization operations are contingent on the degree to which local authorities are willing and able to perform decentralized tasks and responsibilities. As discussed, there are important constraints on local willingness and ability to do this, also when it comes to environmental policy. These can be arguments for refraining from decentralization. However, they can also be arguments for retaining a degree of central guidance to exert influence over local performance. In that case, central policies and the associated reliance on the coordinative model of governance are used to set the conditions required for decentralization.

158   *Making decentralization work*

First, economies of scale imply that central guidance has benefits in coping with knowledge-intensive tasks or implementing common policy formats to perform repetitive government functions. If local authorities are not able to attract sufficient resources, knowledge and expertise to perform a function, decentralization should either be avoided or given central government support. When it comes to tasks that require high levels of technical expertise, decentralization is hardly realistic. Local authorities can hardly be expected to attract the required expertise, even with central government support. When it comes to proactive, integrated and tailor-made approaches, local authorities can also face important constraints. Local authorities can face financial constraints, a lack of staff, limited managerial skills and unfamiliarity with tasks such as proactive and integrated approaches. In such cases, a degree of central support is again a prerequisite if decentralization is to be pursued at all.

Second, external effects undermine the potential for decentralization, given the social dilemma character of issues and the limited sphere of influence of localities over these issues. A common response is to maintain a high degree of central control. This does not deny that the proximity of local authorities to local circumstances and stakeholders involved in producing environmental stress puts them in a good place to contribute to reducing this stress. However, local contributions should be confined to those which are within the local sphere of influence, while there is also a need for persuasive policies to prevent 'free riders' and stimulate local initiatives.

Finally, the weak profile of the environment undermines the position of environmental interests in local governance, both in relation to local political and administrative willingness and in relation to the local balance of power. This calls for a degree of central influence to stimulate local willingness. As I argued, the desire to retain minimum levels of protection for humans and ecosystems against environmental stress can even be an argument for installing generic policies supported by centralized control in order to guarantee that such minimum levels are met. However, a degree of central-government control over local performance is also relevant when it comes to environmental policies in which minimum levels are not considered important. If decentralization in environmental policy is to result in more proactive and integrated approaches, the 'weak profile' of the environment can be an important bottleneck. Instead of commanding and controlling local performance (e.g. through strong coercive instruments), more moderate influence can now be exerted. This can include tools such as financial incentives or penalties, contracts and agreements, flexible regulations or mere persuasive policies. Decentralization is then used to invite and enable localities to develop their own course of action, *within* the frames of reference and stimulus provided by central policy imperatives.

Argument 4: robust policies to enable dynamic approaches

Although central policies and regulations can be important to guarantee, support or stimulate local performance, they can also be important constraints on

Making decentralization work 159

such performance. Decentralized operations can run into problems if localities face too many constraints on their local freedom to deal with decentralized tasks and responsibilities. In addition, central policies and regulations require local resources such as time, budget and staff. These resources are also needed to perform decentralized tasks and responsibilities. This will be problematic if local authorities are faced with fragmented, complex and voluminous central policies and regulations. Instead of a routine implementation of policies and regulations, local units then find themselves in a constant state of crisis management to try to navigate the complexity of the policies and regulations. As a result of decentralization, overstressed local authorities and stakeholders see their workload increase but are hardly, if at all, able to cope.

It is hardly realistic to suppose that local authorities can add value to central policies and requirements when they face too many constraints on local freedom to act or fragmented, complex and voluminous central policies and regulations. Decentralization should then be accompanied by measures to simplify and reduce the regulations facing local authorities, stakeholders and citizens, in order to enable them to cope with their new tasks and responsibilities. It is an effort to develop a robust and well-coordinated set of central policies that can be routinely implemented and enables localities to respond to newly decentralized tasks and responsibilities. I argue that such a robust and well-coordinated set of policies is needed before we proceed with developing a framework of cooperation between local governance and central policies that can stimulate dynamic local approaches to add value to central policies and requirements.

An empirical inquiry

In this chapter I have presented four key arguments that can be used to make choices with regard to advancing the role of the local level in environmental policy. In doing so, I concluded that these arguments should be positioned in the search for an 'optimal institutional arrangement' in which central policies and local governance interact. This provides a nuanced perspective on decentralization that suggests that decentralization cannot just be assumed to result in the envisioned outcomes. Rather, I suggest, we should be well aware of its advantages and disadvantages, while decentralization is also a process that is subject to some important conditions that should be respected if the envisioned outcomes should be realized. Mostly, I argue that the coordinative model should not be considered as a relic of the past. It has important benefits in terms of responding to conditions of limited complexity, or focusing on fulfilling a series of predefined priorities. In addition, it can play a key role in setting the conditions required for decentralization and the associated use of proactive, area-based and integrated approaches in order to realize their envisioned outcomes. Hence, if we are to pursue governance renewal and its associated desire to rely on more dynamic approaches in a local setting, we should still address the question of how the coordinative model can help us to build a robust and stable foundation for these new and dynamic approaches.

160   *Making decentralization work*

In the following two chapters I will analyse two recent decentralization operations in environmental policy that focus on stimulating the development of proactive and integrated approaches to the environment. These operations took place in the EU and the Netherlands. I will compare the approaches chosen by the EU and the Dutch national government to the arguments and conditions discussed in this chapter. I am interested in the extent to which the EU and the Dutch national government have taken these arguments and conditions into account. In addition, I am interested in the consequences of both the success and failure to do so. On the one hand, this will allow for reflecting on the degree to which the arguments I have proposed in this chapter are empirically meaningful, which might also help improve, criticize or nuance them. On the other hand, it will also allow for investigating possible responses to limited successes in the delivery of the envisioned outcomes of the decentralization operations studied.

5   A European focus on the local

Officially, the European Union (EU) has only fairly recently been involved in environmental policy. Up until the signing and ratification of the 1987 Single European Act, environmental issues were not a formal part of the policy agenda of the predecessor of the EU, the European Economic Community (EEC). Unofficially though, EU environmental policies were drafted from the publication of the first EEC Environmental Action Programme of 1973 (OJEC 1973) onwards. The development of EEC/EU environmental policy therefore runs largely parallel to most international environmental policy developments, at least in the Western world. This is also true for the EEC/EU's interest in proactive and integrated approaches to the local environment that started with the 'Green Paper on the Urban Environment' (CEC 1990).

In this chapter I will discuss EEC/EU attempts to stimulate local proactive and integrated approaches to the environment. In this, my focus lies on the 'Thematic Strategy on the Urban Environment' (CEC 2004, 2006). The 'Thematic Strategy' is one of seven thematic strategies proposed and adopted by the EU following its 6th Environmental Action Programme. Each of these strategies proposes a set of ambitions, instruments and strategies for addressing seven environmental priority issues the EU recognizes are difficult to tackle due to their complexity and the diversity of actors concerned. The urban environment was considered one of these themes. The CEC (2004) considered that traditional centralized regulations fell short of coping with the complexity and character of urban environmental issues. Furthermore, previous work on stimulating local proactive and integrated approaches to the environment was considered to have met limited success. With the 'Thematic Strategy on the Urban Environment' the CEC aimed to install legislation for promoting proactive and integrated environmental management at the local level. The 'Thematic Strategy' meant to do so by giving local authorities additional responsibilities and tasks in addressing urban environmental issues.

In this chapter I aim to uncover the practical consequences of the 'Thematic Strategy', while assessing it in a context of existing EU policies affecting the local level in environmental policy. Furthermore, I aim to explain why these consequences have occurred. In doing so, I will also reflect on the four arguments presented in Chapter 4. These arguments meant to help inform attempts to increase the role of the local level in environmental policy, also when pursuing

162   *A European focus on the local*

proactive and integrated environmental policies. I aim to assess whether these arguments can indeed help to do so, by studying the approach taken by the EU.

I begin this chapter by addressing the motives for drafting EU environmental policies and the current structure of EU environmental policies (section 5.1). Second, the increased EU interests in the urban environment are discussed, which culminated in the desire of the EU to promote proactive and integrated approaches to the local environment (section 5.2). The discussion on the rationale and structure of EU environmental policy and interest in the urban environment provide a background of information against which we can better understand the development and content of the 'Thematic Strategy'. I discuss the development, content and implementation of the 'Thematic Strategy' in sections 5.3 to 5.5.

My research into the 'Thematic Strategy' takes into account that the draft 'Thematic Strategy' as proposed by the European Commission (CEC 2004) is rather different from the actual finalized strategy (CEC 2006) and the proposals made in its aftermath (CEC 2007a, 2007b, 2010a) The draft 'Thematic Strategy' was the intended culmination of almost 15 years of work by the CEC on promoting urban environmental governance and expressed the CEC's perspective on its own role in promoting and stimulating local authorities to pursue proactive and integrated policies to the local environment. With the 'Thematic Strategy' the CEC aimed at providing a strong stimulus for proactive and integrated policies to the local environment. Supported by an EU directive, larger EU towns and cities would be obliged by the CEC to draft an environmental management plan, sustainable urban transport plan and environmental management system. Most stakeholders, including member states, non-government organizations and local governments were rather critical towards this 'strong framework of EU policies' (CEC 2004). In response, the CEC dropped most of its ambitions, leaving the finalized 'Thematic Strategy' as a voluntary policy advice. The finalized 'Thematic Strategy' is therefore much 'weaker' than intended by the European Commission (CEC) in its draft and, as I will discuss, has little impact on local environmental governance. The limited impact of the current EU policies in promoting proactive and integrated policies to the local environment can be largely explained by understanding why the draft 'Thematic Strategy' did not make it. It is also therefore that this chapter is focussed on identifying and analysing the reception of the draft 'Thematic Strategy' and its key proposals for obligatory plan and management formats to be drafted by larger EU towns and cities by EU stakeholders.

Section 5.3, addresses the reception of the draft 'Thematic Strategy' and the subsequent discussions and critiques during the consultation rounds with stakeholders and member states. In section 5.4, the ideas as advanced in the draft 'Thematic Strategy' will be compared with urban practices in a series of cities that were studied during one of the projects meant to advise on the development of the 'Thematic Strategy' (i.e. the 'Liveable Cities' project). The finalized 'Thematic Strategy' and its impact will be discussed in section 5.5. Finally, and based on the data discussed in sections 5.3 to 5.5, I will in section 5.6 compare the CEC's approach on stimulating local environmental governance with the conditions and arguments that inform choices regarding decentralization I discussed in Chapter 4. Finally, then, section 5.7 contains the conclusions to this chapter.

5.1 A rationale for EEC/EU environmental policies

The 1957 Treaty of Rome stands as the formal start of the European Economic Community (EEC) that later became the EU. The Treaty of Rome makes no mention of environmental protection, apart from a general indication that a good 'standard of living' (article 2) and 'health' (article 36) are policy objectives. This low priority given to environmental issues is not at all unexpected. Not only did few political agendas touch upon environmental issues in those years, the EEC itself was founded for other reasons. Important was the desire to develop a stable economic basis in Europe with the hope and expectancy that this could prevent the kind of political and military conflicts that had ravaged much of the continent during the first half of the 20th century. The main task of the EEC was the pursuit of economic integration between its members which should result in "a harmonious development of economic activities" (article 2 of the 1957 Treaty of Rome). With economic integration to promote economic growth in Europe as the *raison de être* of the EEC/EU (see also Pelkmans 1997, Urwin 1993), environmental policies were at first of little or no interest to the EEC/EU. This changed during the early 1970s, when two important motives provided the trigger for drafting environmental policies by the EEC/EU (e.g. Barnes & Barnes 1999, Hildebrand 2002, Sbragia 1996 and Zito 2000). The first was to coordinate the development of environmental policies in EEC/EU member states in the face of the protection of an EEC/EU common market. The second was the protection of the environment itself.

Two motives for environmental policies

The creation of a common market requires "the elimination of economic frontiers between two or more economies" (Pelkmans 1997; p. 2). National or regional environmental regulations can result in such frontiers. Environmental regulations can, for example, set restrictions on importing or selling products, on certain production processes or on the use of products and materials. Such restrictions typically differ between countries, which can frustrate trade between countries and cause competitive (dis)advantages. In response, the EEC/EU considered it essential to coordinate the development of environmental regulations in its member states that could result in trade barriers and comparative (dis)advantages. Coordinating this development is called *harmonization* (e.g. Barnes & Barnes 1999, Jordan 2002 and Wallace & Wallace 1997).

During the early and late 1970s most EEC member states frantically began drafting environmental policies (compare Andersen & Liefferink 1997, Sbragia 1996, Van Tatenhove, Arts & Leroy 2000). Differences between various member-state regulations soon created trade barriers and comparative (dis)advantages. The EEC responded by issuing its first 'Environmental Action Programme' in 1973, marking the official start of EEC/EU environmental policies. Soon the EEC also developed policies so as to regulate domestic environmental policies. While this was inspired by the general desire to harmonize regulations, the EEC was also urged by many environmental 'leader states' (such as Germany, Denmark and

164 *A European focus on the local*

the Netherlands) to stimulate the uniformity of environmental regulations in the EEC/EU (e.g. Andersen & Liefferink 1997). After all, these 'leader states' had more advanced and strict environmental policies than many other environmental 'laggard states'. This resulted in economic disadvantages for these 'leaders', inspiring them to demand that environmental 'laggard states' would bring their environmental policies in line with their stricter regulations (e.g. McCormick 2001, Sbragia 1996 and Zito 2000). Harmonization was therefore based upon both an ideological standpoint (protection of the common market) and on an international relations standpoint (comparative advantages).

Environmental protection was the second key motive driving the development of environmental policies in the EEC/EU. Based on international public and political pressure, both individual member states (i.e. the 'leader states' just referred to) and many lobby groups – both inside and outside of EEC/EU institutions – argued for environmental policies (see Andersen & Liefferink 1997, Zito 2000). Although the political climate in Europe and therefore the EEC was fairly receptive to these calls, the 1957 treaty provided hardly any legal frame for developing environmental policies in the EEC/EU. It was only when economic integration was at stake that the Treaty allowed for EEC regulations. It would eventually take until the development and ratification of the Single European Act (SEA) of 1987 for environmental protection to gain a real legal basis for itself. Environmental policies therefore had to be developed in a 'legal vacuum', often related to economic integration arguments. Nevertheless, environmental protection was in practice used as a motive for drafting environmental policies in the EEC, as can, for example, be seen in first Environmental Action Programme 1973 (e.g. Hildebrand 2002, McCormick 2001).

Early development

The 1973 first Environmental Action Programme stands as the formal start of an EEC/EU environmental policy (see also Andersen & Liefferink 1997, Barnes & Barnes 1999, Hildebrand 2002 and McCormick 2001). The Environmental Action Programme was followed by a rapid growth of EEC/EU environmental policies and regulations. Considering the 'legal vacuum' within which these policies were developed, the EEC/EU had to be creative in getting these policies adopted. The EEC/EU therefore relied on Treaty articles that focussed on general EEC/EU goals and on economic integration. It resulted in a creative use of the Treaty articles 100 – on the harmonization of national laws that directly affect the functioning of the Common Market – and 235 – measures needed to obtain a Community objective (e.g. Barnes & Barnes 1999, Hildebrand 2002 and McCormick 2001). So even if environmental protection rather than economic integration was the motive for developing EEC/EU environmental policies, these policies often *could not be separated* from the functioning of the common market; i.e. harmonization. Consequently, relatively uniform policies were developed (see also Barnes & Barnes 1999, Jordan 2002 and Sbragia 1996).

A European focus on the local   165

Quite similar to the early development of national policy schemes, EEC/EU environmental policies were in the 1970s developed ad hoc, sectorally, and relied on the command and control tradition of the coordinative model that still dominated planning practice. From the 1980s onwards, most nation states were able to improve the integration of their sectoral policies and would install policies that relied less in the coordinative model (see also Andersen & Liefferink 1997). The EEC/EU had more difficulties pursuing such integration and innovation. First, the EEC/EU had a feeble legal basis for the development of environmental policies, which meant that environmental policies were developed mostly through an opportunistic process of seizing the opportunity of political willingness. This opportunism reinforced the ad hoc and reactive development of policies (see also Hildebrand 2002, McCormick 2001). The 1976 Seveso chemical spill is a good example of producing such a political window for the development of regulation; i.e. the Seveso Directive on the major-accident hazards of certain industrial activities.

Second, environmental policies were often advanced through the so-called *'Monnet method'*, named after Jean Monnet, one of the founding fathers of the EU (Jordan 1999). The method is based on 'hiding' the detailed consequences of a new piece of policy by using a *technocratic focus*. Using a technocratic focus can make policies less attractive topics in social or political debates, while a technocratic focus also gives policies an aura of scientific fact that can further prevent social or political debate. In other words, the Monnet method proved an excellent strategy for pushing new policies forward without much debate. Given the legal vacuum in which environmental policies had to be developed and the fact that environmental topics already have a relatively technical character, the Monnet method became a popular means for drafting environmental policies by the administrators of the European Commission. Or as Jordan states, "the Commission's tactic was to concentrate on seemingly 'technical' issues such as environmental standards and avoid 'political' debates about the surrender of sovereignty" (1999; p. 3). In such a way, environmental policies could progress "by stealth" (ibid).

Developing and adopting environmental policies based on an opportunistic approach proved remarkably successful, as the rapid and extensive development of these EEC/EU policies during the 1973–1987 period shows (e.g. Hildebrand 2002). This approach, however, did contribute to a focus on technical, regulation-based and uniform policies that also conveniently fitted in with the idea of harmonization. It resulted in a bias towards command and control policies that, despite important changes in recent decades, has only gradually been changing (e.g. Holzinger et al. 2006, Jordan 2002, Knill & Lenschow 2004 and Zito 2000).

Multilevel governance in the EU

As Zito (2000) notes, it would be too simplistic to formulate the rise of EU environmental policies just within the realm of (economic) integration theory or to

166   *A European focus on the local*

consider the astonishing growth of EU environmental policies as just the result of its technocratic focus and the use of the Monnet method. Already during the 1980s EEC/EU environmental policies were becoming more diverse. This was accelerated by an increased acknowledgement of EEC/EU environmental policies by member states and stakeholders as an important part of EEC/EU policies in the 1980s (compare Jordan 1998). This growing acknowledgment allowed for more innovative and far-going environmental policies. Furthermore, it fuelled the idea that environmental protection should be a legitimate EEC/EU objective. Decisions by the European Court of Justice (ECJ) and the acceptance of new Environmental Action Programmes by the Council of Europe further paved the road for a formal recognition of environmental protection. This recognition came with the 1987 Single European Act (SEA), which was confirmed in the 1992 Maastricht Treaty and further expanded with the 1997 Amsterdam Treaty.

Following the formal recognition of environmental protection as an EEC/EU objective, environmental policy soon diversified. Especially during the last two decades, EU environmental decision making has grown more complex and is subject to its own dynamics and rules, producing its own relatively distinct policy outcomes. The notion of multilevel governance is a helpful concept in understanding these rules and dynamics (e.g. Bache & Flinders 2004, Hooghe & Marks 2001, Jordan 1998 and Svedin et al. 2001).

In the early years the EEC resembled many characteristics of an international organization, meaning that its powerbase was dependent solely on cooperating sovereign states. During the last decades, the role of the EEC/EU has however gradually increased in many policy domains. Therefore, the EU became "a truly supranational bureaucracy that can openly seek to influence the policies of its own members" (Porter & Welsh Brown 1991; p. 47). The EU is, however, also still a rather unique institutional body; it is more than an international organization based on cooperating sovereign states, but less than a true federal unity (e.g. Jordan 1998, McCormick 2001 and Pierre & Guy Peters 2000). This unique character is also why traditional theoretical ideas on federalism or international cooperation are considered limited for describing or explaining the EU. Instead, a distinct body of literature arose that focused on the principle of 'multilevel governance' (e.g. Bache & Flinders 2004, Bernard 2002, Hooghe & Marks 2001, Jordan 2000, Kooiman 2003 and Pierre & Guy Peters 2000). Multilevel governance is based on the assumption that the EU has grown into "a far-reaching multi-level governance system in which policy-making powers are shared between supranational, national and sub-national actors" (Jordan 1998; p. 1). This sharing of powers is important in almost all stages of EU policymaking and implementation, resulting in distinct policy outcomes. Multilevel governance, generally not considered as a theory but as a conceptual framework, helps to identify the complex interplay of actors in EU environmental governance.

Mechanisms of decision making

Before 1987 decision making in the EEC/EU was the sole right of the European Council that consists of governmental representatives of all member states, typically ministers or the political heads of state. Environmental policies had to be agreed upon under articles 100 and 235, for which a full majority was needed in the Council. After 1987 full majority voting has been abandoned for many environmental issues. Instead, most environmental policies can now be accepted based on a qualified majority (a 2/3 majority). Furthermore, the European Parliament (EP) gained in status and now has the capacity to reject or amend EU policies and regulations; i.e. a status that is in a formal sense equal to that of the Council. These changes make that as of 1987 articles 100 and 235 are no longer necessarily needed for drafting environmental policies, nor is full majority voting needed. Although established in the 1987 Treaty, most changes eventually took until the mid-1990s to become common practice.

The shift from full majority voting to qualified majority voting is a quite important shift. Before this shift, the representatives of all member states had to agree on new legislation. Zito (2000) shows how this procedure resulted in *common denominator bargaining* (see also Scharpf 1988). That is, bargaining results in policies that *all* member states agree upon i.e. the lowest common denominator. In practice, common denominator bargaining makes it difficult to push for serious innovations in policies. Typically, there will always be one or two member states that find the innovation problematic and use their veto right. Consequently, policies often confirm the status quo and are rather conservative. Full-majority voting also causes a bias to relatively uniform policies. Many differences cause comparative (dis)advantages and, in the face of individual veto rights, these differences are unlikely to be accepted (e.g. Barnes & Barnes 1999).

Individual veto rights no longer exist with qualified majority voting. Still, member states do not want to take a position that puts them in a minority position. Such a position would imply that they become rather isolated and lose influence on the policy outcomes. Member states will therefore search for a coalition of at least 1/3 of member states, which avoids a majority to become a 2/3 majority. In addition, the dominant opinions within the EP also need to be taken into account for any policy to be adopted. Again, there is a push to seek for coalitions that are strong enough to either prevent policies from being accepted or that encourage adoption.

On the one hand, the process of seeking for coalitions supporting alternative proposals can be understood as what Zito (2000) calls *collective entrepreneurship*; i.e. a tactic to "generate, design and implement innovative ideas in the public domain" (p. 37, compare Scharpf 1988). This pushes for more innovative policies that are convincing enough to be accepted by a fairly large group of member states and the EP. On the other hand, finding a 2/3 majority still calls for relatively easily acceptable policies. Some degree of common denominator bargaining is likely to remain, as is the bias to uniform and conservative policies (see also McCorminck 2001). Even with relying on a 2/3 majority, seriously innovating and reorganizing EU environmental policies remains more difficult than it is with national schemes. In the meantime, the continued focus on harmonization of environmental policies and regulations still provides a push for uniform policies. Thus, the kind of diversity in policies that allows for solution strategies based on the 'locally specific problem conditions' – as was linked to advancing proactive and integrated policies in Chapter 2 – remains difficult to embed in an EU frame.

168 *A European focus on the local*

Bargaining before decision making

Not all bargaining takes place in the Council or the EP. There are also informal processes of lobbying and bargaining that occur before decision making takes place. The European Commission (CEC) is the executive body of the EU and is responsible for proposing and implementing policies and regulations. After that, the proposal is sent to the Council and later the EP, who accepts, amends or dismisses the proposal. In practice, most bargaining takes place within the CEC and between the CEC and actors such as lobby groups or member states *before* the Council or the EP discuss them (e.g. McCormick 2001, Zito 2000).

EU environmental policymaking is influenced by a fairly large amount of actors and organizations; i.e. businesses, NGOs, branches of the CEC or domestic authorities (e.g. Bernard 2002, Jordan 2002, McCormick 2001 and Zito 2000). Partly, these actors and organizations exert influence through expert groups and committees, which the CEC sets up to advise them while preparing legislation. The CEC also sets up *informal working groups* in which key actors and organizations are represented. Within these working groups, CEC officers debate and negotiate with the represented experts the policies suggested by the CEC. If working group members, among which the member-state representatives, agree upon CEC policies, the Council will typically accord these policies without much debate. If working-group members do not agree, the CEC usually changes its proposals and, based on that, ensures agreement in the working groups later on. In other words, the advice of the informal working groups is quite decisive in developing EU policies and regulations.

The role of informal working groups is also the reason why bargaining efforts usually take place outside of the formal institutions of the EU (McCormick 2001, Zito 2000). National and non-government parties can influence decision making quite significantly through their participation in the informal working groups. In the meantime, many of the business, non-government and national representatives are *technical experts* and/or administrative staff (Zito 2000). It is not rare that they tend to rely on known strategies, such as regulations and technical solutions, reinforcing the technocratic bias of EU environmental policies and hence, the reliance on the 'coordinative model' (i.e. regulations, directives, standards, etc.).

Current perspective

EU environmental policies are still fairly strongly reliant on a command and control tradition; with its uniform, sectoral and often technocratic character (see Holzinger et al. 2006, Jordan 2002, Knill & Lenschow 2004, Zito 2000 and Zuidema 2005). Still the EU is increasingly allowing for greater autonomy for domestic administrations to decide upon *how* policy outcomes should achieved. In other words, although the EU often sets unambiguous, generic and binding targets, the means of achieving them are often left to the discretion of national or even regional governments. Therefore, despite relying on the command and control of the policy *outcomes* in member states, the EU often refrains from commanding and controlling the exact implementation.

Furthermore, since the 1987 SEA, environmental policies *are* becoming more diversified. Similar to national environmental policies, changes include the use of more argumentative and market-based instruments (e.g. emissions trading, eco-labels, procurement tools, policy recommendations, etc., see also Jordan et al. 2005, Lemos & Agrawal 2006 and Stavins 2003). Among these innovations is the so-called 'urban environmental agenda' (e.g. CEC 1990, 2004, 2006). Pursued from the early 1990s onwards, it is this agenda that stands out as the prime attempt of the EU to promote more proactive and integrated approaches for dealing with local – mostly urban – environmental issues. It is also this attempt that I will continue to focus on.

5.2 Towards a local environmental agenda in the EU

The publication of the 1990 'Green Paper on the Urban Environment' by the CEC was the start of an EU focus on proactive and integrated approaches to the local, and explicitly, urban environment. The 1990 Green Paper was inspired by the 4th Environmental Action Programme (1987–1992), which referred to 'comprehensive environmental programmes in inner city areas', and by several CEC conferences on urban environmental issues. The Green Paper itself played an important role in two ways. It both explained the motives for the EU to become involved in urban environmental matters and paved the road for the EU to stimulate proactive and integrated urban environmental policies.

Subsidiarity and the urban environment

As mentioned in Chapter 2, subsidiarity refers to the idea that decisions should be taken at the lowest possible level of authority so as to be closest to the actual problem (e.g. Barnes & Barnes 1999, Jordan 1999 and Zito 2000). Subsidiarity is often used in reacting to a strong interference of the EU in matters considered domestic. National, regional and local authorities use the idea of subsidiarity for defending their administrative and political powers. In addition, subsidiarity is a strong concept in defending the idea that different areas need different policies; i.e. in line with the arguments favouring decentralization as discussed in Chapters 2 and 4. In recent decades, subsidiarity has also become an important objective in EU policy making, also when it comes to environmental policies.

There is a clear tension between harmonization and subsidiarity as overarching objectives when developing EU environmental policies. While harmonization tends to uniform policies (equity), subsidiarity tends to differentiation. The EU urban agenda largely escapes this tension, as it is meant to be *additional* to existing uniform policies. Hence, it does not directly intervene with the idea of harmonization. Nevertheless, when reflecting on the idea of subsidiarity, the question does emerge whether EU interventions are at all suitable for dealing with local issues. The EU is not the body that is 'closest to the actual problems'. The EU has nevertheless expressed motives why it wants to be involved anyway. In the 1990 Green Paper four motives are mentioned (p. 35): 1) the international implications

170   *A European focus on the local*

of pollution originating from urban areas (cross-border effects), 2) the commonality of problems within the urban environment which calls for cooperation among member states in search of solutions, 3) recognition of a European dimension of the historical and cultural heritage of our towns and cities, and 4) the necessity for considering potential impacts on the environment and in particular in urban areas, of community policy in all sectors.

The rise of the urban agenda

The 1990 Green Paper indicates that "dealing with the problems of the urban environment requires going beyond sectoral approaches" (p. 1). Based on the interrelatedness of problems, the CEC expressed in the Green Paper that it no longer considered sectoral and piecemeal solutions sufficient for coping with urban environmental challenges. The CEC's considerations linked in well with the more general shifts in governance and those in environmental governance more specifically (Chapter 2). Proposals made in the Green Paper include more research into urban environmental problems, subsidized projects, new legislation (e.g. energy conservation, economic instruments), additional financial support and guidelines on the incorporation of environmental considerations into planning strategies (CEC 1990; pp. 54–57). Mostly, the Green Paper paved the road for developing distinct urban environmental policies.

During the 1990s the intentions of the 1990 Green Paper were confirmed by the CEC. These intentions were also supported by the spin-off resulting from the 1992 UN conference in Rio de Janeiro on 'Environment and Development'; the UN action plan for sustainable development, Local Agenda 21 (LA21), which was picked up all over the EU. At the 1994 'First European Conference on Sustainable Cities & Towns' in Aalborg the LA21 ideas were translated into the Aalborg Charter, which later became part of the EU supported 'Sustainable Cities and Towns Campaign'. In 2011, about 5,000 local authorities have worked or are working with LA21 and over 2,500 signed the 1994 Aalborg Charter which was replaced in 2004 by the Aalborg Commitments. Central to each of these initiatives was the intent of cities to be *proactive* and to develop more *holistic* (i.e. integrated) policies. Such initiatives were also supported and employed by the 'EU Expert Group on the Urban Environment'. Set up in 1991, the 'Expert Group' has proven important in confronting the CEC with the realities of local and urban environmental governance by informing the CEC on local issues. One of the initiatives it took is the 'Sustainable Cities' project in 1993 with its focus on holistic urban policies (CEC 1997).

Gradually ideas emerged within the CEC to design a formal set of policies that could stimulate and accommodate local authorities to develop more proactive, holistic and integrative approaches to the environment. The 1990 Green Paper indicated that improved urban environmental governance should involve both vertical integration and horizontal integration (p. 32). Vertical integration implied that different levels of authority should draw policies that mutually support each other. Such an integration would then create vertical (top down) policy structures that make the desired horizontal integration on a local scale easier. The 1997 CEC report 'Towards an Urban Agenda in the EU' indicated that this was still the idea seven years later. In this report it was concluded that "policy efforts in Europe already address many of the problems affecting European cities; but these efforts have often been piecemeal, reactive and lacking in vision" (p. 3). The CEC (1997) argued that the effectiveness of existing policies should be stimulated through "a more focused approach using existing instruments at national and community level and enhanced co-operation and co-ordination at all levels" (p. 3). Thus, it was not the existing regulatory approach or even sectoral approach that was to be altered, but rather its implementation in practice. In other words, amendments to the coordinative model were preferred over real renewal operations.

To the Expert Group the conclusion drawn by the CEC in 'Towards an Urban Agenda in the EU' was less evident. In its reaction to the report, the Expert Group stressed that "the sectoral approaches of the European and national levels impede attempts by the local level to achieve integration between different levels (vertical) and between sectors (horizontal)" (1998; p. 4). In the eyes of the Expert Group the 1990 Green Paper and the 1997 'Towards an Urban Agenda in the EU' should have 1) resulted in more coherent EU environmental policies and regulations and 2) stimulate and support urban environmental governments for producing these improvements. The Expert Group argued that such results were not visible. In fact, the CEC even continued issuing new and often ill coordinated sectoral regulations during the 1990s and 2000s.

New sectoral regulations

In the late 1990s three important sectoral environmental directives were issued that directly affect urban environmental governance: the IPPC directive (96/61/EC) on integrated pollution prevention and control, the Seveso II Directive (96/82/EC) on the control of major-accident hazards involving dangerous substances and the updated EIA (Environmental Impact Assessment) Directive 97/11/EC. These directives were all drafted within one year. The CEC did make efforts for a better coordination between these directives and was successful in avoiding explicit inconsistencies between the directives, while the competent authorities were enabled to combine the procedures of these directives during implementation. Still, urban authorities had to make serious efforts figuring out how to make such combinations (IMPEL 1998). In practice, these directives essentially remained sectoral policies that were rather ill coordinated (IMPEL 1998).

Subsequently, two other sectoral policies have been issued that also affect urban environmental governance: the new air pollution directive (1999/30/EC) on nitrogen oxides, sulphur oxides, lead and small particles in ambient air and the noise directive (2002/49/EC) on the assessment and management of environmental noise. The air quality directive sets explicit environmental limit values that must be met all over the EU. It states that if limit values are not met, or are likely not to be met in the near future, action plans must be developed for dealing with air pollution in these areas. Making and implementing these action plans is largely

172　A European focus on the local

the responsibility of regional or local authorities. In practice, many urban areas in the EU are at risk of not meeting the standards and thus have to draw up air quality action plans. In larger EU cities, the noise directive also calls for the development of action plans on noise abatement. Importantly, both action plans – although obligations coming out of the same directorate general of the CEC and issued within three years of each other – are *not* coordinated at all. Just as the IPPC, Seveso II and EIA directives, these two directives represent yet again two sectoral obligations. The EU might have recognized the need for more coordinated and integrated action at a local level; it has not recognized what this means for its own sectoral policies. Instead of reducing the fragmentation of its existing regulations, it is instead increasing this fragmentation.

Towards a 'thematic strategy'

Throughout the 1990s, the EU addressed urban environmental issues mostly with voluntary advice and projects (for an overview see CEC 2004). Examples include the Sustainable Cities project supporting the implementation of Local Agenda 21 and the greening of the Structural Funds (see also CEC 1997). Essential building blocks in these voluntary attempts were the exchange of best practices and the increase of the financial opportunities for innovative urban environmental management projects.

In 2001 the process was accelerated by the choice of the 'urban environment' as one of the so-called 'Thematic Strategies' supporting the implementation the 6th Environmental Action Programme. These 'Thematic Strategies' addressed environmental priority issues characterized by their complexity and diversity of actors concerned. The urban environment was assigned as one of these issues, alongside thematic strategies on air quality, natural resources, pesticides, soil, the marine environment and waste and recycling. The CEC initially aimed for six thematic strategies. Only through pressure from the European Parliament was the urban environment added; i.e. their political will created the political mandate for a 'Thematic Strategy'. Still, the 'Thematic Strategy' was developed against a background of widespread support from national, regional, city and other stakeholders for a proactive and integrated approach to the urban environment (e.g. CEC 2005, EEB 2004, Enviplans 2006, Eurocities 2004b and ICLEI 2004).

In 2004 the CEC produced the draft proposal 'Towards a Thematic Strategy on the Urban Environment' (CEC 2004). This proposal was informed by various informal working groups, who were coordinated by the 'EU Expert Group on the Urban Environment'. The working groups consisted of both member state representatives and other stakeholders. The draft 'Thematic Strategy' conveyed the vision of the CEC on encouraging proactive and integrated approaches to the urban environment. It became the start of fierce debates between stakeholders, lobby groups and member states.

In 'Towards a Thematic Strategy', the CEC confirmed its commitment to improve the integration of policies at the local level. The CEC stated that "active and integrated management of environmental issues for the whole urban area is

A European focus on the local   173

the only way to achieve a high quality and healthy urban environment" (2004; p. 9). To achieve this integrated management, the CEC concluded that the voluntary approach of the past had not succeeded in seriously enhancing an integrated approach (CEC 2004). The CEC subsequently concluded that "a stronger framework at the European level is therefore necessary to revitalize and generalize the environmental management of Europe's largest towns and cities" (CEC 2004; p. 8). So in contrast to earlier attempts, the CEC concluded in the draft 'Thematic Strategy' that additional powers at the EU level are needed.

In 'Towards a Thematic Strategy' (CEC 2004), the CEC maintained a focus on the sharing of best practices and the creation of financial opportunities for cities. In addition, it proposed a 'stronger framework' that should go beyond voluntary measures and that should "strengthen institutional arrangements on sustainable development, including at the local level" (CEC 2004; p. 9). To facilitate such a stronger framework, the Commission considered that "explicit environmental targets, actions and monitoring programmes that link environmental policies to economic and social policies are required", which implies that "urban municipalities therefore need to put in place an environmental management plan. To ensure its implementation and monitor its progress, they need to adopt an appropriate environmental management system" (2004; p. 9). Finally, given the urgency of transport issues, "these towns and cities should furthermore develop and implement a sustainable urban transport plan" (CEC 2004; p. 5). To the CEC, 'urban municipalities' are those in urban areas with over 100.000 inhabitants; i.e. the largest 500 EU cities.

Local environmental management and sustainable urban transport plans were to set clear targets while they were also supposed to "increase the co-operation between different levels of government (local, regional and national), between different departments within local administrations, and between neighbouring administrations, as well as increasing citizen and stakeholder participation" (2004; p. 10). In addition, these plans were meant to uncover "gaps and defects in current environmental management" and support the development of "a coherent sustainable development policy" (p. 10). Consequently, these plans addressed the fragmentation of power and policies in urban governance so as to improve urban environmental governance (CEC 1997, 2004).

The 2004 draft proposal remained relatively vague on the ways in which it was to deliver on its promises. What was clear was that the delivery of the 'stronger framework' relied on legislation for ensuring that the plans and management systems were installed locally. This legislation was not meant to "dictate the solutions and targets that they [cities] should adopt, since no two urban areas are the same" (2004; p. 4). The CEC thus seemed to recognize that common policy formats were constrained in their ability to respond to detailed local circumstances. Instead, the CEC argued for strengthening the role of environmental policies in urban governance and a 'better management of the urban environment'. Instead of prescriptions, the Commission would issue "recommendations, guidelines, indicators, data, standards, evaluation techniques, training and other actions of a more technical support nature to help towns and cities assess and manage different

174   *A European focus on the local*

aspects of their environment" (CEC 2004; p. 5). In doing so, however, the CEC surprisingly opted for proposing what can, by all means, be considered common policy formats.

5.3 The 'Thematic Strategy' under debate

The opinions of NGOs, cities, regions and member states were collected during the development of the draft 'Thematic Strategy'. Nevertheless, the wider consultation process (2004–2005) started *after* the publication of the draft. The 2004–2005 consultations consisted of direct liaisons between the CEC and member states, three hearings in the 'Extended EU Expert Group on the Urban Environment' and of work conducted in two distinct working groups of the 'Extended EU Expert Group on the Urban Environment' (on environmental management and on sustainable urban transport plans). Although the 'Thematic Strategy' was developed against a background of widespread support from national, regional, city and other stakeholders for a more proactive and integrated approach to the urban environment, important criticism did emerge during the 2004–2005 consultations. It resulted in large differences between the draft and finalized 'Thematic Strategy'. This section investigates why these differences were considered needed.

The data used for the analysis made in this section is based on two sources. The first is the reception of the draft 'Thematic Strategy' in the 'Extended EU Expert Group on the Urban Environment'. This group was based on the 'EU Expert Group on the Urban Environment', which was founded in 1991 for advising the CEC on urban environmental matters. In 2004 the original 'EU Expert Group on the Urban Environment' was expanded to inform the CEC specifically on the development of the 'Thematic Strategy'. A group of more than 100 experts from governments (member states, regions and cities) and non-governmental agencies and groups (e.g. lobby groups, scientific organizations) met three times in 2004 and 2005. Stakeholders commented on the proposed 'Thematic Strategy' and could criticize and endorse ideas the CEC put forward. The author was one of the participating experts, which allowed for the collection of stakeholder opinions, the most prominent debates and doubts and hence, for an assessment of the stakeholders' response to the proposed 'Thematic Strategy' (see also Zuidema 2011). Next to making personal notes during these sessions, the minutes of the meeting assisted in collecting stakeholder options. The consultation meetings were public and minutes are also publicly available (DG Environment 2004a, 2004b, 2005). The second source of data consists of position papers and public statements made by various stakeholders prior to or in the aftermath of each of the sessions. If used, references to these papers will be made in the text below.

The arguments used by stakeholders in supporting and criticizing the draft 'Thematic Strategy' can be categorized in four main groups of arguments. The first category of arguments focuses on the idea of a stronger (i.e. obligatory) framework for local plans and management systems to be installed at the EU level. The second category is about the capacity of the proposed instruments (i.e. the plans and the management system) to improve local policy coordination and integration. The third category addresses the relation between these new instruments and the existing environmental policies cities and towns already have in place. Finally, and related, the fourth category addresses local time, resources and skills for coping with the tasks envisioned to be decentralized, also related to support by the EU and EU policies and regulations.

A stronger framework

Several member states, cities and lobby groups, warmly welcomed strong obligatory measures. Specifically, Mediterranean and new accession states were relatively positive about such obligatory measures (see also CEC 2005, Enviplans 2006). They considered obligatory measures to compensate for the relative lack of institutional frames already in place for supporting local environmental governance in their countries (see minutes of the consultation sessions, DG Environment 2004a, 2004b, 2005). In addition to these states, various NGOs welcomed obligatory measures as they believed these measures could help to push local environmental governance beyond mere rhetoric (i.e. compensate for the relative 'weak profile' of the environment). The European Environmental Bureau, for example, collected various responses from NGOs and concluded that "voluntary approaches do not appear to deliver. If delivery is left at the aspirational level, there is therefore serious danger that nothing will happen" (EEB 2004; pp. 6–7). Hence, the European Environmental Bureau endorsed the obligatory nature of the plans, but later did suggest this was most relevant for new accession countries, as it is mostly in these countries where experience and skills for working with local environmental management are limited (see also TCPA 2004). Finally, the two main organizations representing local authorities in the EU (Eurocities and ICLEI) were supportive of obligatory measures (Eurocities 2004b, ICLEI 2004).

Despite a serious number of supporters, the majority of stakeholders rejected obligatory measures or only supported them under certain conditions (see also CEC 2005; p. 5). In the first place, rejections followed political motives, most notably the subsidiarity principle. The European Council of Ministers was one of the first stakeholders to question the obligatory 'stronger framework' the CEC proposed in its draft 'Thematic Strategy' (Council of the European Union 2004). To the Council, the proposed strong EU involvement in local matters was at least questionable in light of the subsidiarity principle. In addition, various nation states (e.g. Spain, Sweden, the Netherlands and the United Kingdom) were reluctant towards a high degree of involvement from the EU in local matters (taken from author notes and the minutes and notes of the first consultation meeting, 7 April 2004, DG Environment 2004a). Rather, as the Association of London Government expressed, "a more suitable approach would be the creation of a more flexible strategy, which offers various options and affords each local authority the autonomy to determine local needs and local solutions" (2004; p. 2). The EU was not commonly considered the 'lowest level of authority possible' to cope with local environmental issues.

176 *A European focus on the local*

In addition to political motives, stakeholders also had practical motives for objecting to the obligatory measures proposed. Many considered the proposed instruments as having little or no added value over existing policies and plans. The following two subsections address these doubts in more detail. Unless other sources are mentioned, data is drawn from the minutes and author notes taken during the consultation meetings with the 'Extended EU Expert Group on the Urban Environment' of 7 April 2004, 24 September 2004 and 17 May 2005 (DG Environment 2004a, 2004b, 2005, Zuidema 2011).

Coordination and integration

To the CEC the 'Thematic Strategy' was to "increase the co-operation between different levels of government (local, regional and national), between different departments within local administrations, and between neighbouring administrations, as well as increasing citizen and stakeholder participation" (2004; p. 10). In addition, the CEC aimed at identifying "gaps and defects in current environmental management" and the development of "a coherent sustainable development policy" (p. 10). To the CEC, sustainable urban transport and environmental management plans supported by environmental management systems were considered appropriate for delivering a coherent framework and its related increased cooperation. Member states, cities and NGOs endorsed the need for improved coordination and integration. Most of them, however, doubted whether the proposed instruments would do the job.

A first category of doubts surrounded the area or jurisdiction that the plans and management system would cover. To the CEC (2004; p. 5), the two plans and the management system should be developed in "capital cities and urban agglomerations of more than 100,000 inhabitants (i.e. the EU 25's largest 500 towns and cities)," while, "as urban areas often extend beyond the municipality's administrative boundaries, the plan would apply to the whole urban area and may require cooperation between neighbouring administrations" (CEC 2004; p. 10). Both the 100,000 threshold and urban fragmentation in the face of external effects of environmental issues proved causes for fierce debate.

The 100,000 threshold was a problem for many countries and cities, as it was not just considered arbitrary, but also too high. Latvia, for example, noted that such a threshold would mean that only one or maybe two cities in Latvia would draft a plan. Others also noted that the 100,000 threshold would exclude many relevant urban areas (e.g. Spain, EEB 2004). The European Environmental Bureau proposed lowering the threshold to 50,000 inhabitants (EEB 2004). In the meantime, the debate on the fragmentation of urban areas was an even bigger issue during the consultation meetings with the 'Extended EU Expert Group on the Urban Environment'.

Many urban areas consist of several neighbouring municipalities in one functional urban area. Many central cities are even home to only the minority of the population of the entire urban area, some reaching levels below one quarter of this population (e.g. Katowice, Lille, Manchester and Porto). Several other urban areas are themselves composed out of many separate municipalities of comparable size or even of distinct cities that grew together (e.g. Athens, Brussels, London, Paris and Ruhrgebiet). This fragmentation is a key barrier for formulating policies covering the entire urban area. In the meantime, many environmental issues such as congestion, air pollution, background noise, sewage, etc. cross jurisdictional boundaries and manifest themselves on at least the level of the urban region. In the face of these external effects, coordination between policies is needed, which is hard to arrange given the fragmented jurisdictional layout of these urban areas. For many stakeholders during the consultation meetings of the 'Extended EU Expert Group on the Urban Environment', overcoming the problems of fragmentation and related external effects was a paramount issue for the 'Thematic Strategy'. Furthermore, the Council of Europe for also highlighted fragmentation as one of the key themes to address before accepting the need for a 'Thematic Strategy' (Council of the European Union 2004), whilst many NGOs expressed a similar message (e.g. COSLA 2005, Eurocities 2005, ICLEI 2004, but also Deutscher Städte- und Gemeindebund and Dutch Society for Nature and Environment).

Towns and cities hoped that the EU could address this issue (see section 5.4). This could have involved modest stimuli such as knowledge dissemination of best practices, policy advice and financial aid for pioneering projects. Some stakeholders also hoped for a clear definition of what constitutes an urban area. Not only could this become an argument for central cities to urge adjacent municipalities to cooperate, it might also imply that there would be planning obligations following the 'Thematic Strategy' to cooperate. Despite the considered importance of urban fragmentation, reading the draft 'Thematic Strategy' shows that this issue was hardly addressed (CEC 2004). The CEC did ask the 'Extended EU Expert Group on the Urban Environment' for advice. An Expert Working Group on environmental management plans and systems was asked for this advice. They noted that such a plan or management system "should follow a 'functional area' (rather than 'administrative boundary') approach, as this would involve more people and stakeholders in the plan, and would address the city's overall environmental impact on wider areas" (EGUE 2005; p. 29). In addition, it noted that "Member States are best placed to define the urban areas that should adopt EMPs, and the Competent Authorities (CAs) responsible for preparing the EMPs" (ibid; p. 39). Many stakeholders expressed during the consultation meetings of the 'Extended EU Expert Group on the Urban Environment' that letting member states decide was not satisfactory as these stakeholders considered it as simply passing the buck. Fragmentation remained on the agenda during subsequent meetings of the 'Extended EU Expert Group on the Urban Environment' without a solution.

A second category of doubts surrounded the degree to which stakeholders believed the 'Thematic Strategy' would actually help deliver a more integrated approach. The CEC had chosen for the use of environmental management and sustainable urban transport plans and an additional management system (CEC 2004). Many stakeholders during the consultation meetings of the 'Extended EU Expert Group on the Urban Environment' were not convinced about the CEC's choice. First, they expressed that the approach essentially consisted of two separate plan

178   *A European focus on the local*

obligations, focusing on a relatively isolated set of themes. Hence, they were considered sectoral, rather than holistic. Many stakeholders argued during the consultation meetings of the 'Extended EU Expert Group on the Urban Environment' that such a sectoral approach did not help in delivering an integrated approach. Various member states argued during the consultation meetings that working with separate environmental management and sustainable urban transport plans could even undermine an integrated approach (e.g. Spain, UK). In addition, many stakeholders expressed during the consultation meetings that they were disappointed by the CEC's focus on the 'environment' instead of 'sustainability' in its environmental management plans (e.g. EEB 2003, Eurocities 2004a, 2004b and ICLEI 2004). A focus on the environment alone, as these stakeholders also argued during the during the consultation meetings, could undermine the establishment of links with social, economic and spatial planning policies (e.g. EEB 2003, Eurocities 2004b; p. 4, Council of European Municipalities and Regions, CEMR 2004). Many NGOs, member states and cities instead opted for a more holistic approach during the consultation meetings. In one of its position papers, Eurocities noted that "many cities are . . . concerned that the environmental focus of the proposed plans should not detract from the need to ensure a more integrated approach, bringing together environment, transport, health, social, employment and economic policies" (2004a; p. 1). In addition, ICLEI (2004) suggested that a focus on 'quality of life' rather than the environment would be better adapted to the needs and desires of cities.

The Expert Working Group on Environmental Management Plans also addressed the choice in favour of a more holistic approach, with a focus on sustainability instead of environmental management. They noted that an environmental management plan "should aim at environmental sustainability, through a more comprehensive or holistic approach to urban policy (public and private), so tackling the contradictions due to compartmentalization of sectoral policies" (EGUE 2005; p. 16). To this end, they recommended "that the proposed urban EMP be part of an over-all 'sustainability plan' for cities" (2005; p. 28). Importantly, the Expert Working Group did not propose an actual 'sustainability plan' in addition or as alternative to the environmental management plan. Not only was such a proposal outside of the Expert Working Group's terms of reference; DG Environment could not really respond to such a request. DG Environment has little competence and even less power when it comes to economic and social affairs. A focus on sustainable urban management plans would, however, seriously affect these terrains. Support and collaboration from other DGs would be required, as would support from the EP and member states. Even if this support would be raised, the spatial planning consequences of a directive focusing on sustainable urban management plans would be a second key challenge. A sustainable urban management plan would, under normal circumstances, include spatial planning consequences. The EU, however, has no competences in spatial planning. This would imply that a directive on sustainable urban management plans should be agreed by full majority by all EU member states, or should be altered in such a way that it would have no spatial planning consequences. Both requirements were considered unlikely to be met. Summarizing, the competences and institutional makeup of the EU prevented DG Environment pursuing sustainable urban management. Still, most stakeholders found a focus on sustainability the most preferable, if not the only acceptable way forward.

A third and final category of doubts related to the degree to which the draft 'Thematic Strategy' was adequately paying attention to the coordination deficits between governments and non-government stakeholders and the public. Much in the same way as discussed in Chapters 2 and 4, moving beyond the public–private divide was something most stakeholders during the consultation meetings of the 'Extended EU Expert Group on the Urban Environment' considered essential, but still hard to put into practice. Various towns and cities and their representing agencies noted that towns and cities often lacked experience, skills and resources for doing so. In the draft 'Thematic Strategy', the CEC did suggest that "public participation in decision making is recognized as a prerequisite for achieving sustainability" (2004; p. 40). The acknowledgement of 'public participation' as a priority in the draft 'Thematic Strategy' was also welcomed by most stakeholders during the consultation meetings. There were, however, doubts concerning how the CEC planned to deal with public participation.

During the preparation of the draft 'Thematic Strategy', the European Environmental Bureau (2003; p. 1) collected opinions of many NGOs and noted "a concern over the lack of community participation, environmental justice and social inclusion issues" in the CEC discussions. Despite these early critiques, the draft 'Thematic Strategy' made little progress in addressing participation and partnerships. The words 'participation' or 'partnerships' were only mentioned as 'good ideas' or in relation to the consultation process of the 'Thematic Strategy' itself (see CEC 2004). The European Environmental Bureau found that the CEC "should . . . be more concrete about how public participation would be included in the management and transport plans" (EEB 2004; p. 8). In addition, Eurocities stressed "the need to address the issue of public consultation and participation in a more comprehensive manner than has been done so far, possibly by introducing a requirement on cities to report progress to their citizens" (2004b; p. 5). To Eurocities, consultation and participation should not just include more public involvement, but should also address "new and innovative ways of governance, such as the concept of multi-partite agreements proposed in the Commission's White Paper on Governance of July 2001, which facilitate partnerships between public actors in the area concerned" (2005; p. 6). The draft proposal's lack of attention to these topics, however, shows that the CEC had little interest in helping cities overcome the public–private divide. Consequently, many doubts regarding the contents of this draft were expressed by stakeholders during the consultation meetings.

Additional policies or duplication?

There was not just a widespread lack of confidence in the added value of the proposed measures in the draft 'Thematic Strategy' amongst of many stakeholders

180   *A European focus on the local*

during the consultation meetings. In addition, many stakeholders expressed that they feared these measures could duplicate or frustrate existing local policies and initiatives. As they expressed during the consultation meetings, the proposed measures (i.e. the two plans and the management system) could therefore be counterproductive by diverting resources away from local initiatives that already performed well. These warnings against duplication or frustration of existing local efforts were relatively widespread among the stakeholders. Based on the idea that "many cities across Europe already have their own strategic, environmental and transport plans" (Eurocities 2004b; p. 6), many stakeholders during the consultation meetings expressed doubts on the added value of the obligatory plans and management system. Indeed, research by the CEC (2005) revealed that in some EU member states up to 90% of the largest cities (>100.000 inhabitants) have their own environmental management plans, while even in the least progressive member states about 30% do. These plans are often supported by environmental management systems, whilst some cities use such a system instead of environmental management plans. In addition, strategies for managing urban transport are present in about 15% of the larger cities in less performing countries to up to 95% in other, better performing, countries (CEC 2005).

Member states, cities and NGOs raised the issue of duplication during the consultation sessions (e.g. Belgium, Sweden, the Netherlands; the cities of Birmingham, Helsinki and London; Eurocities; Deutscher Städte-und Gemeindebund). A key argument that these stakeholders expressed during the consultation meetings was that new obligatory measures should 'fit in' with existing local initiatives (e.g. Czech Republic, Eurocities, Expert Working Group on Environmental Management Plan and ICLEI). In one of its position papers, Eurocities argued that "any new legislative requirements should recognize these [existing local] plans and make room for them, ensuring that they are integrated and built upon" (2004b; p. 6). In addition, ICLEI suggested that "any legal obligation would need to respect existing plans and systems at the local level which already serve the purpose of sound environmental and transport management, but which are called differently" (2004; p. 7). Finally, the Expert Working Group on Environmental Management Plans advising the CEC, also stated that "the proposed urban EMPs must not replace existing plans that meet the specific objectives proposed by the Working Group, and must have visible benefits for citizens and contribute to (at least) EU environmental planning" (2005; p. 39).

Although it was quite clear that new EU measures should 'fit in' with existing local initiatives, many stakeholders expressed during the consultation meetings that the CEC was not convincing in how it would do so. Eurocities, for example, noted that the draft 'Thematic Strategy' hardly addressed the question how existing local, national and EU policies and regulations could be coordinated. In addition, ICLEI noted that the focus in the draft was on proposing new plans and hardly on the process of positioning these plans in urban governance for delivering the desired change with these plans. Hence, as the London Government Association (LGA) posed during the consultation meetings, it was simply not clear *how* the new obligations would fit in with existing governance structures (see also cities of Brussels and Birmingham).

Given the risk of duplication and frustration and the CEC's lack of attention to responding to this risk, various parties were quite explicit in their critiques on the draft 'Thematic Strategy' during the consultation meetings. Representatives from Belgium, for example, warned that new obligations should "not destroy existing legislation and those plans that work". The Deutscher Städte- und Gemeindebund suggested that "there already is so much action, do not frustrate this by new EU policies". The Council of European Municipalities and Regions (CEMR) added to this that it feared the obligatory plans could 'cut across' current measures. Finally, the Association of London Governments concluded that "given the strides being made by local authorities in sustainable urban management the LGA does not support the Commission's proposal to impose a requirement for mandatory urban management plans" (2004; p. 2).

Limits to local abilities

While duplication and frustration of local initiatives were highlighted by stakeholders during the consultation meetings, many also highlighted the importance of the 'finite resources' of local governments (e.g. EEB 2004, Eurocities 2004b, ICLEI 2004, TCPA 2004, also UK and Spain). To the Expert Working Group on Environmental Management,

> The main perceived obstacles regarding urban EMPs are related to the extra work-load (in time and man-power) at both national and local levels. This is particularly relevant in those Member States that already have planning commitments and legal requirements for the management of the environment.
>
> (DG Environment 2005; p. 21)

In addition, the Town and Country Planning Association summarized that the 'Thematic Strategy'

> should advocate using existing systems of spatial planning, urban management and traffic management, if they are suitable. The creation of parallel systems will consume scarce resources unnecessarily and divert them from the substantive aim of not creating systems but improving the urban environment.
>
> (TCPA 2004)

Thus, as again the London Government Association summarized, "the proposals from the Commission to impose new obligations on local authorities are therefore met with caution as they could require changes to the distribution of resources that may not be appropriate for all urban areas" (2004; p. 4).

Stakeholders also expressed during the consultation meetings that local problems of responding to new obligations should be linked to existing coordinative challenges related to existing sectoral EU policies and regulations. Stakeholders claimed that the fragmented and bureaucratic nature of EU (environmental) policies and procedures was considered a cause for local fragmentation and excessive workloads. In addition, they claimed that this fragmentation frustrated

182 *A European focus on the local*

the development and delivery of proactive and integrated environmental policies, as was, for example, stated by stakeholders such as the Council of European Municipalities and Regions (CEMR) and Eurocities (see also EGUE 1998, section 5.2). Given these complaints, deregulation and simplification of EU policies was a topic addressed during the consultation meetings and especially in various position papers. Eurocities (2004b; p. 4), for example, called for a greater emphasis of the CEC on the need to develop European actions for supporting the integration of the EU's own policy areas into a coherent structure that supports sustainable urban environment. Eurocities in particular mentioned improved integration of the EU's transport policy, climate policy, waste policy, environmental health care, cohesion policy and state aid policy. The Council of European Municipalities and Regions also referred to the increasingly complex and bureaucratic procedures associated with application to EU structural funds (CEMR 2004).

The 'Thematic Strategy' does not address the fragmentation and bureaucratic complexities of CEC policies (CEC 2004), nor does the CEC use other strategies for improving the 'internal integration' of its policies. Instead, as discussed in section 5.2, the EU continues to develop sectoral policies that are not actively coordinated to each other. They are signals of the fact that deregulation and the simplification of environmental policies are no issues for the CEC. Little surprisingly, Eurocities again concluded in 2006 that "EU environmental legalization is very complex and fragmented. Eurocities calls for more harmonisation as regards timeframe and requirements as this legislation contradicts other EU environmental, social and economic objectives" (2006; p. 8). Local authorities are clearly not actively enabled by the EU by a reduction of potentially unnecessary pressures on their workloads.

Based on the problems with finite resources and staff, many stakeholders argued for additional EU support for the development and delivery of the two plans and the management system during the consultation meetings. In one of its position papers, Eurocities also noted that

> The development of sustainable transport and management plans and their implementation through management systems will certainly put a further strain on already limited local authority budgets, therefore the question of how these additional responsibilities will be paid for is an extremely relevant one to address at the outset of the process.
>
> (2004b; p. 7, see also European Environmental
> Bureau 2003, 2004)

In addition, ICLEI agreed to binding requirements, but only "*upon condition* that national governments and the European Commission agree to provide a sound and supportive framework to accompany local action in line with the Thematic Strategy" (2004; p. 3).

There was little confidence among stakeholders during the consultation meetings regarding the CEC's proposed means of support for cities in implementing the two plans. The CEC proposed two main categories of supportive mechanisms in the draft 'Thematic Strategy' (see CEC 2004). The first was the promotion and continuation of financial support for the exchange of good practices and the training of city staff. These supportive mechanisms were highly welcomed by many stakeholders during the consultation meetings (e.g. Council of European Municipalities and Regions [CEMR], Eurocities, ICLEI, London Association of Government, and Expert Working Group on Environmental Management Plans). However, these mechanisms were not really different from what was already being done. In addition, they were voluntary mechanisms requiring subscriptions and tenders, so only a limited number of towns and cities could use them. What was new was the second means of support proposed: the obligatory environmental management system that should support the implementation of environmental management plans. Again, stakeholders during the consultation meetings clearly expressed that they welcomed the *potential* added value of such a management system. However, as the Expert Working Group on Environmental Management Plans (2005) indicated, for example, the obligatory status of these management systems was less widely accepted during the consultation meetings. To the Expert Working Group, "it was clear . . . that there is some reluctance regarding EMS adoption, due to lack of consensus among Local Authorities on the need formally to execute all steps foreseen in the most commonly used EMS". Therefore, the Expert Working Group did not recommend any *specific* EMS, rather it "recommends the adoption of a formal, possibly certified set of procedures for the establishment of the urban Environmental Management Plan", allowing cities more discretion for finding their 'own best ways' of coping with their challenges.

In the meantime, the draft 'Thematic Strategy' was not suggesting any additional financial support for cities in the development and delivery of the obligatory measures (see CEC 2004). Consequently, several stakeholders during the consultation meetings expressed that they considered the 'Thematic Strategy' and the proposed obligatory measures simply as additional burdens that were hard to meet given existing budgetary constraints. For example, the city of Birmingham noted a likely mismatch between the aim of the obligatory plans and the available competences, resources and instruments available locally for implementing and enforcing them. ICLEI (2004) added that current conditions (i.e. poor financial situations, lack of human resources and training) in many local authorities did not allow for what they called 'reckless actions' of adding additional obligatory policy formats as well. The Council of European Municipalities and Regions (CEMR) also worried that the obligatory plans "could . . . risk creating extra administrative and financial strain rather than providing added value" (2004; p. 1).

Letting go of obligatory measures

Summarizing then, "many experts expressed the view that obligations were not appropriate citing concerns over duplication of efforts, bureaucracy, cost and the subsidiarity principle" (minutes from the meeting of the Extended Expert Group

184   *A European focus on the local*

on the Urban Environment, 7 April 2004). In response, the CEC dropped the proposed obligatory measures and stated that

> given the diversity of urban areas and existing national, regional and local obligations, and the difficulties linked to establishing common standards on all urban environment issues, it was decided that legislation would not be the best way to achieve the objectives of this Strategy.
>
> (2006; p. 4)

In section 5.5 I will discuss how the finalized 'Thematic Strategy' did aim to respond. Before that, I will first asses how the ideas and tools proposed by the CEC in the draft 'Thematic Strategy' compare to the practice of urban environmental governance as analysed during the 'Liveable Cities' project.

5.4 Liveable cities

The 'Liveable Cities' project (2005–2007) was one of the projects the CEC used for assessing the reception of its central ideas proposed in the draft 'Thematic Strategy' (CEC 2004) to practice. These central ideas were an obligatory environmental management plan, sustainable urban transportation plan and environmental management system for larger EU towns and cities. DG Environment selected various projects as 'testing grounds' to put these ideas to practice. The 'Enviplans' project investigated the design, development and implementation of environmental management plans (Enviplans 2006), the 'Pilot' project addressed sustainable urban transport plans (Pilot 2007) and the 'Managing Urban Europe 25' project assessed existing and possible new environmental management systems (MUE 25, 2008). The 'Liveable Cities' project also focussed on the proposed environmental management plan, but had a somewhat broader focus than 'environmental management'. The project focussed on making 'sustainable urban management plans'. This broader focus was based on the idea that 'sustainable development' was favoured as a guiding principle over 'environmental management' for pursuing the desired 'integrated approach to the urban environment' that the CEC aimed for. The project partners were the University of Groningen; the Dutch Ministry of Housing, Spatial Planning and the Environment; Eurocities; the Sustainable Cities Institute (University of Northumbria); Ethics etc. . . . consultancy and the cities of Aalborg, Bourgas, Bristol, Copenhagen, Lille Métropole, Malmö, Rotterdam, The Hague and Venice.

The 'Liveable Cities' project had two main objectives: the development of sustainable urban management plans in all nine partner cities *and* the development of a so-called 'model plan' for sustainable urban management. The model plan would explain the structure of a sustainable urban management plan and would advise on how such a plan could be developed and implemented. This 'model plan' eventually became the 'Guidance Document for Sustainable Urban Management' (Creedy et al. 2007). The author was part of a four-member research team that conducted the research fuelling the development of the 'Guidance

A European focus on the local   185

Document'. As such, the author was directly involved in collecting and analysing data during the 'Liveable Cities' project. Within the context of the 'Liveable Cities' project, the aim was to write a 'Guidance Document for Sustainable Urban Management' (Creedy et al. 2007). The author also contributed to writing the 'Guidance Document'. The 'Guidance Document' expresses the project partners' conclusion regarding their response to the proposals made in the 'Thematic Strategy'. It reflected the state of the art of sustainable urban management of a series of EU towns and cities. By doing so, it meant to identify good practices of sustainable urban management, key problems towns and cities face in sustainable urban management and possible means for responding to them.

During the 'Liveable Cities' project, the status of environmental governance in the participating cities was studied as was the feasibility of the proposed measures of the 'Thematic Strategy'. The analysis of this section is based on the data produced during the 'Liveable Cities' project and interviews conducted by the author in the partner cities (Zuidema 2011).

The 'Liveable Cities' project

The research during the 'Liveable Cities' project was based on two complementary research agendas. The first research agenda during the 'Liveable Cities' project consisted of a selection of eight case studies of sustainable urban management in all participant cities (except for Rotterdam). Each case study was presented in a 'case report' (Liveable Cities 2005a, 2005b, 2005c, 2005c, 2006a, 2006b, 2006c. 2006d, 2006e, see also Creedy et al. 2007). These reports started with a general introduction to urban environmental governance as conducted in the partner city, while addressing the institutional structure of the partner city and the contexts of national policies and regulations. In addition to that, a key example of urban environmental governance (the actual case) was presented so as to show how urban environmental policies and strategies were applied in practice in the partner city. Finally, the 'case report' contained a reflection on the case and, in doing so, also on the environmental policies and strategies in the partner city. This reflection was based on 1) a desk study of relevant policy documents, 2) discussions with local government officials, 3) site visits to the case during a 'multi-day workshop' where the 'Liveable City' partners would be invited to see and discuss the case, 4) discussions with stakeholders, 5) the use of two 'reference cases' of sustainable urban management in other EU towns or cities and 6) a peer review of the case by a selected group of 'experts' from the participating cities.

The second research agenda during the 'Liveable Cities' project consisted of 'expert workshops'. These 'expert workshops' took place during the 'multi-day workshops' when the 'Liveable Cities' partners would come together to discuss the eight cases. Early on during the 'Liveable Cities' project, the focus of the 'expert workshops' was on either the lessons that could be learned from the case that was visited, or on the brainstorm sessions regarding key themes of sustainable urban management. These themes were decided upon by the research team. These 'expert workshops' provided a basis for the structure and contents of the

186   *A European focus on the local*

'Guidance Document'. Later 'expert workshops' would reflect on first drafts of the 'Guidance Document' and allowed the 'Liveable Cities' partners to provide comments, suggestions and in the final stages, to endorse the finalized 'Guidance Document'. The final version of the 'Guidance Document' was endorsed by all project partners. The 'expert workshops' were visited by the 'Liveable Cities' partners (local government officials), and by experts on sustainable urban management that were specifically invited. In total, more than 100 experts from EU towns and cities, national and regional governments, the academic world's sister projects of the 'Liveable Cities' project and various non-government organizations contributed during the 'expert workshops'. In addition, a total of 12 cities also presented cases of sustainable urban management during the workshops in addition to the nine partner cities. More details on the contents of the 'expert workshops', the participants, the means of data collection, the role of the author in the research process and the accessibility of the data can be found in Zuidema (2011), while the results are presented in the subsequent case reports (Liveable Cities 2005a, 2005b, 2005c, 2005c, 2006a, 2006b, 2006c. 2006d, 2006e).

A broader focus

A first key lesson emerged early on in the project when vast differences between the participating cities became evident. Next to well-anticipated geographical, economic and physical differences, vast institutional, political and cultural differences were also encountered as soon as the first workshop discussions (The Hague, 31 March 2005). Workforce, administrative routines, legal competences, expertise, etc. all varied greatly, as did social and political tendencies and the allocation of competences. Given the resulting differences in the structures and processes of planning and policymaking in the participating cities, the project partners agreed that there is no 'one best way' of organizing urban sustainable management (Venice Workshop, 2 December 2005). Rather, it was concluded that "although European cities and towns have similar problems, each requires a unique set of solutions – Sustainable Urban Management must be 'tailored' to the local situation" (Creedy et al. 2007; p. 4). As was discussed during the Venice workshop, such a 'tailored' governance approach was considered not only contingent on the circumstances surrounding the specific issue faced. It was also considered crucial that it was adapted to the cultural, political, economic, organizational, etc. conditions. Different political, policy-making and partnership strategies are required for responding to different socio-political *and* administrative realities. The project partners eventually concluded that there is no one governance blueprint for delivering sustainable urban management to be used in all towns and cities; i.e. a single 'model plan' for sustainable urban management simply doesn't exist. It is in line with the message of Chapters 3 and 4 that common policy formats are not likely to fit in with the diversity of local circumstances.

Instead of developing a single 'model plan' for sustainable urban management, as proposed by the European Commission, an alternative approach was adopted. During the 'Liveable Cities' project, all partners contributed to the development

A European focus on the local 187

of a 'Guidance Document' for sustainable urban management (Creedy et al. 2007). While doing so, the partners rejected a sole focus on a sustainable urban management *plan*, despite the project's agenda (Liveable Cities 2006a, also 2005a and 2005b). Rather, the partners emphasized the *instrumental* role of plans in the more encompassing processes of planning and policymaking. In other words, plans, considered as legally framed documents that record the agreed objectives, targets and time frames for achieving them, are only among the tools to be used in planning and policymaking. The original objective within the 'Liveable Project', however, was for as many cities as possible to develop and agree on a 'sustainable urban management plan'. The assumption underlying this objective was that such a plan would be adequate for improving sustainable urban management in these cities, as the CEC expressed in the draft 'Thematic Strategy' (CEC 2004). During the project it was agreed that the narrow focus on only making plans potentially ignores the importance of other urban management aspects or tools, such as political bargaining, process management, partnerships, programmes, implementation, etc. (Liveable Cities 2005a, 2006a).

The cities also explained why making a sustainable urban management plan would not necessarily add value, both during the interviews with city officials (see also Zuidema 2011) and during several of the workshop sessions (Liveable Cities 2005a, 2005b, 2006a, 2006b and 2006c). First, most cities already worked with one or more environmental plans and strategies (Aalborg, Copenhagen, The Hague, Rotterdam and in development in Bourgas), while in the other cities sustainable policies were integrated in more overarching strategic plans (Bristol, Venice, Lille). In addition, most cities worked with LA21 or other operational plans or programmes on sustainable development (see also Creedy et al. 2007). Hence, most cities expressed that they saw little added value in adding an extra plan to this set of policies and plans (Liveable Cities 2006a, see also Liveable Cities 2005a, 2005b). Furthermore, cities as diverse as Bristol, Malmö and Venice indicated that they worked with way over 50 plans and policy documents in governing their cities. As was expressed during the interviews with city officials (Zuidema 2011), most city professionals are not aware of all these plans. Fewer people have read them all, let alone those professionals who can relate all these documents to each other. With the exception of Bourgas, which was still developing many of its policies, the cities were not keen to draft 'yet another plan' (Liveable Cities 2006a). For example, "for Copenhagen, written plans have not been effective in the past. A new approach was needed, involving practical targets with resources and an action plan focus" (Liveable Cities 2006a). Similarly, "we [Rotterdam] did not believe that making a SUM [sustainable urban management] 'plan' would be a useful process for us" (Liveable Cities 2006a). Eventually, early and detailed discussions between the project partners concluded that a single new 'sustainable urban management plan' was not considered something that would add value to existing management efforts and might even be counterproductive.

In response, the cities rejected the original project objective of developing a 'sustainable urban management plan' in their own city (Liveable Cities 2006a). Instead, some chose for restructuring their existing or newly drafted plans and

188　*A European focus on the local*

strategies according to the learning taking place during the project. The other cities used this learning to improve the overall management capacity in sustainable governance. Although plans were still considered important elements in such a strategy, the cities agreed they were only a part of a more encompassing governance effort (Liveable Cities 2005b). Hence, it was also agreed that the project output would not solely provide guidance on a 'sustainable urban management plan' (i.e. the 'model plan'). Instead, a 'broader focus' was called for, which would reflect the diversity of governance practices and instruments used in sustainable urban management. This 'broader focus' was already called for during the Aalborg en Venice workshops in late 2005. It gradually developed during the project and was presented and endorsed during the Bristol workshop (6 September 2006, Liveable Cities 2006c). It is also reflected in the guidance document 'Towards Liveable Cities and Towns' (Creedy et al. 2007) that replaced the initial 'model plan' for sustainable urban management.

More than a plan

The 'broader perspective' argued for was meant for addressing "the difficulties caused by thematic, departmental, and sectoral specific initiatives" while recognizing "that partnerships between governments and the civil society are the essential lubricating oil for Sustainable Urban Management" (Creedy et al. 2007; p. 4). The main messages the cities aimed to express is, first, that political leadership and commitment are crucial for both the development and delivery of sustainable ambitions (Liveable Cities 2006b, 2006c). Second, a strong mixture of instruments can be used in developing and delivering sustainable ambitions (see Creedy et al. 2007). They can involve many procedures, process management schemes or even entire systems for environmental management that cities use for dealing with thematic and sectoral fragmentation (including EMAS, ISO 14001 and specific local initiatives such as used in Aalborg, The Hague and Rotterdam). Furthermore, they can involve a wide variety of communicative or partnership arrangements. In the 'Liveable Cities' project, they included partnership constructions with other municipalities (e.g. Lille, Rotterdam, Copenhagen, Malmö), with private stakeholders and the general public (e.g. Bristol, Copenhagen, The Hague and Venice) or with local subcontractors for performing part of the municipal tasks (e.g. Bourgas, Copenhagen, Venice). Among the initiatives are the Danish and Swedish Green Cities / DOGME 2000 project (Malmö and Copenhagen), the British Local Strategic Partnerships (Bristol) and regional cooperative agencies (e.g. Lille, Rotterdam and The Hague).

With this mix of instruments used, the cities confirm that "although hard management tools (i.e. plans, permits, laws and procedures) are highly important, these are only the means of power and the framework within which the actual work needs to be done" (Venice workshop). This actual work consists of debating, connecting, integrating, deciding, doing, etc., to be conducted by and with city departments, politicians and other stakeholders. All project partners agreed that any approach to improve urban planning and policymaking should react to

Box 5.1   Seven key elements of sustainable urban management (Creedy et al. 2007)

The 'Liveable Cities' discussions produced seven elements to be addressed in sustainable urban management:

- Principles; the essential aspects of sustainable development and urban management required for liveable cities
- Politics; the nature of political involvement and commitment that is required to support liveable cities
- Partnership; the governance models and responsibility agreements for working with stakeholders that are needed for liveable cities
- Processes; for agreeing visions and objectives for managing and monitoring the delivery of actions for liveable cities
- Policies; meeting local needs, respecting local traditions, satisfying national and European regulations
- Plans; the role and integration of spatial, thematic, agenda 21 and other plans
- Programmes; of action to achieve the agreed objectives for liveable cities.

all these activities (Liveable Cities 2006c). Thus, the 'Liveable Cities' guidance is based on agreeing that in addition to plans, politics, processes, programmes of action, partnerships and process management also matter (Box 5.1).

Guidance for 'sustainable urban management'

The choice in favour of the desired 'broad perspective', which was already called for from the second workshop onwards (Aalborg workshop), allowed for the subsequent work in cities and workshop discussions to address the full range of sustainable governance in the participating cities. First, sessions during the multi-day workshops were no longer focussed on just the development of a sustainable urban management *plan*. Instead, they focussed on sustainable urban management and included discussions regarding each of the 'key elements' as summarized in Box 5.1. These sessions generated much data on the role of each of these elements, rather than on the role of plans alone. Second, the case studies were restructured so as to better respond to the broader focus. Finally, the work conducted in the partner cities was no longer dominated by the need to develop a sustainable urban management *plan*. Instead, this work was inspired by all seven elements of sustainable urban management. Cities assessed their current plans, policies and management operations for identifying 'gaps' in their sustainable governance. This was partly inspired by a benchmark of relevant sustainable and

190 *A European focus on the local*

environmental policies and operations, as presented in the 'Guidance Document' (see Creedy et al. 2007; p. 58). The assessment helped show the current status of the partner cities' sustainable urban management. It produced data on the main gaps, barriers and problems that these cities face in developing and delivering their sustainable urban policies (as expressed in the case reports). It also provided indications of cities' current willingness and ability to pursue proactive and integrated approaches to the environment. As such, the project highlighted which problems cities mainly face and what kind of support or persuasion from central governments they might need or can respond to.

Out of the data generated during the case studies, workshops and interviews, three main issues can be distilled that the cities are struggling with. First, cities struggle with widespread coordination and integration deficits they recognize between the cities' existing plans, policies and departments. In relation to this, the cities also have a desire to better link policies and administrations with the social and political dynamics surrounding urban governance; i.e. the coordination between politics, government bureaucracies, the market and the 'civil society'. Third, serious problems were uncovered in the actual delivery of sustainable policies. All three problems will shortly be addressed here.

Poor policy coordination

The coordination deficits identified during the 'Liveable Cities' project were quite similar to those highlighted during the consultation process of the draft 'Thematic Strategy' (section 5.3). Deficits mostly related to fragmented urban jurisdictions, the coordination between various levels of authority, the communication and integration between fragmented and competing city departments and the administrative officers' skills and competences.

The *fragmentation of urban areas* is a prominent issue in more than half of 'Liveable Cities' partner cities. The example of Lille was already touched upon in section 5.3. Lille itself has less than a quarter of a million inhabitants, while the entire urban area has more than one million inhabitants. The creation of 'Lille Métropole' was needed for coordinating the policies and programmes of all the 85 municipalities making up the urban area (Liveable Cities 2006e). Cities such as Copenhagen, The Hague, Rotterdam and Bristol face similar issues to Lille. All these cities are home to only about half the population of their entire urban areas. This can cause serious problems (Creedy et al. 2007). In Bristol there are, for example, complaints about the Greenfield developments going on in adjacent municipalities undermining Bristol's own attempts at reducing urban sprawl. One example is the large indoor mall at the Cribbs Causeway junction outside the city, which was developed in an adjacent municipality in spite of Bristol's attempts to stop it.

Coordination deficits also relate to the fragmentation of formal governments themselves. On the one hand, fragmentation has to do with the development of policies by various levels of authority (i.e. *vertical coordination*). A mismatch was repeatedly noted between city responsibilities and the competences and resources

A European focus on the local   191

the cities are given (Liveable Cities 2006a, 2006b). Bourgas, for example, indicated that it gained much in responsibilities through extensive decentralization. However, Bourgas had not gained much in legal competences, funding or staff for dealing with these new responsibilities. Many important competences still originate at a regional level (e.g. some possibilities in issuing permits or fines), whilst many funding opportunities are kept centrally. Bourgas indicated they had little 'power to prevent' unwanted developments and faced increased instability in their urban development. In addition, in Venice,

> competences on environmental pollution issues are very much spread, firstly between different levels such as state, region, province and urban, and also on a municipal level. It is sometimes not ever clear to us as environmental offices where all competences originate, which makes management difficult.
> (Interview with city official, 30 November 2005)

On the other hand, fragmentation occurs within municipal governments (i.e. *horizontal coordination*). First, cities noted that horizontal coordination was frustrated by the fragmented nature of many (inter)national policies and regulations (e.g. Liveable Cities 2005a, 2006a). Experts from Malmö, for example, expressed during the interviews that "much national legislation got in the way of integration" (interview with city officials, 14 February 2006 and Liveable Cities 2006a). Second, horizontal coordination deficits are also caused by the organization of municipal agencies and policies themselves. Venice is most illustrative in this (see Liveable Cities 2005b). In Venice several environmental themes run through many distinct departments (i.e. on water and environmental information). In addition, Venice municipality works from two distinct offices (in the towns of Mestre and Venice). These two offices have their own geographical territory in the municipality for several environmental themes. However, responsibility for some other environmental themes resides solely in one of the two offices. Not surprisingly, during the Venice peer review it was also concluded that "the fragmented plans on different themes makes it difficult to achieve an integrated approach" (Liveable Cities 2005b). Furthermore, experts from Malmö noted that they worked with a total of 72 municipal plans and programmes, which undermines policy coordination (interview with city officials, 14 February 2006). Although less extreme in other cities, it is hardly surprisingly that many city partners agreed that "experience shows that environmental officers often run into plans made by other departments, which are already developed into quite a significant degree; this can result in being too late for the environmental interest to be well integrated in these plans" (Liveable Cities 2005b).

In addition to high coordination demands, cities noted that the sectoral organization of their policies and administration also resulted in large differences in *organizational attitudes and working cultures* (see Liveable Cities 2005a, 2005b). For example, in Aalborg developing a sustainable urban management strategy met much administrative resistance. Aalborg experts noted that a common response from representatives of departments within the city administration to

192   *A European focus on the local*

developing such a strategy was 'mind your own sector!' (Liveable Cities 2005a). Various departments simply see any attempt for sustainable urban management, in Aalborg conducted by spatial planners and environmental representatives, as mingling with their own competences. In Aalborg such 'mingling' was rejected; "don't intervene in our work and don't tell us to do our job differently" (ibid). It confirms "strong sectoral interests and politics in organisations" and that "people hold on to their own organisational topics" (ibid).

Related to 'sectoral attitudes' are the 'cultural differences' noted between various departments.

> Splitting an organization into distinct departments can make integration quite difficult as each works in its own way, as each department has its own understanding, interpretation and language and techniques of making these known. Differences of opinion and clashes between and within organizations within municipal government are thus natural.
>
> (Liveable Cities 2005a)

Cultural differences mean that "it is crucial to develop routines and mechanisms that facilitate officers to communicate and mutually adapt and stimulate to avoid ad hoc communication" (Liveable Cities 2006a). To improve communication, suggestions included rotating vertically and sectorally organized policies through project working, informal interdepartmental working, steering groups and formal routines of consultation. Also, the "emancipation of both environmental and spatial officers is crucial; both need to respect each other's objectives and tasks" (Liveable Cities 2006a).

Finally, developing and delivering proactive and integrated approaches to the urban environment is also determined by the current *managerial skills of most environmental experts* and *the available resources.* It was concluded during the workshop discussions in Aalborg that in coming to proactive and integrated approaches, "it is crucial to convince non-environmental officers, politicians and possible related stakeholders of the potential contribution an environmental officer can have; presenting the environmental officer as a constructive partner". The same workshops discussions showed, however, that many environmental officers are not used to a constructive and strategic attitude. The City of Malmö expressed, "as many of them are used to be [technical] experts, it is difficult for them to make compromises and think outside their box" (Liveable Cities 2005a). In response, cities such as The Hague and Maastricht (a city contributing an example case) expressed that they work with a separate unit of environmental officers addressing permits and control and a unit with more strategic and integrated officers. Consequently, these cities state, staff can be hired that have the right skills for functioning as intermediaries between environmental experts, other municipal departments and private partners. They help environmental issues to be "presented in a more constructive way than in a deconstructive way; i.e. 'we will help you overcome costs and barriers' instead of 'we will check you and set crucial barriers'" (Liveable Cities 2005a). Despite some good suggestions for

A European focus on the local 193

improving municipal officer skills and competences, many cities expressed they face a lack of funding and time to act. As indicated during the Bourgas workshop, officers in most cities already lack the time to even read documents and proposals issued by other departments! Also, experts from Bristol indicated that the municipal allocation of budget is based on budgets per department, preventing real integrative efforts (interviews with city officials, 3 September 2006). In the meantime, political and administrative leadership for overcoming sectoral differences is also needed, bringing us to the socio-political realities in cities.

Socio-political realities

To begin with, the cities argued that a relatively limited commitment from politicians and heads of department to environmental ambitions is a major limit for pursuing proactive and integrated environmental policies (Liveable Cities 2006c). Often, as was noted during the Bourgas workshop, 'ownership' of sustainable urban management tends to be a problem, where neither environmental officers nor the entire municipal administration seem to be correct agents. Rather, the cities agreed that politicians or the entire urban society should be involved in sustainable urban management. However, environmental issues typically involve technical data such as decibels, concentrations of small particles and toxics or risk assessments. For political representatives and social or market stakeholders, environmental issues can thus be difficult to understand, referring to the 'weak profile' of the environment (Chapter 4). Cities thus noted a "gap between politicians and administration" (Liveable Cities 2006c), as discussed during the Bristol workshop, resulting in a lack of expertise in the political arena. Cities also pointed out that "political knowledge and understanding is a policy in its own right", whilst "politicians need to be helped to vision" (ibid). Without help by administrators, politicians might well have a lack of ambition in environmental and sustainable initiatives and a bias towards minimum requirements and rhetoric rather than genuine actions (ibid). Sustainable urban management thus requires that politics will not be overlooked. The respondents however noted that their traditional technical environmental bureaucracies still tend to do so (ibid).

For engaging the urban society, project partners suggested that "sustainable urban management must be based on open, inclusive, transparent, and accountable decision-making", as it "is a social activity for the whole urban community protecting the democratic value of policies, legitimising decisions, stimulating stakeholder responsibility and encouraging changes in behaviour" (Creedy et al. 2007; p. 7). For the city partners, it was clear that such engaging goes beyond informing and educating people, but it also involves partnerships for clarifying and empowering stakeholders' roles and contributions. Cities thus considered partnerships important for linking societal and market processes with traditional formal government structures (Liveable Cities 2005b, 2006c).

On the one hand, cities celebrated partnerships for their potential contribution to democratic legitimacy (Liveable Cities 2006c). Consequently, most partner cities were keen to engage the public in decision making and policy delivery

194   *A European focus on the local*

(explicit examples were found in Aalborg, Copenhagen, Malmö and The Hague, but also in the example cities of Antwerp, Apeldoorn and Breda, see Creedy et al. 2007). Despite this commitment, these cities face serious difficulties in engaging the public. These difficulties were well illustrated in the case of Copenhagen. The city had installed eight separate Local Agenda 21 offices, which meant to "anchor democracy and citizen involvement in local areas and by this raise the local awareness for sustainability and local commitment" (Liveable Cities 2006d). Based on grassroots working, the offices engaged in dialogues and actual initiatives with citizens and businesses. The approach used in Copenhagen was, for example, adopted in the urban renewal project in Nørrebro Park where "it has been a very central idea that the whole renewal project is based locally with much of local involvement from the residents" (Liveable Cities 2006d). Copenhagen, despite its efforts and successes, still stated that it "has to be admitted that also in this case, as in many other cases, the percentage of local residents involved in the project is low" (Liveable Cities 2006d). Similar experiences were reported in other cities (Liveable Cities 2006a, 2006b). For example, Malmö faced problems engaging ethnic minorities, whilst Bourgas had to train people regarding how they could participate.

On the other hand, the cities also recognized that they needed the support and resources of private parties (Liveable Cities 2005a, 2006c). A lack of financial leverage of cities and a lack of municipal land ownership were mentioned by various cities as key barriers for delivering sustainable policies. Aalborg, for example, indicated that "[land] ownership is crucial for both initiating projects and to set environmental demands that go beyond formal law" (Liveable Cities 2005a). The Hague also learned during the implementation of LA21 in the 1990s that intensive cooperation between crucial stakeholders in and outside of government was needed for achieving good results (interview city official, 18 February 2005). In Sweden, a lack of national policies and funding on soil remediation was highlighted as an example of why public–private partnerships were essential to almost any brownfield development. Cities simply need financial aid from landowners of polluted sites and the developers for clean ups (Liveable Cities 2006a).

All cities confirmed the need to strengthen their relations with public *and* private partners (Liveable Cities 2006c). The cities' ambition is not just to consult and inform these partners, but also engage these partners in networks of communication, collaboration and negotiation (ibid). However, cities note that they still struggle with engaging in such networks. The gap-analyses in the cities indicated that the involvement of NGOs, businesses and the public is only just starting up in most cities. Although good examples can be found (e.g. Bristol, Venice and the example cities of Bilbao, Breda, Drachten and Maastricht, see also Creedy et al. 2007), most cities indicate that involvement is often still little more than informing and educating (e.g. Liveable Cities 2005b, 2006c). In fact, most 'Liveable City' partners claimed to often lack the experience, ability and political and administrative support for setting up convincing partnerships.

Ambitions revisited

The 'Liveable Cities' project shows that the *delivery* of environmental and sustainable ambitions is another key problem in most urban areas (as can be seen in all case reports). City representatives noted that legal requirements and legal environmental standards (limit values) are often still the most important drivers in sustainable urban management in their cities (Liveable Cities 2005a). As the project cities are in an EU project on sustainable environmental management, they can be considered relatively proactive in environmental management. If even proactive cities fail to do much more than meeting minimum requirements, there is little reason to believe this is much better in most other EU cities.

The dominance of legal requirements and standards in sustainable management came forward in most cities. In Venice it was noted that "urban planners working on strategic planning do not consider environmental variables as such, except for rules that have been set such as zones around busy roads" (interview with policy officer, 30 November 2005). In Malmö it was added that "strategic policies are traditionally not as popular as restricting policies as it is harder to get them done" (interview with policy officer, 14 February 2006). Bourgas noted that it mostly acts in line with National and EU rules that are already tough enough. In Aalborg it was also noted that the most power for acting in the face of economic and social interests came from environmental standards (interview with policy officer, 9 September 2005). Finally, Bristol noted that "unless they are obligatory, international obligations get overlooked" (interview with policy officer, 4 September 2006).

Most city representatives were critical of their cities' focus and reliance on legal standards alone, as these provide only 'modest' stimuli for a true improvement of urban environmental policies (Liveable Cities 2005a). City officials interviewed in Aalborg suggested that "the focus of national standards results in too little focus on which issues and aspects are not covered by these standards". In response, cities *try* to "move from legal compliance to visionary planning using new ways of working that deliver ambitious sustainability targets appropriate for the challenges" (Creedy et al. 2007; p. 45). Thus, instead of relying on legal limit values as quality ambitions, these limit values are only considered the minimum ambitions. The real ambition is for "towns and cities to go beyond this minimum to deliver urban areas that are as good as is reasonably achievable" (Creedy et al. 2007; p. 45). Many cities have endorsed plans containing strategic and even more operational environmental and sustainable ambitions that are additional to minimum requirements. As the dominance of legal requirements and limit values illustrates, however, cities often get stuck in good intentions.

One major cause of limited commitment to ambitious environmental policies turned out to be the difficulty of moving from the political rhetoric surrounding concepts such as sustainability, quality of life and liveability to actual local political and social commitment and workable policies. The 'fuzzy' or 'ambiguous' character of these concepts make them hard to operationalize. Various respondents

196 *A European focus on the local*

indicated they often got stuck in good intentions to put concepts to practice (e.g. interviews in Bristol and Malmö).

Another major cause of limited commitment that reoccurred during various workshops and peer review sessions was time (e.g. Liveable Cities 2005a, 2006b and 2006e). Social, partnership and political commitment often fades after a plan or policy is agreed. As expressed during these workshops, decision makers tend to assume that government bureaucracies will simply 'deliver' their decisions. Doing so overlooks the importance of translating decisions into day-to-day commitment of all key stakeholders. To prevent implementation deficits, it was stressed during the Bristol workshop to always ask the question "how do resources follow political commitment and how do actions follow resources?" Programmes of action that institutionalize the contribution of political, social and market parties are possible answers to this question as is recording the support from partnerships, social and market parties in covenants and contracts (Liveable Cities 2006c, see also Liveable Cities 2005a, 2005b).

To generate commitment for sustainable policies, various cities noted that support from (inter)national governments was rather crucial (compare section 5.3). This was most clear in Bourgas, where ongoing decentralization from the central state towards the local level was causing serious problems (see Liveable Cities 2005b). Although decentralization was in itself welcomed, the absence of additional financial, legislative or human capacity for municipalities undermined the potential for dealing with the decentralized responsibilities. Hence, city officials from Bourgas noted that decentralization contributed to "local authorities facing problems of instability, lack of capacity and financial resources" (Liveable Cities 2006b). Other cities also noted similar problems. In Malmö the lack of Swedish legislation on soil issues and safety risks posed problems (Liveable Cities 2006a), while Aalborg pointed at themes such as sustainable construction and mobility that were not included in national policies (Liveable Cities 2005a).

Some conclusions on local environmental governance

The data generated during the 'Liveable Cities' project was especially helpful for understanding the state of the art of urban environmental governance in the EU, including cities' main problems and needs. In this, cities confirmed many of the problems that were discussed by the 'Extended EU Expert Group on the Urban Environment'.

In the first place, cities are struggling with the 'public–private' divide and the associated problems of power dispersal and democratic legitimacy. This both follows the need and desire to involve local citizens, but also to respond to power dispersal and the possible conflicting interests between key stakeholders. Especially, it turns out that the development of partnership models for involving, empowering and negotiating with private stakeholders and NGOs is considered as important. In the meantime, experience, knowledge and resources for developing convincing partnerships are often limited. Second, cities widely acknowledge the need for more proactive and integrated policies to the environment. Many cities had various policies, plans and programmes in place to this cause. Cities widely recognized the interrelatedness of problems and the policies, but stated that inter-relatedness was not always anticipated in integrated plans or cross-departmental working. In fact, as was just noted, various cities expressed that they struggled with *too many* coordinative instruments such as plans and programmes. Consequently, cities argued that promoting local proactive and integrated approaches to the environment should move beyond the installation of two plans and a management system. Rather, they endorsed a strategy which is based on seven distinct elements of urban management as highlighted in Box 5.2 (i.e. politics, partnerships, principles, process, policies, plans and programmes). Their hopes and ambitions were to use these elements to cross-reference current urban policies and to reveal gaps in the urban governance capacity for sustainable urban management (see Creedy et al. 2007). Finally, cities noted problems with gaining and keeping political commitment. The 'weak profile' of the environment was widely noted, even within cities that, confirmed by their participation in the 'Liveable Cities' project, can be considered relatively strongly committed to sustainable and environmental priorities. Political support and the implementation of well-meant policies were considered key bottlenecks in pushing local environmental governance forward.

5.5 The thematic strategy and its impact

The EU adopted the final 'Thematic Strategy on the Urban Environment' in January 2006. Despite the CEC's choice for obligatory measures in the draft 'Thematic Strategy' of 2004, the final 'Thematic Strategy' contains mere voluntary measures. Many of the proposals included in the finalized 'Thematic Strategy' are continuations of existing policies. These proposals include the exchange of good practices, financial support through the IMPEL network (Implementation and Enforcement of Environmental Law), the Cohesion and LIFE+ (the EU's financial instrument supporting environmental and nature conservation projects) funds and increasing the availability and visibility of information on environmental policies. These activities existed before the 'Thematic Strategy' was issued and are only updated based on the 'Thematic Strategy'. For example, the CEC has responded to its recognition that "many local authorities have expressed the need for specific skills to adopt an integrated approach to management involving cross-sector cooperation and training on specific environmental legislation, effective public participation and encouraging citizens' behavior" (CEC 2006; p. 7). This response is to alter the LIFE+ regulations, which means that two of the nine LIFE+ themes are now related to integrated policies (i.e. urban environment, quality of life, environmental management).

The only real new element in the 'Thematic Strategy' is the promise of two 'technical reports' that mean to advise towns and cities on environmental management. They are guidance reports on 'Integrated Environmental Management' (CEC 2007a) and on the preparation of 'sustainable urban transport plans' (CEC 2007b). Interestingly, the finalized 'Thematic Strategy' does not even mention environmental management *plans* anymore. It only addresses integrated

198 A European focus on the local

environmental management and sustainable urban transport plans. The report on sustainable urban transport plans highlights the general structure and steps for producing such a plan. In addition, it highlights the existing possibilities for funding transportation projects, especially in relation to improving environmental conditions and sustainable targets. The 'technical report' on 'Integrated Environmental Management' is more extensive and meant as a guidance confirmed by its subtitle: 'Guidance in relation to the Thematic Strategy on the Urban Environment'.

The 2007 guidance on 'Integrated Environmental Management' summarizes and structures the existing opportunities for local authorities to apply for funding on urban environmental issues. Furthermore, the guidance report advises on improving local environmental governance. Interestingly, the guidance bears witness to many of the points raised during the consultation rounds and the 'Liveable Cities' project. The guidance highlights the importance of political support (p. 10, p. 12), the need to incorporate more than just environmental issues but also other aspects of sustainable development (p. 12), the involvement and sharing of tasks and responsibilities between stakeholders (p. 12) and between various sectors (p. 14). While the guidance has thus picked up on the learning during the consultation rounds and projects such as 'Liveable Cities', it provides little clear advice. Rather, the guidance gives some examples on the side and a general description of integrated environmental management. It proposes a rather common and fairly simplistic five-step approach from a base line review, to target setting, approval in a political domain, implementation and reporting and evaluation. A series of steps at least most towns and cities are already well aware of.

First reactions

The finalized 'Thematic Strategy' did not nearly generate the number of reactions as the 2004 draft did. This is of course hardly surprisingly. Only Eurocities (2006), the European Parliament (2006) and the European Environmental Bureau (2006) came with a formal response, leaving many of the stakeholders represented during the consultation sessions with the 'Extended EU Expert Group on the Urban Environment' without a response. This can be explained by the voluntary measures promoted by the 'Thematic Strategy'. The finalized strategy therefore has no impact for cities and member states if they have no desire for responding to it. This voluntary nature also generated most comments from stakeholders. Most prominent were reactions by Eurocities, the European Environmental Bureau and the European Parliament who all endorse and support the intentions of the finalized 'Thematic Strategy' but consider the strategy to be relatively 'weak'.

The voluntary basis is considered by Eurocities to undermine the potential of the 'Thematic Strategy' to be implemented. Hence, Eurocities suggests that "the proposal by the commission on implementation of the TSUE [Thematic Strategy of the Urban Environment] requires further precision and development" (2006; p. 3). The European Environmental Bureau (EEB 2006) is even more explicit and simply calls the strategy 'too weak' (p. 1). The European Environmental Bureau suggests that "by not defining specific (mandatory) measures to promote integration of policy formulation and implementation the Thematic Strategy on the Urban Environment is unlikely to meaningfully contribute to this aim" (p. 1). The European Parliament also moves along a similar line as the European Environmental Bureau. On 29 September 2006 it adopted a resolution with a vast majority (2006/2061(INI)) in which it states that the current 'Thematic Strategy' "is not sufficient to achieve the goals set in the 6th Environmental Action Programme". It thus "regrets that, contrary to the intentions of the 6th Environmental Action Programme, there are no legally binding measures and deadlines proposed by the Commission to address any of the goals set by the 6th Environmental Action Programme". Consequently, the European Parliament asks commission and member states for increasing and intensifying policies and actions as compared to the current strategy, which it judges is not going 'far enough'.

The European Parliament suggests that "the Commission, in cooperation with the national authorities, should encourage every agglomeration having more than 100.000 inhabitants to establish a Sustainable Urban Management Plan (SUMP) and a Sustainable Urban Transport Plan (SUTP)". The parliament thus suggests that a broader focus (i.e. sustainability instead of environmental) is preferable. In the meantime, it continues to point at obligatory measures, just as the draft proposal suggested. The parliament does highlight that such measures need to be accompanied by financial aid. Hence, it calls on the commission "to identify sources of additional specific funding to meet the new challenges arising from the strategy for the urban environment". Finally, the parliament highlights that for the current strategy to be implemented, concrete projects on best practices, mid-term reviews and domestic encouragement is needed.

The European Environmental Bureau concludes that

> Based on the proposed Thematic Strategy on the Urban Environment, urban environmental and development policy will continue to be developed and implemented on a sector basis, which to date has been the prime reason for the deterioration of the urban environment and increasing impacts on the wider environment.
>
> (2006; p. 2)

They consider the current 'Thematic Strategy' to have little added value compared to previous policy attempts. In response, the European Environmental Bureau states: "the Commission should propose directives making environmental management plans and systems as well as sustainable urban transport plans mandatory for all cities above 50.000 inhabitants" (p. 2) and that "these plans should contain binding EU objectives on for example on decreasing CO_2 emissions, ecological footprints and the share of automobiles in the urban transport modal split as well as establishing a clear link to existing EU environmental objectives" (p. 2).

Eurocities also argues for stronger CEC support. Instead of obligatory measures it argues for "stronger support for implementation, such as strong political commitment, extra funding, annual Member State reports on management plans, the launch by the Commission and Member States of a European project to track

200   *A European focus on the local*

the implementation of the TSUE in cities" (2006; p. 3). Linked to this, Eurocities considers "more streamlined EU environmental legislation" and "funding through simplified environmental funding programmes with simpler bureaucratic processes for applications and reporting" crucial (p. 3).

Expectations, results and follow-up

The doubts regarding the likely impact of the 'Thematic Strategy' are best elaborated by the European Environmental Bureau:

> This Strategy will mean that good urban environment practitioners, as in the past, will continue to improve their performance with the added aid of voluntary support mechanisms, whereas poor urban environment performers will have too little incentive to significantly change their urban management and development approach.
>
> (2006; p. 2)

While member states are invited to implement the 'Thematic Strategy', they are not subject to evaluation by the CEC. Using the strategy is therefore completely up to individual member-states and cities.

A short internet scan of the use of the 'Thematic Strategy' in member-state policies did not reveal much news (Google, 11 November 2009). From the sample of member states (selected for language reasons), no evidence of a real use of the 'Thematic Strategy' could be found. Only a few even mention the strategy on national websites. In the Netherlands, the strategy is made available on the website of the Ministry of Housing, Spatial Planning and the Environment, with additional guidance reports. In Flanders (Belgium) the Ministry of Environmental Management leaves it with a short PowerPoint presentation on their website. In Ireland, no references were found, although Dublin City Council did refer to it as a reference document in its vision for the Dublin Bay (Dublin City Council 2007). In the United Kingdom, Austria, Germany and Estonia, no references could be found to the strategy. Use of the two 'technical reports' (CEC 2007a, 2007b) is even more dubious. Two years after they were published, the two 'technical reports' are hardly mentioned on the internet; i.e. the one on Integrated Environmental Management got 27 hits, of which 26 were in the first year after publication and the one on sustainable urban transport plans only 20, all in the first year after publication (Google, 11 November 2009, in English, Dutch and German). By contrast, the draft 'Towards a Thematic Strategy on the Urban Environment' still got 221 hits after being published over five years ago. Unsurprisingly, the World Health Organization (2010) deemed the results of the finalized 'Thematic Strategy' not just as disappointing, but also as inadequate.

In the years following the adoption of the 'Thematic Strategy' and the related 'technical reports', the EU did continue its work on promoting sustainable development, also related to urban areas. Nevertheless, this work does not seem to rejuvenate the attempts of the 2004–2007 period. On the one hand, the EU has continued with its urban environmental agenda by relying on its previous approach of giving voluntary policy advice. In 2010 DG Environment of the CEC issued 'Making Our Cities Attractive and Sustainable' (CEC 2010a). It highlights ongoing environmental challenges in Europe's urban areas and does highlight the relevance of environmental management plans. Apart from some good examples and a clear list of the existing financial possibilities cities might use, it is, however, not more than well-meant advice. On the other hand, with the urban environmental agenda remaining voluntary, also other EU policy efforts continued that also have an urban impact. For example, in 2007 its member states, for example, signed the 2007 Leipzig Charter on Sustainable European Cities, expressing domestic political support for more integrated urban policies, while mostly taking a social sustainability perspective. In being guidelines to inspire urban practices in its member states, it is up to member states to act in practice without actual EU pressure on implementation. More important in recent years has been the development of Europe's growth strategy called 'Europe 2020' that was adopted in 2010 (CEC 2010b).

Europe 2020 aims at 'smart, sustainable, inclusive growth', making sustainable development among its prime inspirational frameworks. Inspired by the 2008 economic crisis, the strategy is developed in a context of economic decline and stagnation, unemployment and budget cuts. It provides an arena where political concerns focus specifically on growth and jobs. Europe 2020, therefore, is also clearly focused on these ambitions, with setting clear targets on increasing employment rates, research and development, levels of education and decreased poverty rates. The 'green side' of sustainable development is not left out, but seems focused especially on carbon dioxide. The strategy sees the reduction of greenhouse gas emissions as a key target by at least 20% as compared to 1990, aims for improved energy savings and means to boost the share of renewable energy to 20% in 2020.

Translated into national targets, the EU itself remains a supervisor of domestic progress and has also installed a wide set of projects and specific policies to assist the implementation of the strategy. Urban sustainability is part of that, sometimes going beyond a focus on greenhouse gas. But again the approach remains based on largely voluntary tools (e.g. CEC 2009). Urban sustainability is largely addressed by holding on to networks of exchanging 'best practices' and existing or new funding opportunities the EU has on issues varying from air pollution, sustainable transport and energy efficiency (for an overview see Committee of the Regions 2012; pp. 37–40). Furthermore, linked to one of Europe 2020's flagship initiatives, 'Innovation Union', are the so-called 'European Innovation Partnerships'. These partnerships, much like the idea of learning in 'niches' by experimenting, aim at contributing to meeting the Europe 2020 targets by promoting the adoption of novel technologies. Most relevant for urban areas is the 'Smart Cities and Communities' initiative (see CEC 2012). Supported by almost 4000 mayors confirming their towns and cities' commitment to reduce greenhouse gas emissions, it is about sharing novel practices on energy, ICT and transport to be supported by EU funding (Committee of the Regions 2012). These practices are, however,

202 *A European focus on the local*

largely confined to technological innovations and not necessarily with new governance formats and strategies.

In pursuing its Europe 2020 strategy, the EU acknowledges the need and relevance of experimenting with novel practices. It might seem a slight turn back to what the 'EU Expert Group on the Urban Environment' opted for back in the 1990s (EGUE 1996). Nevertheless, with these experiments still focussing on the voluntary sharing of 'best practices', it is not a real shift away from the past voluntary approach that characterizes the urban environmental agenda. Furthermore, as the experimentation with innovative technologies also shows, many of these policies are also developed largely disconnected from the urban environmental agenda of DG Environment. Finally, in focussing strongly on novel technologies, the current focus is also no real follow-up to what the Expert group meant back in the late 1990s, where they stated how

> sustainable urban management . . . requires a range of tools addressing environmental, social and economic concerns in order to provide the necessary basis for integration, as well as a reconsideration of the governance, institutional arrangements and capacities of different levels of government, and partnerships between public and private sectors, NGOs and citizens.
>
> (EGUE 1998; p. 4)

It has meant that, following Europe 2020, the EU promotes urban sustainability by largely focussing on efforts at economic and social development, while the environmental dimension is confined largely to the reduction of greenhouse gas emissions not in the least based on adopting innovative technologies.

The 'Thematic Strategy' did not only have a marginal impact within Europe's towns and cities. Also, it had a marginal impact in pushing the EU bureaucratic system forwards in dealing with many of the issues that were raised by experts when developing the 'Thematic Strategy'. The urgency of the economic crisis of 2008 might well account for a political reframing the focus on urban sustainability in terms of 'smart growth' that is more economically driven. Whatever the exact cause, what is clear is that the urban environmental agenda, up until a few years ago a key building block for the EU in promoting urban sustainability from a 'green' or environmental perspective, clearly has been marginalized. Maybe the most convincing indicator of this is that in 2015 the 'Thematic Strategy' still features prominently on the website of DG Environment.

5.6 Reflecting on the 'thematic strategy'

In Chapter 4 I presented four arguments that can help inform attempts to increase the role of the local level in environmental policy, also regarding the pursuit of proactive and integrated environmental policies. In this section I will return to all four arguments. I will reflect on these arguments based on the data collected regarding the approach taken in the EU for advancing the role of the local level in environmental policy. On the one hand, I hope to establish whether these

A European focus on the local   203

arguments make sense. While doing so, on the other hand, I also hope to better explain the consequences of the approach taken in the EU for advancing the role of the local level in environmental policy.

Argument 1: complexity as criterion

The first argument I discussed in Chapter 4 was that the degree of complexity of the issues faced can be a key argument in support of an increased role of the local level. It was an argument that the EU also supported and used. After all, the pursuit of an urban environmental agenda in the EU during the last two decades was a response to the EU's conclusion that relying only on sectoral and piece-meal policies is insufficient for coping with urban environmental challenges. The CEC (2004) also explained that it considered such common policy formats too piecemeal, reactive and lacking in vision for dealing with urban environmental issues. Instead, they urged for 'active and integrated management' of the urban area where local authorities were to take the lead, since 'no two urban areas are the same' (CEC 2004). The CEC thus concluded that that existing common policy formats issued by central governments are limited in their capacity to respond to the complexities associated with unique local circumstances. It provides support for the claim that complexity of the issues faced can be a key argument in support of an increased role of the local level.

This claim was further supported during my empirical inquiry, most notably during the research conducted during the 'Liveable Cities' project. The 'Liveable Cities' project showed that cities rely on a wide array of different measures and policy instruments. They are expressions of a comprehensive set of governance formats that address aspects such as political bargaining, process management, partnerships, programmes, implementation, etc. The exact mix of these governance formats turned out to be dependent on geographical, economical and physical differences, and on institutional, political and cultural differences. Various cities are facing different environmental challenges that are closely related to unique local circumstances. In the meantime, also aspects such as workforce, administrative routines, legal competences, expertise, etc. all vary in EU cities and towns, as do social and political tendencies and the allocation of competences. Given these different circumstances, the conclusion was drawn and supported by cities that there is no one best format for approaching proactive and integrated approaches to the local environment. It indeed illustrates that common policy formats are ill adapted to the complexities associated with unique local circumstances.

As the EU considered that local authorities should take the lead in developing and delivering local proactive and integrated approaches to the environment, it explicitly said that it did not aim to 'dictate solutions' (CEC 2004). Surprisingly though, the draft 'Thematic Strategy' (CEC 2004) did propose a common format for an environmental management and sustainable urban transport plan and for an environmental management system. While the EU thus accepted that a more integrated approach was needed for dealing with the complexity of urban environmental issues, it still chose for common policy formats to install such

204   *A European focus on the local*

approaches in a local realm. In Chapter 4 I argued that such common policy formats would not go well with the desired pursuit of local integrated policies. The structural configuration as proposed (i.e. relatively centralistic common formats) does not go well with the functional ambitions (i.e. specific integrated and area-based). This contingency between function and structure was not taken into account by the CEC in its draft 'Thematic Strategy', which had important consequences.

The CEC's choice for common policy formats was not well received by most stakeholders (section 5.3 and 5.4). Rather, most stakeholders and cities rejected the policy formats proposed by the draft 'Thematic Strategy', with the general exception of those coming from new accession states and some Mediterranean states (see section 5.3, see also Enviplans 2006). The approach advocated in the draft 'Thematic Strategy' was considered as too rigid, uniform and to interfere with local initiatives that are tailored to the local circumstances (section 5.3). Instead, as was well supported by arguments during the 'Liveable Cities' workshops (section 5.4), pursuing integrated approaches would benefit from a locally contingent 'mix of governance formats'. It provides further support for the argument that common policy formats are indeed ill adapted to cope with the complexities associated with unique local circumstances; i.e. structural configurations are contingent on the functional ambitions.

In response to the critiques by stakeholders, the finalized 'Thematic Strategy' no longer contains obligations for installing common formats in its larger towns and cities. Instead, it has shifted focus to voluntary measures that advise and persuade cities to pursue proactive and integrated approaches to the environment. This shifted focus implies that the EU does not dictate solutions, but instead merely provided cities with policy advice. This advice is nevertheless still based on the idea of installing common policy formats for sustainable urban transport and environmental management, in addition to minor financial incentives and a focus on training and the exchange of best practices. In the following sub-sections I will discuss how these 'fit in' with the most prominent constraints on local willingness and ability. I will start with seeing how the EU responded to the idea of retaining generic and common policy formats for meeting minimum levels of protection in a local realm.

Argument 2: a focus on protection

In adopting a post-contingency approach, I argued that a reduced capacity to meet single fixed goals is one of the dominant consequences of decentralization. In the meantime, the desire to install minimum levels of protection against environmental stress in a local realm can be an important motive for setting such single fixed goals. The EU has also chosen to retain such minimum requirements. In the draft proposal for its 'Thematic Strategy' the CEC also notes

> Many of the problems facing Europe's towns and cities are common ones and there are clear opportunities at the European level to develop, share and

facilitate the implementation of appropriate solutions. A wide range of Community policies, actions and funding programs are tackling these problems.

(2004; p. 4.)

These consist of the kind of sectoral policies that are used for addressing environmental issues where the protection of human health is at stake and where issues have a common manifestation. The 'Thematic Strategy' is not meant to replace these minimum requirements, but rather, was to function as additional to these minimum requirements. After all, the CEC also noted that these generic and sectoral policies function "mostly as separate exercises without considering the specific needs of the urban context or the potential synergies between them" (2004; p. 4). In response, the CEC calls for a 'strong framework' for solving urban environmental problems *in addition to* existing sectoral policies. It was with proposing an obligatory environmental management plan, sustainable urban transport plan and environmental management system that the CEC hoped to install such a 'strong framework'.

The EU wanted a stronger framework to revitalize and generalize the environmental management of urban authorities and to add value to minimum requirements. This was inspired by the CEC's conclusion that the existing policies have not prevented serious environmental problems in most urban areas with adverse consequences for the local 'quality of life' and 'health problems' (CEC 2004). The 'Thematic Strategy' was, at least partly, meant to address current adverse health consequences of environmental stress. As I claimed in my theoretical inquiry, such a focus on protection indeed requires a fairly strong governance framework. In the draft 'Thematic Strategy' the CEC does make such a choice, albeit without setting clear targets or objectives. Inspired by widespread criticism, the CEC has withdrawn the proposal for a strong framework to be replaced by only voluntary policy advice. It is a rather radical shift from a focus on fairly strong policy instruments towards rather weak instruments. The current voluntary policy approach provides hardly any stimuli for improving urban environmental conditions, other than some well-meant advice. Based on my theoretical claims, this would imply that it should not be expected to structurally (i.e. in all EU towns and cities) result in improved health conditions. This is also why the WHO (2010), European Environmental Bureau (2006) and European Parliament (2006) have called the 'Thematic Strategy' 'too weak' and 'inadequate'. The limited use of the 'Thematic Strategy' and its proposed tools suggests that voluntary measures are indeed incompatible with the desire to structurally improve conditions in EU towns and cities. This directly brings us to assess the most common constraints on local willingness and ability to pursue proactive and integrated approaches to the environment: economies of scale, external effects and the weak profile of the environment.

Argument 3: part 1, responding to economies of scale

Economies of scale relate to both the possibility to benefit from routinely implementing common policy formats and the possibility to attract technical and possibly

206   *A European focus on the local*

managerial expertise. As I just argued, proactive and integrated approaches to the local environment can hardly benefit from the efficiency of routinely implementing centrally developed common policy formats. They do, however, involve knowledge-intensive tasks, both regarding technical and managerial skills. Given their capacity to attract more qualified staff and hire specialists, I argued in Chapter 4 that larger organizations have economies of scale over smaller ones. This begs the question whether local authorities are able to attract sufficient numbers of skilled employees and if they are not, if and how central guidance can respond to correct these constraints.

The research findings suggest that there are few problems to attract sufficient technical expertise in the cities studied. Technical expertise was hardly a theme during the consultation sessions in the 'Extended EU Expert Group on the Urban Environment', nor was it a dominant theme during the 'Liveable Cities' workshops. This suggests that economies of scale with regards to technical expertise seem to have little or no impact on the capacity of EU town and cities to pursue proactive and integrated environmental policies. We should, however, take into account that the 'Thematic Strategy' is focussed on larger towns and cities (i.e. >100.000 inhabitants) that usually are able to pursue a functional specialization in response to different environmental themes. Smaller towns and cities might face more difficulties. There are also indications of this (ICLEI 2004).

The research did suggest that problems with limited funding, staff and managerial and strategic planning competences were common in EU towns and cities. These problems can, first of all, be related to the *managerial skills* needed for pursuing proactive and integrated approaches. Many towns and cities indicated that their organizations and the existing managerial skills are not well adapted to develop convincing partnerships, or to develop cross-sectoral and integrated approaches. Local administrations were often noted to be biased towards regulatory and technocratic municipal policies, also regarding the competences and routines of local staff. On the one hand, as the 'Liveable Cities' project most vividly revealed, traditional sectoral policies have manifested themselves in inward-looking departments and sub-departments, which rely on their own routines, cultures and preferences. The result is a lack of experience in entering into cross-sectoral working, requiring more visionary and conceptual policies and communicative skills for engaging in more integrative and collaborative working. Common issues were communicative problems between departments ('speaking the same language') and competition among municipal departments over budget, policy outcomes and interests. On the other hand, towns and cities noted a lack of experience and resources to enter into bargaining, negotiations or public–private partnerships for dealing with coordinative and financial challenges that accompany the fragmentation of power. Cities recognized the importance of partnerships for engaging citizens or local stakeholders such as business, housing corporations or project developers in policy development and delivery. However, they also noted that they often lacked experience and competences to do so.

Given the problems they encountered, towns and cities asked for support and guidance regarding communicative skills, integrated working and for pursuing

A European focus on the local   207

convincing partnerships with non-governmental stakeholders. Such support and guidance is not easy to provide by the EU as they are hardly in a good position to judge the detailed circumstances in various towns and cities that should be taken into account in improving local integrated working or for developing partnerships. The kind of support and guidance hoped for was therefore not a direct EU involvement. Instead, during the consultation sessions in the 'Extended EU Expert Group on the Urban Environment' and policy statements by city stakeholder groups, various calls were made upon the EU to expand research into these areas, the training of municipal staff, the dissemination of good practice examples, process management tools for improving coordination, or (financial) stimuli for cities for pursuing partnerships. The EU has not responded to these calls. The issue of partnerships was hardly even mentioned in both the draft and finalized 'Thematic Strategy'. Integration and collaboration was also not discussed in terms of competences or skills. Instead, the 'Thematic Strategy' responded by proposing new coordinative instruments (i.e. an environmental management system and sustainable urban transport plan). These instruments were however received with much scepticism. Not just because they were not the support desired (i.e. improving managerial skills), but also since these proposed instruments were already used by most EU towns and cities.

Problems with local ability can also be related to the existence of *overstressed municipal departments*. There were many remarks about limited financial resources and overstressed municipal staff was frequently mentioned. This was also well illustrated by the 'Liveable Cities' project, where respondents noted they often work with over 50 separate plans and policy documents, each focussed on different local policy agendas. It contributes to conditions that can best be described as those of *management overload*, where incoherencies and even conflicting policies are no exceptions. In the face of such conditions, time and resources are scarce and that also manifests itself in a reluctance to cope with additional planning obligations as proposed by the CEC in the draft 'Thematic Strategy'. Stakeholders repeatedly argued that towns and cities needed additional resources for implementing additional obligatory plan formats. In addition, calls were made in favour of a focus on sustainable urban management tools that the EU could advise on, which indicate how cities might overcome existing administrative fragmentation. The CEC, however, had no choice but to refrain from a focus on sustainable urban management (section 5.3). In the meantime, the CEC also assumed that cities would be able and willing to bear the costs of working with additional policy formats themselves and refrained from installing additional financial support. Faced with overstressed administrations, the size of the additional work proposed by the draft 'Thematic Strategy' and the supposed limited added value of the obligatory measures, most stakeholders rejected the proposed measures.

In Chapter 4, I argued that central policies can be important means for supporting local ability in the face of limited resources, funding and knowledge for pursuing proactive and integrated approaches to the environment. My empirical inquiry supports this claim and suggests that many towns and cities are at least partly

208　*A European focus on the local*

unable to pursue proactive and integrated approaches, as mostly became clear during the 'Liveable Cities' project (section 5.4). The EU hardly responded to the main problems that cities face; i.e. managerial skills and limited time, staff and funding. It does not allow us to assess whether central policies *could have been* important to support local ability to pursue proactive and integrated approaches to the environment. However, it is clear that without such support, the EU has not really influenced this local ability. Furthermore, many towns and cities expressed during the consultation sessions with the 'EU Expert Group on the Urban Environment', various position papers (CEMR 2004, Eurocities 2004b, ICLEI 2004), the interviews with city officials and the 'Liveable Cities' workshops that central support would have been welcomed and could have made a difference.

Argument 3: part 2, responding to external effects

External effects are a next constraining factor for local performance in environmental policy that was discussed in Chapter 4. External effects were also encountered during the empirical inquiry. The biggest problem that towns and cities encountered was the fragmentation of urban areas in many individual jurisdictional units. This requires coordination between various urban jurisdictions when it comes to issues that manifest themselves on an urban scale, such as congestion, localized air pollution, background noise, sewage, etc. Such coordination is however neither evident, nor easy. Rather, as the 'Liveable Cities' project confirmed, neighbouring municipalities compete for resources and developments, which can easily come at the cost of the overall environmental quality in an urban region. I also found examples of this during the 'Liveable Cities' project (section 5.4). Towns and cities hoped that the EU was willing to pursue possible responses to these issues (see section 5.3). However, the draft and finalized 'Thematic Strategy' only *suggest* that plans and policies are to be made at the level of the 'functional urban area'. There is no mention of how this can be done in either the draft or finalized 'Thematic Strategy', nor in the technical reports. It is important to note that the EU might not be the right institutional level to propose how best to cope with such jurisdictional fragmentation. Neither is it evident what such a proposal should entail. Still, it is in the meantime evident that many member states, cities and non-governmental stakeholders did call upon the EU for guidance on urban fragmentation. It suggests that central policies might have made a difference.

External effects also interact with the weak profile of the environment, especially where it involves issues whose effects manifest themselves on even higher spatial scales. This is, for example, the case with issues such as background air pollution and sustainable development (e.g. energy savings, biodiversity, sustainable construction, etc.). The 'costs' of environmental measures are often local and short term, while the 'benefits' are often long term and spill over jurisdictional borders. It requires politicians, administrators and stakeholders to move outside of the 'comfort zone' of ordinary planning and existing legislation. It calls for coordination and partnerships across borders and for ambitious policies that can be financially unpopular or long term. Most cities indicated serious problems to

A European focus on the local 209

do so without central policy stimuli as they face both a lack of widespread political, administrative and public support and are constrained to pursue ambitions due to local power dispersal. It would imply that the 'Thematic Strategy' could provide such stimuli, bringing us to the EU response to the weak profile of the environment.

Argument 3: part 3, responding to the weak profile

At first sight there seems to be a positive attitude of most EU towns and cities towards proactive and integrated approaches to the environment. This is, for example, illustrated by the large number of towns and cities that work with environmental management or sustainable management plans and systems (section 5.3). In addition, many towns and cities have joined initiatives such as the Aalborg Charter, Aalborg Commitments, LA21 and many other (inter)national schemes, suggesting a positive attitude towards proactive and integrated approaches to the environment. This positive attitude can however not hide the fact that setting ambitious objectives and especially, getting them implemented, is less evident amongst even the more proactive 'Liveable Cities' partners. Hence, conclusions drawn by, for example, the European Commission (CEC 2004) and the European Environmental Bureau (EEB 2006) are supported by my research: proactive and integrated approaches to the environment find it difficult to move from rhetoric to action. The data also shows that the 'weak profile' of the environment is constraining local performance in proactive and integrated approaches towards the environment.

In the first place, political will and administrative enthusiasm for ambitious and proactive environmental or sustainable policies was found limited (see section 5.4). Legal standards are still the main drivers in environmental governance, even among the relatively ambitious 'Liveable Cities' partners. Except for urgent or highly visible local issues, additional environmental ambitions are not a common priority for most politicians, municipal departments or urban stakeholders. This was related to the lack of visibility and knowledge of many environmental issues, especially due to the technical and specialist information surrounding these issues. Political representatives, administrative heads of department and many other urban stakeholders find it difficult to understand and judge the consequences of environmental problems and qualities. Hence, environmental policies are often considered the domain of technical experts in environmental departments who check if developments are within the legal limits.

Second, as was shown during the 'Liveable Cities' project (section 5.4), the weak profile of environmental issues shifts the balance of power away from environmental and sustainable priorities. Even if environmental issues are high on political or social agendas, implementation proves difficult. Associated with the supposed unbeneficial cost–benefit ratio of achieving environmental ambitions, environmental officers find competing over resources and policy outcomes with non-environmental departments and stakeholders difficult. This is especially problematic as many resources needed to add value to minimum requirements are

210 *A European focus on the local*

at the disposal of non-governmental stakeholders such as housing corporations, businesses and project developers. Local authorities recognize they need support from private parties, especially in the face of limited land ownership and financial resources. Doing so is problematic due to limited experience and expertise in setting up convincing partnerships and also, the weak profile of the environment.

Given the weak profile of the environment, I have argued in Chapter 4 that support from higher levels of authority can enable and stimulate local authorities to pursue more ambitious environmental policies. In the draft proposal of the 'Thematic Strategy' the CEC indicated that it also considered such support necessary, given the lack of success of the previous voluntary proposals. To this end, it proposed the installation of a 'stronger framework at the European level', expressed in obligations for an environmental management and sustainable urban transport plan and the related environmental management systems. Given the resistance to these obligatory formats, the CEC dropped them in the finalized 'Thematic Strategy'.

In the finalized 'Thematic Strategy' the CEC shifted quite radically from a focus on obligatory measures towards voluntary and persuasive policies. Consequently, the support given by the EU to its towns and cities is fairly weak and does not offer the kind of stimuli and 'checks and balances' as discussed in Chapter 4. Although sectoral environmental policies do continue to function as safeguards, there are hardly any real stimuli for cities adding value to them through proactive and integrated policies. The current strategy relies on the continuation of financial support for local pioneering experiments, the exchange of good practices for producing mutual learning and of the use of training programmes. Furthermore, to facilitate these experiments and programmes, current financial support systems such as the Structural Funds, the LIFE+ funds and the 7th Framework research funds are continued and partly altered. There is, however, no expansion of these experiments and programmes or of the financial means available for cities to participate in them. The only new support provided by the 'Thematic Strategy' are the two 'technical reports' meant to advise and persuade cities to develop integrated environmental management systems and sustainable urban transport plans. These 'technical reports' can hardly be considered real incentives though. Not only are they voluntary, they are also written for government officials, most notably environmental experts. It is of course doubtful whether environmental experts are the only or even most relevant actors in urban governance pushing proactive and integrated approaches to the local environment forward. Crucial actors such as politicians, powerful stakeholders and non-environmental officers are largely neglected by the 'technical reports'.

The EU uses central guidance and the associated coordinative instruments when it comes to issues of limited complexity and to maintain minimum levels of protection against environmental stress. It also focused on coordinative instruments for pushing local proactive and integrated approaches to the environment forward; i.e. the obligatory plans. Since these were not accepted by most stakeholders, the EU radically shifted to a reliance on voluntary policies. It did not pursue alternative possibilities as discussed in Chapter 4, which rely on a hybrid

A European focus on the local   211

combination of central guidance and coordinative policies with financial or persuasive elements. Examples include flexible regulations, contracts, voluntary agreements or eco-labels, financial aid, education and training or the increased dissemination of knowledge, etc. The consequence is that the EU now fails to provide any real support or incentives that push proactive and integrated approaches to the local environment further. In Chapter 4 I argued that this can easily result in disappointing results in practice, which is indeed what happened (section 5.5).

Argument 4: robust and dynamic

Finally, then, I also argued in Chapter 4 that local willingness and ability to perform tasks in environmental management depend on a robust and well organized set of central policies which can be routinely implemented. It is hardly realistic to suppose that local authorities can add value to central policies when they face too many constraints on local freedom to act or when they face fragmented, complex and voluminous central policies and regulations. The collected data suggests that EU environmental policies are certainly not exempt from such problems (see section 5.2 and 5.3). In the meantime, as was most vividly expressed by the 'Liveable Cities' partners, European towns and cities are facing serious challenges to coordinate the many policy ambitions they are dealing with. National policies and local administrative fragmentation are among the causes of these problems (section 5.4). The fragmented and bureaucratic nature of EU (environmental) policies and procedures was also considered as a cause (see also section 5.3). Examples included poor coordination between EU transport policy, climate policy, waste policy, environmental health care, cohesion policy and state aid policy, while many policies and procedures were considered overly complex and bureaucratic. As was discussed in section 5.2, EU policies stimulate and reinforce sectoral fragmentation in local administrations and constrain local efforts to develop and deliver proactive and integrated environmental policies.

Given the complaints about fragmentation and excessive bureaucracy, deregulation and simplification of EU policies is relevant to enable local authorities to cope with new or additional tasks. There were also various calls for such deregulation efforts during the consultation sessions with the 'Extended EU Expert Group on the Urban Environment' (section 5.3). The 'Thematic Strategy' does not, however, address the fragmentation and bureaucratic complexities of sectoral policies, nor does the EU use other strategies for improving the 'internal integration' of its policies. Instead, the EU continues to develop sectoral policies that are not actively coordinated with each other (section 5.2). The 'Thematic Strategy' was meant to help respond to sectoral policies by improving the integration of policies in a local realm. To this end, towns and cities urged for a focus on sustainable urban management due to its holistic focus (sections 5.3 and 5.4). Hence, the 'Thematic Strategy' could potentially be a means for engaging various administrative departments and local stakeholders in a more encompassing governance effort addressing local issues. Unfortunately, the competences of the DG Environment of the EU prevented such a choice. The result is a focus on planning

212   *A European focus on the local*

formats that still have a rather sectoral focus, where an integrated approach is to be stimulated by a focus on environmental policies and transport policies. As the 'Liveable Cities' partners indicated, for example, there is limited confidence that environmental or transport departments are in a strong position to push an integrated approach forward. The 'Thematic Strategy' proposed administrative planning formats that were not considered to contribute to a streamlining of local governance efforts. In fact, as was also expressed during the consultation rounds, the two plan formats are even considered to potentially cut through local integrated policies. As there has been no real attempt at reducing the fragmentation and bureaucracy of EU regulations, cities are still constrained by these regulations in pursuing proactive and integrated approaches to the environment.

5.7 Conclusions

This chapter aimed to investigate the practical consequences of the EU approach to increase the role of the local level in environmental policy and, also, to explain these consequences. In doing so, I reflected on the theoretical arguments I put forward in Chapter 4 that meant to help inform attempts to increase the role of the local level in environmental policy. In this concluding section, I return to both ambitions.

Consequences investigated

The 'Thematic Strategy on the Urban Environment' was meant to be a culmination of over 15 years of work on improving environmental conditions in EU towns and cities. Initially, the EU aimed to move beyond the 'voluntary approaches' it had relied on in doing so during most of the 1990s. This was also clearly expressed in the 2004 draft 'Thematic Strategy'. The main aim was promoting a 'strong framework' of EU policies that stimulated more ambitious, proactive and integrated approaches to the local environment. In practice, this meant that the CEC proposed three legal obligations to be part of an EU directive: an 'environmental management plan', a 'sustainable urban transport plan' and an 'environmental management system'. Most stakeholders, including member states, NGOs and city governments, however, rejected the proposed means of installing the desired 'strong framework'. It was considered to be too rigid, uniform and to interfere with local initiatives in environmental governance. In response, the CEC radically changed course. The finalized 'Thematic Strategy' (CEC 2006) relies solely on voluntary approaches. The second main consequence uncovered was that this finalized 'Thematic Strategy' has not resulted in much impact and is criticised for being 'too weak' to do so.

Explaining the consequences: a post-contingency approach

In Chapter 4 I have used a post-contingency approach to develop theoretical arguments, which help inform attempts to increase the role of the local level in

A European focus on the local   213

environmental policy. With my study into the practical consequences of the draft and finalized 'Thematic Strategy', I have been able to reflect on these arguments. In the first place, conditions of complexity indeed provide an argument for decentralization and against the use of central policy formats that should be generically implemented. Second, decentralization can result in disappointing results in the face of constraints on local willingness and ability, such as economies of scale, external effects and the weak profile of the environment and by possible local problems to routinely and efficiently implement central policies and regulations. My empirical inquiry shows that each of these constraints is real and, albeit with differencing degrees, undermines local willingness and ability to pursue proactive and integrated approaches to the environment. In Chapter 4 I claimed that central policies can be a crucial means for supporting and stimulating local authorities to pursue proactive and integrated approaches to the environment. Not just if there is an ambition to meet single fixed goals such as minimum environmental ambitions, but also for helping local authorities to respond to the aforementioned constraints. With its 'Thematic Strategy' the EU has however not responded to these constraints. It means that I could not assess whether central policies would have helped local authorities to respond to the EU's request for local proactive and integrative policies. The calls for such support by cities and NGOs and the lack of follow-up to the 'Thematic Strategy', however, do suggest that the absence of central support and stimuli is a key contributing factor to this lack of follow-up.

What about the 'thematic strategy'?

In developing the draft and finalized 'Thematic Strategy', the CEC has largely overlooked and ignored the main challenges its towns and cities face regarding their pursuit of proactive and integrated approaches to the environment. Instead, the CEC relied on the assumption that the administrative process of making plans and carrying them out with an associate management system would be sufficient for installing proactive and integrated approaches to the environment at a local level. It also assumed that such an administrative process could be stimulated by relying on EU-issued policy formats that were to be applied generically in larger EU towns and cities. Most member states, NGOs and cities did not support this assumption. I have also seen some of the main reasons why. Not only are there doubts as to whether the EU should be explicitly involved in local matters such as urban environmental governance. The research also shows that proactive and integrated environmental approaches require a 'mix of governance formats', which is both diverse and tailored to the local circumstances. Such a 'mix' should ideally also address aspects such as political bargaining, process management, partnerships, programmes, implementation, etc. It is also exactly in developing and working with such a 'mix' that EU towns and cities seem to have problems, related to constraints on local willingness and ability such as economies of scale, external effects and the weak profile of the environment. If the EU aims to truly invest in improving urban environmental governance processes, these constraints should also be addressed. The finalized 'Thematic Strategy' largely overlooks

214   *A European focus on the local*

these issues, as do its supporting 'technical reports' on 'Integrated Environmental Management' (CEC 2007a) and 'sustainable urban transport plans' (CEC 2007b).

In the meantime, the rejection of the proposals made in the draft 'Thematic Strategy' point towards a lack of support for a strong influence of the EU in pursuing proactive and integrated approaches to the local environment. It is, however, too simplistic to assume that such influence would also be rejected if the CEC would have altered its approach. In addition to obligations for installing the kind of coordinative instruments the CEC proposed in the draft 'Thematic Strategy', voluntary approaches are not the only alternative. Options such as moderate coercive tools, subsidies, awards, contracts and (voluntary) agreements or eco-labels are also possible. In addition, using central policy expertise or funding for research, knowledge dissemination and training of municipal officers might also support local willingness and ability to pursue proactive and integrated approaches to the environment. Clearly, many of these options would require more financial commitments, such as is the case with training, education, subsidies, covenants, etc. In addition, they might also involve political courage to pursue a more ambitious approach to urban environmental governance focused on urban sustainability, spatial planning and stimulating regional cooperation. It is not my aim or position to directly advise the EU on the path it might pursue next. I do, however, know that a post-contingency approach tells us that even pursuing a functional ambition such as proactive and integrated approaches can benefit from the support and stimuli provided by central policies.

6 Beyond the minimum in the Netherlands

Traditionally, Dutch environmental policies have been strongly dominated by the coordinative model of governance with the central state in charge. The prime responsibility of municipalities was to implement and enforce national regulations, largely based on the use of permits and licences. During the last 20 years some important changes have occurred within Dutch environmental policy. Many of these relate to the national policy ambition to improve the integration of Dutch environmental policies with other policies, most notably with spatial planning. To the Dutch Ministry of Housing, Spatial Planning and the Environment (VROM), these changes are to contribute to the creation of an 'optimal environmental quality' on a local level (VROM 2001). This optimal environmental quality is considered dependent on local circumstances. Given their proximity to local interests and circumstances, municipalities are given a leading role in pursuing such an optimal environmental quality.

Following the recent changes, the Dutch national government takes responsibility for protecting a so-called 'base quality' in environmental policy that is inspired by minimum legal requirements (VROM 2004b). Municipalities are subsequently responsible for a quality that 'goes beyond the minimum' (VROM 2004b) as they are considered to be in the best place to judge the most urgent and relevant local priorities and strategies. They can harvest local opportunities for proactive environmental policies and balance local priorities and interests in integrated approaches. Hence, based on the local circumstances, proactive and integrated approaches are to add value to minimum requirements.

The approach taken by the Dutch national government differs from the EU's approach as pursued with its 'Thematic Strategy'. Through decentralization and deregulation, the Dutch national government removed several strict national targets so as to allow local authorities to set their own ambitions based on local priorities. In this chapter I aim to uncover the practical consequences of the Dutch approach for advancing the role of the local level in environmental policy and to explain why these consequences have occurred. In doing so, I will again reflect on the four arguments I presented in Chapter 4. These arguments are meant to help inform attempts to increase the role of the local level in environmental policy, also when it comes to the pursuit of proactive and integrated environmental policies. I aim to assess whether these arguments can indeed help to do so, by studying the approach taken in the Netherlands.

216　*Beyond the minimum in the Netherlands*

This chapter starts with explaining the development of Dutch environmental policies and the approach used for pursuing proactive and integrated approaches to the local environment. The idea is to explain how and why recent changes in Dutch environmental policy have been pursued. It both means to identify the main motives used and the national policies and regulations that continue to influence local performance. It thus provides an overview of the kind of national policies that are in place that can both stimulate, enable or constrain local efforts to pursue proactive and integrated approaches. In section 6.1, the development and main changes in Dutch environmental policy until the more recent focus on proactive and integrated approaches at the local level are discussed (i.e. until approximately 1990). More recent developments are discussed in sections 6.2 and 6.3, on respectively the connection between environmental policies and alternative policy fields (6.2) and on the decentralization and deregulation operations pursued (6.3). Special focus in these sections will be on the envisioned role of municipalities in the pursuit of proactive and integrated approaches to the local environment. In section 6.4, the current organization of Dutch national environmental policies is analysed, while focusing on five distinct policy themes regarding environmental health and hygiene policies.

The final sections of this chapter focus on the *performance* of Dutch municipalities in developing and delivering proactive and integrated approaches to the environment and for achieving an environmental quality that 'goes beyond the minimum'. Section 6.5 addresses the level of success encountered, while section 6.6 explains the causes for success and failure. Section 6.7 will assess how the approach taken by the Dutch national government to the arguments and conditions for advancing the role of the local level in governance compares to the arguments proposed in Chapter 4. Section 6.8 contains the conclusions to this chapter.

6.1 A start for Dutch environmental policies

Just as in most other Western countries, modern-day Dutch environmental policy finds its origin in the 'environmental revolution' of the late 1960s and early 1970s. Located in the densely populated and heavily industrialized delta of the Rhine, Scheldt and Meuse rivers, the Netherlands was one of the first countries that felt the consequences of the rapid economic growth in western Europe. Starting in the early 1970s, the Dutch response to the environmental problems encountered was fast and well structured. This swift response even gave the Netherlands a reputation of a European frontrunner in managing environmental problems (e.g. Liefferink & Van der Zouwen 2004, Weale 1992).

Expressed most explicitly in the 1972 Emergency Memorandum on Environmental Hygiene (Urgentienota Milieuhygiëne, VoMIL 1972), the Dutch chose for a rapid development of environmental legislation and policies. The Dutch set clear limitations on social and economic behaviour with a *reactive* and *compartmentalised* approach that meant to 'clean up' the three identified compartments: air, water and soil (De Roo 2003). After a few years of debate and institutionalization, the 1976 'Policy Document on Ambient Environmental Standards' (VoMil

Beyond the minimum in the Netherlands 217

1976) confirmed the approach chosen. The document contained a series of rather ambitious and strict environmental standards that dictated maximum levels of environmental stress tolerated in air, water and soil. To ensure that these maximum levels were not breached, many emission, immission, product or process related standards were introduced (see also De Roo 2003, Van Tatenhove 1993). These standards were informed by scientific research on the tolerable exposure to pollutants by humans and, to a lesser degree, by ecosystems. Meant to be 'strict but realistic' (De Roo 2003), these standards were also generic, as all people should in principle be protected to a similar degree. Therefore, standards were applied uniformly in the entire country and were decided upon by the state. Provincial and municipal governments only shared executive responsibility with the national government. Clearly, Dutch environmental policy was modelled after the coordinative model; i.e. based on a top-down organization, within a 'command-and-control' tradition that relied on a sectoral and compartmentalised approach.

Fragmentation

The Dutch did not escape from the typical problems of installing regulatory, sectoral and compartmentalised policies. These problems mostly became visible during the 1980s. The sectoral and compartmentalised approach produced relatively isolated pieces of policy which were often poorly coordinated. This lack of coordination was only aggravated by the fact that many environmental policies and regulations in the Netherlands were developed within different ministries. For example, issues related to water were dealt with by the Ministry of Transport, Public Works and Water management, issues related to mining, energy and radiation by Economic Affairs and issues on pesticides by Agriculture. This resulted in what Van Tatenhove (1993) calls a 'fragmented institutionalization' of environmental policies. It meant that variations in the organization, procedures and mechanisms of enforcement of these policies were common, while multiple organizations and levels of authority were involved in their development and implementation.

Around 1980 policy fragmentation caused serious coordination problems. One of the consequences was the transfer of pollution from one compartment to another (e.g. from water to soil, from soil to air, etc.). Waste that could no longer be stored in landfills (soil pollution), for example, had to be burned (air pollution), while pollution disposed of in running waterways (water pollution), for example, also needed to be burned (air pollution) or stored on land (soil pollution). Other consequences included incoherencies, overlap and conflicts between policies, which undermined the effectiveness and efficiency of Dutch environmental policy. Eventually, "it had become almost impossible to take into account the connections both between the 'sectors' of environmental policy and with other policy fields such as physical planning, water management or agricultural policy" (Liefferink 1997; p. 216, compare De Roo 2003, Van Tatenhove 1993). In reaction to these problems, 'integration' became the new buzzword in Dutch environmental policy. From 1983 onwards, integration was seen as "a precondition for an

218   *Beyond the minimum in the Netherlands*

effective and efficient environmental policy that takes account of the effectiveness and efficiency of the whole range of environmental policy" (VROM 1983; p. 6).

Internal integration

In the mid-eighties the newly established Ministry of Housing, Spatial Planning and the Environment (VROM) prioritized what they called the 'internal integration' of environmental policies (between separate sectors in environmental policy). The Ministry's approach was explained in the 1983 'Plan Integration Environmental Policies' (PIM, VROM 1983) the 1984 'More than the Sum of the Parts' (VROM 1984a) and the annual 'Indicative Multi-Year Programme for the Environment' (VROM 1984b). The ambition: "to develop a cohesive and consistent policy for all government activity relating to the environment" (De Roo 2003; p. 175).

Internal integration was first pursued through two complementary lines of policy: 'effect-oriented policies' and 'source-oriented policies'. Effect-oriented policies focussed on a series of thematic priority issues, which the national government considered to be poorly addressed by existing sectoral and compartmentalised approaches. In response, it proposed these issues to be addressed based on a tailor-made package of solutions. These themes included issues such as acidification, eutrophication, the diffusion of substances, climate change and disturbance (i.e. noise, odour, vibrations, safety risks, local air pollution). Source-oriented policies were already common in the Netherlands, focussing on, for example, the emission of stress from machines or on end-of-pipe measures. The novel feature in the 1980s was to address the sources of pollution by focusing on the main target groups that were involved in causing this pollution, such as the transport and energy sectors, households and agriculture. Coherent strategies were developed to alter target group behaviour and the use of materials or products by them.

With 'internal integration' the Dutch national government meant to improve the functioning of the existing environmental policies and regulations; i.e. to improve the functioning of the coordinative model. Deregulation was used for streamlining and combining existing regulations, procedures and licence structures, while new framework laws were developed to better coordinate and combine existing sectoral legislation. These changes were not meant to challenge the sectoral organization of Dutch environmental policy, nor did they challenge the reliance on environmental standards. The 'internal integration' operations did, however, involve the introduction of some new governance instruments. These included financial instruments, such as subsidies, fines, the polluter-pays principle and deposits on cans or bottles (see also Winsemius 1986). Furthermore, communicative, persuasive and collaborative policies were among the new approaches used, as illustrated by large government campaigns regarding target group behaviour and increased stakeholder involvement. Finally, and in addition to the increased use of new governance instruments, 'internal integration' resulted in a slight move away from reactive policies and clean-up operations, towards preventive or proactive policies (VROM 1983).

Beyond the minimum in the Netherlands 219

A more proactive approach was apparent in the focus of the Dutch national government on improving the 'internalization' (Dutch: verinnerlijking) of environmental values in the decision making of economic and societal actors (Winsemius 1986). In addition to the focus on 'internal integration', this 'verinnerlijking' aimed for environmental values to function as guidelines for social and economic behaviour. The focus on target groups was a first means to do so. There were, however, also extensive persuasive campaigns highlighting moral responsibilities of citizens and businesses, while management systems and accountancy systems were introduced in businesses and agriculture to shift responsibilities away from the central state. Finally, inspired by the Dutch neo-corporatist tradition (Liefferink 1997), negotiation and bargaining in wider policy networks became increasingly important. It resulted in a relatively extensive use of covenants between government agencies and businesses and the use of large advisory commissions to advise government policies as compared to other Western countries. It illustrates that Dutch environmental policies gradually shifted away from a full dependence on the coordinative model of governance towards a more diversified body of policies. This shift, according to Liefferink, "reflected the tendency towards declining confidence towards the state's problem solving capacity" (1997; p. 217). Still, most of the new instruments introduced did not *replace* existing coordinative instruments and the associated dominance of the central state, but were *additional*.

At the end of the 1980s, the 'internal integration' of environmental policies was regarded as a success (De Roo 2003). The 1989 first Dutch 'National Environmental Policy Plan' (NMP1, TK 1989) therefore summarized how internal integration was to be concluded and maintained. With the problems of fragmentation considered under control, the NMP1 was to take Dutch environmental policy to the next level. Following the idea of 'verinnerlijking', it aimed to show how the environment could function as a guideline for general social and economic behaviour. The move towards proactive environmental policies to be integrated with other governance activities was continuing.

The NMP1

The NMP1 was internationally celebrated for being one of the most daring and innovative policy documents on environmental management in the world until then (see also De Roo 2003, Liefferink 1997). It showed the Dutch desire to move away from 'damage control' and 'cleaning up'. In 1992 Weale, for example, suggested that the NMP1 is "perhaps the most serious attempt to integrate environmental concerns into the full range of public policy" (p. 125) and similarly a "union of economy and ecology" (p. 135). This ambition of the NMP1 drew from the then still recent 1987 Brundtland report (WCED 1987) and its claim for sustainable development; i.e. with the ambition to pursue the integration of environmental concerns into all relevant governance activities and general social and economic behaviour.

The NMP1 was developed against growing international doubts regarding the effectiveness and efficiency of environmental standards and other

220 *Beyond the minimum in the Netherlands*

regulatory instruments in environmental management (e.g. Milbrath 1989). Dutch environmental management, however, still enjoyed much support from scientists, politicians and social pressure groups for strong state intervention and the associated regulatory tools (e.g. standards). In 1989 Glasbergen and Dieperink doubted whether deregulation, decentralization and privatisation should be used in renewing environmental policies at all. In a similar stance, Mol (1989) stressed that it is essentially centralized government intervention that is needed in environmental management. Based on this continuing confidence in central government control and regulatory policies, the NMP1 did not take the gradual move away from a regulatory and coordinative mode of governance that started in Dutch environmental policies in the 1980s much further. Rather, it reinforced what was done in the 1980s regarding the source- and effect-oriented policies. The continued overall deterioration of many aspects of the Dutch environment in spite of past legislation even meant that additional regulatory measures were considered needed. With the internal integration now almost finished, the NMP1 considered the Netherlands was ready for even more ambitious environmental regulations to be directed by the state. Although these could be developed in a more integrated institutional setting, they remained focussed on fulfilling strict environmental regulations. Environmental standards remained the key instruments, while financial, persuasive and communicative strategies were also instigated and supported by state regulations. The NMP1 remained regulatory based, relying largely on the logic of the coordinative model in which the central state pulls the strings (see also De Roo 2003).

6.2 Towards external integration

The NMP1 was both ambitious and optimistic. It proposed policies that should ensure that all key environmental issues were resolved or, at least, controllable within one generation. The optimism of the NMP1 received some serious blows in the early 1990s. It was especially the 'external integration' of environmental policies with other policies that contributed to these blows.

A link with spatial planning

The external integration between environmental policies and other government policies had partly been facilitated with target group policies. Through contracts, covenants and persuasion, environmental values were at least partly 'internalized' in the decision making of, for example, businesses, farmers and citizens. The external integration which began in the early 1990s was however focused on the integration between environmental policies and other government policies addressing the physical environment, such as policies focussed on housing, infrastructure, water management, nature, urban development, etc. Most notably, it was focussed on the link between environmental policies and spatial planning.

Spatial planning in the Netherlands traditionally dominates planning for the physical environment. Spatial planning seeks to locate human activities and land use functions for creating the highest possible quality of the physical environment. Before the 1970s this quality was based mostly on social welfare, economic development and aesthetic motives, which were all embedded in spatial planning policies. With the rise of environmental policies, creating a high quality of the physical environment was now also to be defined in terms of environmental health and hygiene. Environmental policy had been developed in parallel to existing spatial planning policies. Following their separate development and focus, spatial planning and environmental policy were two bodies of policies that functioned in their own distinct way (e.g. De Roo 1993, Miller & De Roo 2005). Environmental policy was organized top-down, hierarchically, and used many generic and sectoral regulations. Spatial planning relied mostly on indicative state policies to be translated into detailed and more holistic or integrated policies by municipalities, which held the most powerful planning instruments (i.e. more decentralized). The differences between spatial planning and environmental policy also resulted in conflicts, especially in implementing them in a local realm (e.g. De Roo 1993). Generic environmental targets did not always go well with the integrated ambitions that were common for local spatial planning strategies. Improving the integration of environmental policies and spatial planning was therefore seen as important ever since the late 1970s. The integration of the Ministry for Housing and Spatial Planning (VRO) and Public Health and Environmental Hygiene (VoMil) into one ministry in 1982 was meant to smooth the relation between spatial planning and environmental policies. However, it did not remove the differences between them and the associated conflicts. Around 1990, it was apparent that pursuing this integration should especially focus on the local realm where most problems in combining these policies took place (VROM 1989).

From the start of the development of environmental policies, the Dutch relied on physically separating environmentally intrusive functions ('sources') from environmentally sensitive functions ('exposure'). Such a separation between functions is inspired by the idea that many forms of environmental pollution or stress are subject to a 'tapering effect'; i.e. the level of pollution or stress becomes less when distance to the source of pollution or stress increases. Good examples are noise from busy roads or railways, safety risks around fuel stations or odour from factories. Maintaining a certain distance from the source of pollution or stress can ensure that the levels of exposure or risk become low enough to be tolerated. The distances needed for reaching these tolerable levels can be calculated for various sources of pollution or stress. Based on such calculations, a buffer zone can be installed in order to ensure that sensitive functions, such as housing, would be outside of the zones with intolerable levels of exposure. These zones can be translated into spatial plans and can help when issuing permits to either new urban developments or companies that produce environmental pollution or stress. The use of environmental zones was confirmed by the wide use of the 'Companies and Environmental Zones' report of the VNG (1986). This report listed *indicative* zones that should be taken into account for many hundreds of different installations or functions producing environmental pollution or stress, depending on the kind and level of pollution or stress produced.

222   *Beyond the minimum in the Netherlands*

Until the early 1990s, environmental zones were the main link between spatial planning and environmental policy. In most situations, the indicative distances such as expressed in the 'Companies and Environmental Zones' report could be installed and maintained. In some places, the indicative distances could not be implemented easily or without severe measures. Sometimes, zones overlapped with existing residential areas or with areas designated for new developments. Although such cases were also not uncommon, their visibility and the urgency attributed to them remained limited until the focus on external integration in the early 1990s. The need for a focus on the local realm in further pursuing the external integration between environmental policies and spatial planning resulted in a focus on what was called an 'area-specific approach', that was to function next to existing effect-oriented and source-oriented policies (TK 1990). It was in pursuing this area-specific approach that conflicts between environmental policies and spatial planning gained in visibility and urgency.

Reality bites back

With an area-specific approach environmental priorities should be combined (i.e. integrated) with other policy priorities in an overall strategy based on the desired or existing qualities of a specific area (see also Chapter 2). Combining these priorities proved more difficult than expected. The idea with the area-specific approach is that local parties can do the balancing and combining of various priorities, given their proximity to the specific circumstances. In the early 1990s, local parties were, however, still confronted with strict environmental standards that they had to implement. These standards dictated generic environmental standards, focussed on one specific environmental stressor (i.e. noise, air quality, odour, etc.). As such, local authorities were faced with several isolated environmental standards that should be met a priori to any attempt to balance priorities. That is, area-specific approaches that relied on a structure of decentralized working had to be pursued within a framework of national generic standards that thus relied on a functional focus on single fixed goals. It is, following the discussion on the contingency between function and structure in Chapters 3 and 4, a rather awkward combination. Indeed, practice also showed that this combination proved hard to pursue.

Two main projects were instigated in the early 1990s to experiment with the area-specific approach: the Integrated Environmental Zoning (VROM 1989) and the 'Spatial Planning and Environment' or ROM (TK 1990) project. Both took a different approach and were received quite differently in practice. Learning from these differences inspired many changes to Dutch environmental policies from the mid-1990s onwards (De Roo 2003).

The Integrated Environmental Zoning project was received with much criticism. It was this approach that proved exemplary for showing the consequences of aiming for an area-specific approach with a framework of strict sectoral environmental standards (Box 6.1). The project, first of all, showed occasions where levels of environmental stress were so high and so hard to reduce, that meeting standards was impossible within reasonable technological, financial or temporal frames. Environmental zones would continue to overlap with existing urban areas or with areas reserved for future development. Possible solutions included large restrictions on future development, the closure of industrial estates or roads or the demolition of large city parts. The project thus illustrated the consequences of relying on strict standards: "environmental policy dictates the consequences for spatial planning, so that spatial planning can no longer weigh up alternatives" (Borst et al. 1995; p. 77).

Second, the project showed that it was not always evident who was responsible for coping with the consequences encountered. After all, who is responsible for traffic noise or the background level of air pollution? Also, what if a factory was built before the houses adjacent to it were built? In short, the Integrated Environmental Zoning project illustrated the administrative and political difficulties local authorities encountered in implementing environmental policies imposed on them by the central state.

Box 6.1   Integrated Environmental Zoning

The Netherlands, 3 April 1989. Confidence was high when the Ministry of VROM called for the beginning of the Integrated Environmental Zoning (IEZ) approach (VROM 1989). The approach consisted of a rather innovative way for dealing with environmental problems in urban areas. An overall environmental load was calculated by cumulating different environmental stressors (i.e. noise, odour, safety risks, soil, etc.). The calculated cumulative load could belong to six 'environmental stress classes' running from little to high environmental stress. The levels of stress measured or projected were categorized in various classes of environmental stress, which each were translated into an environmental zone; i.e. based on the logic that stress became less with increased distance to the source. Then, these environmental zones were implemented *directly* into spatial plans including the necessary land use implications. All areas belonging to the zone with the highest stress class would have to be demolished, for example, or extensive reductions at the source should be realized. Classes facing lower levels of stress would have less severe consequences, such as prohibitions for new developments or more modest reductions of stress.

The IEZ approach was meant to become part of Dutch environmental law. The first experimental projects, however, proved that much of the high confidence was misplaced. In more than a few occasions, the direct translation of environmental standards into spatial plans would have devastating consequences. For example, large parts of cities like Arnhem, Dordrecht or Maastricht would have to be demolished (Borst et al. 1995). These consequences were considered too harsh to be accepted.

224　*Beyond the minimum in the Netherlands*

While the problems with a strict implementation of environmental standards were explicitly recognized during the Integrated Environmental Zoning project, the 'Spatial Planning and Environment' (ROM) project had quite different outcomes. The ROM project was based on developing integrated and more participative approaches to environmental issues in 11 priority areas in the Netherlands (see also De Roo 2003). These areas included both areas with high levels of environmental stress and areas that are highly sensitive (i.e. nature reserves and protected landscapes). These areas were chosen for their complex mix of land uses and the number of interests and stakeholders involved. The Spatial Planning and Environment approach allowed for a realistic and pragmatic step-by-step implementation of environmental ambitions, while simultaneously promoting economic and possibly social development. It thus reduced the direct impact of a functional focus on single fixed objectives. Implementation would follow a pragmatic approach, by, for example, taking more time to carry out. In addition, the possibility for local and regional stakeholders to be directly involved in both deciding and implementing the strategies proved popular. Increased public and stakeholder support, the occurrence of mutual learning and a pragmatic step-by-step approach in coping with existing environmental standards resulted in a general feeling of success. Given the problems with using strict environmental standards as illustrated by the Integrated Environmental Zoning project, the message of the more flexible Spatial Planning and Environment approach was heard by many.

Change

Based on the experiences of these two projects, scholars and practitioners reconsidered the role of strong regulatory instruments and state intervention in Dutch environmental policy (e.g. De Roo 2003, Liefferink & Van der Zouwen 2004 and Van Tatenhove et al. 2000). These considerations also affected dominant ideas on the role of the local level in environmental policy. Setting strict a priori targets was increasingly considered to 'suffocate' local authorities in their attempts to produce area-specific policies that would both be tailored to the local situation and would facilitate the desired external integration between environmental policies and other government policies. In reaction, increasing the room to manoeuvre locally by reducing the impact of a priori targets became popular among scholars and practitioners. Decentralization, deregulation and the use of market-based instruments, already pursed widely in other Dutch policy fields (e.g. Nelissen et al. 1996), now also began to affect environmental policy.

6.3 Flexibility and decentralization: new policies and different ambitions

In the second National Environmental Plan (NMP2) the Ministry of VROM already sought for a "shift of emphasis from top-down governance to self-governance within parameters" (TK 1993; p. 42). Increasingly the central government considered itself unfit to bear all responsibilities in environmental policies (i.e. a rejection of a full reliance on the coordinative model of governance). One of the responses by the national government was to increase the role of regional and local authorities in the development and implementation of environmental policies (in line with the idea of a subsidiary).

During the 1990s local and regional governments, partly stimulated by the national government, already began experimenting with new instruments and policies in environmental management. Several provinces began with the development of new Provincial Integrated Area Plans (POP), which combined three traditionally separate provincial plans for the physical environment; i.e. on water management, environmental health and hygiene and spatial planning. In addition, integrated plans and visions were also developed by various individual regions and municipalities (see De Roo & Schwartz 2001). With these experiments, regional and local authorities were seeking for means to integrate environmental policies and water management into the traditionally already more integrated spatial planning policies (see Humblet & De Roo 1995, De Roo et al. 2012). In addition to regional and local experiments, structural changes to Dutch environmental policy took place during the 1990s. Through the reduced impact of sectoral regulations local authorities should be enabled to produce area-specific policies.

Structural changes

First, decentralization and deregulation affected existing sectoral regulations, most notably within soil, odour and noise policies (see also De Roo 2004), to be followed by external safety policies (Flameling 2010). These regulations were previously dominated by 'strict and ambitious' environmental standards. During the 1990s and early 21st century, they however became more flexible and less ambitious regulations. National limit values were made less ambitious (noise policies), became indicative (odour policies) and there was a great increase in the room for interpreting limit or indicative values based on the local priorities and circumstances by local authorities. I will explain these policies in more detail in the following section. For now it is important to note that the result of these changes was an increase in the power of local authorities to make decisions within noise, odour and soil policies.

In addition to the changes made to sectoral environmental policies, the 1990s also saw other structural changes to Dutch environmental policy. Among the prime examples are the so-called City and Environment approach and the Investment Budget for Urban Renewal. The City and Environment approach (see also VROM 2004a) is established in a national interim law. The approach allows for local authorities to deviate from generic national environmental standards if there are pressing reasons for doing so. Deviation is only possible for noise, odour, ammoniac, soil policies and air quality, if not influenced by EU legislation. In addition to a clear motivation for deviating from national standards, local authorities should also pursue compensation measures. For example, allowing for more noise outside can be compensated by reducing the noise indoors from neighbours. The City and Environment law requires that compensation must contribute to

226 *Beyond the minimum in the Netherlands*

improving the overall quality; i.e. any deviation from legal standards should be overcompensated. The flexibility as installed with the City and Environment law makes it possible to adopt an area-based approach and integrated policies even if sectoral legislation would initially prevent developing such approaches or policies. Despite a good reception by stakeholders, the City and Environment use has only been fairly limited. The increased flexibility in sectoral policies on noise, soil and odour seems to have seriously decreased its relevance.

The Investment Budget for Urban Renewal was installed in 2000. Before its installation, municipalities received financial support for urban renewal projects directly linked to various policy priorities such as subsidies for soil remediation, noise abatement or other non-environmental issues such as social housing or mobility. With ISV, subsidies have been united into one 'bulk sum', where municipalities have more discretion to spend it in their urban renewal projects as they seem fit. As the amount of money is based on contracts between the state and local authorities, the state continues to have a degree of influence on municipalities. The state can decide to include provisions on how money should be spent locally. Nevertheless, the Investment Budget for Urban Renewal has given local authorities greater responsibilities and flexibility. As environmental subsidies are now part of a more embracing effort for improving the 'liveability' of these neighbourhoods, the integration of environmental policies with other policies is theoretically also easier to achieve. Evidence is, however, not convincing that this integration is also achieved in practice (Kamphorst 2006).

Fourth national environmental plan

While structural changes to Dutch environmental policies were taking place, the Dutch national government issued its Fourth National Environmental Plan (NMP4) in 2001. In the NMP4, which is still applicable today (2016), the Dutch national government confirms its belief that granting local authorities with new responsibilities and greater freedom helps to develop tailor-made and integrated policies. Decentralization in environmental policy is therefore believed to "result in improvements to the quality of the living environment" (VROM 2001; p. 68, translation CZ). Therefore, the NMP4 states that local government "must be afforded greater freedom and as much integrated responsibility for the local living environment as possible" (VROM 2001; p. 68, translation CZ).

The renewal of various sectoral policies and initiatives such as the City and Environment approach and the ISV already contributed to the desired 'freedom' at a local level. The NMP4 continues by expressing the respective responsibilities of different levels of governments in pursuing the desired improvement of the 'quality of the living environment'. It begins with the national government, which "will set rigid minimum standards for environmental quality and monitor those limits in conjunction with the lower levels of government during implementation and enforcement" (p. 69, translation CZ). This 'rigid minimum' is called the 'base quality' (see also VROM 2004b), which the national government guarantees and safeguards. This 'base quality' is inspired by "minimum requirements where it

Beyond the minimum in the Netherlands 227

concerns the protection of the health and safety of people and of nature" (2001; p. 329, translation CZ). Typically, such requirements are generic environmental standards. However, the 'base quality' is only a *minimum* requirement. The Ministry therefore has a second ambition, which is to pursue "a quality that goes beyond the minimum" (2001; p. 333, translation CZ). The Ministry believes that "the quality of the living environment differs from place to place, from person to person and also from period to period" (2001; p. 328, translation CZ). Given these variations, the Ministry considers a quality that 'goes beyond the minimum' to be tailored to the local circumstances. Hence, local authorities are to take the lead in pursuing a quality that 'goes beyond the minimum', as they are considered to be in the best place for judging local circumstances (VROM 2001).

Intentions with the NMP4

The Ministry aims for local authorities to be allowed to formulate their own ambitions and strategies when it comes to a quality that 'goes beyond the minimum'. In its NMP4, the Ministry of VROM, however, remains unclear about the status or contents of a quality that 'goes beyond the minimum'; i.e. is it optional or more obligatory? The Ministry did try to provide such clarity, based on agreements with the Association of Provincial Authorities (IPO), the Association of Netherlands Municipalities (VNG) and the Union of Waterboards. The idea was to develop both clarity regarding the responsibilities of each of the actors involved, including between the different levels of authority, and regarding the status and contents of a quality that 'goes beyond the minimum'. A policy advisory was to be written to produce the desired clarity.

Initially, the plan was for this policy advice to be 'legally robust' (VROM 2001) and be a fairly strong stimulus for local authorities. The plan of the Ministry in the NMP4 was to agree upon indicative environmental standards that were decided for various environmental themes (e.g. odour, noise, soil, etc.) based on different area types (e.g. central city, suburban, industrial, rural, etc.). These would be nationally agreed indicative standards that were more ambitious than the minimum legal requirements. Local authorities, as was the initial plan, would then be able to deviate from these indicative standards under certain circumstances. The exact status of the indicative standards and the criteria that would have to be met if deviations were allowed were still under debate at the time of writing of the NMP4. The main idea of the Ministry was however clear: a combination of national indicative standards and motivated deviation possibilities were considered a good stimulus for local authorities to indeed pursue a quality which 'goes beyond the minimum'.

The policy advice announced in the NMP4 is the so-called MILO advice, which is an abbreviation for 'Milieu in de Leefomgeving' ('Environment in the Living Environment', VROM 2004b). Although it was supposed to be based on legally robust agreements between the relevant actors and government levels, it has turned out to be simply a voluntary piece of policy advice. It does not set nationally agreed indicative standards, but is focussed on the process of integrating

228   *Beyond the minimum in the Netherlands*

environmental ambitions into spatial planning. In practice, MILO is joined by multiple other policy advisories and informative websites that call upon municipalities to 'go beyond the minimum' (e.g. VROM 2003, 2006, 2007b). Neither of these tools is however an answer to the original intent to develop policy advice that is 'legally robust' or provides a strong stimulus for local authorities. The Ministry only *asks* municipalities to develop their own strategic environmental policies, such as environmental policy plans. There are no national requirements for municipalities to do more than what is legally required: 'going beyond the minimum' is fully optional.

Outcomes investigated

Decentralization and deregulation have paved the road for local authorities to find their own best way for dealing with their environmental problems. This process is supported by the national government's idea that municipalities are in the best place for developing and delivering integrated policies so as to go 'beyond the minimum requirements' in environmental policy. This was explicitly expressed in the NMP4, and confirmed in subsequent policy statements formulated by the various political administrations after 2001 (e.g. VROM 2002, 2006). The national government chose to encourage local integrative policies with merely persuasive policy recommendations. In other words, the development and delivery of these integrated policies and the associated ambition to go 'beyond the minimum requirements' relies on the local willingness and ability to do so.

In the following sections (6.5 and 6.6) I will pursue the question whether local willingness and ability are sufficient to do so. In section 6.5 I will address levels of local performance, specifically focussing on the degree of success of municipalities for delivering a 'quality of the living environment that goes beyond the minimum requirements'. Subsequently, I will in section 6.6 discuss the main explanatory factors of success and failure I identified on the basis of my research. Before that, I will begin with an inventory of the state of the art (2009–2010) Dutch environmental policy. It is based on assessing five relevant environmental themes, where the role of municipalities has been altered since the early 1990s. My aim is to see how changes in national policies and regulations influence municipalities in accommodating proactive and integrated approaches to the environment. The themes chosen are soil remediation, air quality (only on small particles and nitrogen oxides), noise abatement, energy conservation and odour nuisance. Data is based on a study of relevant legislation, national policies and available policy recommendations (references in the text of section 6.4).

6.4 Dutch environmental policy: state of the art

During the 1980s the internal integration of Dutch environmental policy was high on the national policy agenda. As concluded in the 1990 NMP1, the Dutch considered this internal integration to have become a success. Policies and regulations were supposed to be well streamlined, clearly organized and well coordinated (De Roo 2003). During the 1990s and early 21st century, the main national

Beyond the minimum in the Netherlands 229

policy focus therefore shifted to external integration, decentralization, deregulation and on improving linkages with businesses and citizens (see also VROM 2006). Among the outcomes is an increased responsibility and autonomy for local authorities. In the meantime, these changes also altered the landscape of Dutch national environmental policies more generally.

Various sectoral environmental policies were altered before the NMP4 and its focus on a distinction between a 'base quality' and higher ambitions, such as the examples of soil, noise and odour just discussed. These alterations were, however, largely pursued within these individual thematic policy fields without focussing on the relationship between these thematic policy fields. According to various authors (e.g. Baartmans & Van Geleuken 2004, De Roo 2004, De Zeeuw et al. 2009), this has resulted in a reduced coherence and increased fragmentation of Dutch environmental policies. This begs the question whether internal integration has been maintained or that, in contrast, conditions of 'management overload' might prevail for those who have to deal with these policies. In this section, I will respond to this question by addressing the organization of five Dutch environmental policy themes with a prominent role at a local level: soil remediation, air quality (only on small particles and nitrogen oxides), noise abatement, energy production and use, and odour nuisance. The objective is to understand how these five thematic policy fields combine or differ with regards to contents and organization (i.e. the allocation of responsibilities). The research is based on analysing the main policy documents, national policy recommendations and legislation on each of these themes. In addition, I conducted a series of interviews (11 in total) to gain a better understanding of the current organization of Dutch environmental policy, all with experts working for either the Ministry of VROM or the Dutch Association of Municipalities (VNG). The national government also uses websites for improving the accessibility of these policy documents, national policy recommendations and legislation, which can be used to verify the data presented in this section or to find more details. These websites are www.rijksoverheid.nl, which features all national policies and legal regulations; www.infomil.nl, which is focussed on environmental policies and regulations and www.compendiumvoordeleefomgeving.nl, which is focussed on using and implementing environmental policies and regulations. In later phases of the research, the response to these national policies by municipalities is addressed.

Contents: limit values and ambitions

The recent renewal of Dutch sectoral policies is inspired by the idea that the national government is responsible for guaranteeing and safeguarding a 'base quality' (VROM 2004b), inspired by 'rigid minimum standards' (VROM 2001). In addition, the national government aims for 'a quality that goes beyond the minimum' expressed as an 'optimal living quality' (e.g. VROM 2003) or as 'quality of the living environment' (VROM 2006). This 'higher' quality is considered a responsibility of local authorities, based on proactive and integrated strategies that are tailored to the local situation.

230   *Beyond the minimum in the Netherlands*

In theory, the Dutch already worked with a somewhat similar distinction in their environmental policies. Even before the 1990s, a difference existed between so-called 'grenswaarden' (limit values) and 'streefwaarden' (ambitions values). Limit values are the legal limits that may not be exceeded. Ambition values are considered levels that express a genuinely 'good' quality and are the ideal ambitions. Formally, municipalities are obliged to *meet* limit values and *aim for* quality levels in between ambition and limit values, which are in Dutch referred to as 'richtwaarden' (see the 1993 Environmental Protection Act). The status of these so-called 'richtwaarden' remained somewhat vague, as did the means to decide on their exact value. Nevertheless, these differences clearly seem to be reflected in the desired distinction between the standards dominated 'minimum requirements' and the higher ambitions that 'go beyond the minimum' that the Ministry refers to in the 2001 NMP4. One would expect that the policy innovations during the 1990s and 2000s would have therefore further clarified the statuses and meanings of these different levels of ambition.

Ten years after the publication of the NMP4, clarity regarding the differences between minimum requirements and higher ambitions for the five themes studied is largely absent and arguably less than during the 1990s. Rather, each thematic policy field is now organized in its own way. Currently, real differences between minimum requirements and higher ambitions *only* exist in noise and air policies, and even there they are not clear at all. To begin with *noise policies*, the existing 'strict and ambitious' generic standards of the early 1990s have been replaced by a system of two types of limit values. First, there are so-called 'preferable limit values' (between 50 and 57 dB (a) depending on the circumstances) that function as a generic 'limit' value. They are also in line with the minimum legal values that existed before the alterations during the 1990s. The competent authority (i.e. the municipal council) can, however, decide to *exceed* the preferable limit value if there are good reasons to do so. Although the competent authority is obliged to motivate a decision to exceed the preferable limit value, it is essentially free to make such a decision. Now a second limit value comes in, which is the absolute limit value set by the state (between 65 and 70 dB (a) depending on the circumstances). These absolute limit values cannot be exceeded at any time. Third then, there are also (quantitative) ambitions for noise that are higher than the preferable limit values, expressed as ambition values ['streefwaarden']. The status of these ambition values is, however, dubious. Although relatively well known amongst practitioners, they are not widely used in practice and for many practitioners, hardly considered relevant (section 6.8). Concluding, it is difficult to establish what the actual 'base quality' for noise nuisance is. After all, is it the preferable limit value or the absolute limit value?

Air quality policies are formally guided by a difference between limit values and ambition values. In practice air quality policies are, however, fully dominated by the EU-defined limit values (see also Zandvoort & Zuidema 2007). Not a single example or intent of aiming for higher ambitions than the minimum requirement could be found in national reports and policies studied, or during the expert interviews or the case studies. In the meantime, the actual EU-defined limit values

Beyond the minimum in the Netherlands 231

are subject to a degree of flexibility. The Dutch currently work with a system that makes it possible to slightly increase air pollution locally, even to levels higher than what the limit values would permit, only as long as this produces an improvement of the air quality in the direct vicinity of this increase inspired by public health motives. As a consequence, the limit value is subject to some leverage and the definition of the 'base quality' is subject to different interpretations.

All other policy themes that were investigated lack any real difference between minimum requirements and higher ambitions. *Soil policies* do point towards such a difference. Here a difference exists between 'intervention values' (imminent danger is uncovered), 'indication values for serious pollution' (remediation is required before reuse) and 'ambition values' (the soil is clean). Intervention values function as a benchmark for immediate action in the face of public health risks that rarely occur. On the other extreme, if all ambition values are met, the soil is considered clean. In between these two extremes, things become more blurry. On the one hand, for establishing whether remediation is required (i.e. whether there is serious pollution), one has to look at (1) the number of pollutants found in the soil which exceed the ambition values, (2) the type of pollutants that exceed the ambition values and (3) the degree in which they do so. On the other hand, it also depends on the future land use to decide on the degree and kind of pollution that is tolerated. Although this is a very pragmatic approach, the approach does make it hard to clearly establish the minimum requirements, as these requirements should be interpreted within the context of the future use. Hence, there is no real way to establish what the 'base quality' is for soil.

For odour and energy policies, it is even harder to identify the quality criteria used. In *odour policies*, national generic limit values have been dropped altogether. Instead, there is a national frame of reference ('toetsingskader'), which issues indicative limit values. These indicative values are supposed to be interpreted within the local context, inspired by the ALARA principle ('As Low As Reasonably Achievable'). It is up to the competent authority (either the province for large odour producing installations or municipalities for smaller installations) for deciding whether or not to use the national indicative values. While such decisions are made, competent authorities do have to comply with the rather vague national objective of preventing 'serious nuisance'. This concept is not truly defined and thus hard to make operational in practice. All in all, both the minimum requirements and the idea of higher ambitions have no real definition in odour policies. Finally, when it comes to *energy policies*, there is the EPC system: 'Energy Performance Coefficient'. The EPC expresses the kind of energy efficiency standards that new houses need to comply with (i.e. minimum requirements). There are no requirements for existing houses. Other energy policies are mostly related to sustainable construction, but are voluntary and only just developing. Thus, there is no real base quality, nor a set of ambition values.

Faced with the results of the scan of these five thematic policy fields, we can conclude that the difference between minimum requirements and higher ambitions is hard to see within Dutch environmental policies. This difference is not expressed at all (energy), is not used (air) or is at least hard to recognize and define (soil,

232 *Beyond the minimum in the Netherlands*

odour and noise). A 'base quality' or a quality that goes 'beyond the minimum' is hard or impossible to define, even when it comes to individual environmental policy themes. Municipalities are faced with unique regulatory structures for each different policy theme. They are thus forced to cope with their environmental issues within these sectorally organized frames and rationales. While this already undermines the implementation of a 'base quality', municipalities also find little guidance as to which higher ambitions they should aim for. The ideas of an 'optimal living quality' and of a quality that goes 'beyond the minimum', are therefore also difficult to translate into practical actions and objectives.

Organization: allocation of responsibilities

To start with, the allocation of responsibilities for *noise policies* is the clearest, albeit mixed. Local authorities (municipalities) are responsible for most of the choices in noise policies. They issue permits to most companies and enforce national legislation when (re)developing. In addition, local authorities are the competent authority to decide whether or not to deviate from the preferable limit values. The national authority, in this case the Ministry of Housing, Spatial Planning and the Environment, is responsible for enforcing the 'absolute limit values'. In addition to these fairly clearly separated responsibilities, there are many exceptions in dealing with Dutch noise policies. Provinces are, for example, responsible for enforcing legislation and issuing permits for many larger installations (e.g. larger factories) that cause noise nuisance. In addition, there are specifically designated areas that are managed by the national government. Noise originating from national infrastructure is a responsibly of the national government, residing with the Ministry of Infrastructure and the Environment. Noise in areas in the Dutch 'National Ecological Network' is a responsibility of the Ministry of Economic Affairs, Agriculture and Innovation. Finally, airports and air strips are all subject to their own specific noise abatement legislation and are a responsibility of again, the Ministry of Infrastructure and the Environment.

In *air quality* matters, municipalities play a paramount role in ensuring that (EU) limit values are met. Municipalities issue most permits *and* are responsible to take the lead in solving cases where limit values are not met. If EU limit values cannot or might not be met, municipalities take the lead in developing and implementing the EU required air quality action plans, which mean to ensure that limit values will be met in the future. Municipalities, however, share the responsibility for air quality action plans with the Ministry for Infrastructure and the Environment. In addition, the Dutch national government has decided to allow for some deviations from exiting EU limit values. As said earlier, Dutch legislation allows a slight increase in local air pollution, even beyond limit values, as long as this produces an overall improvement of the air quality in the direct vicinity of this increase. Inspired by public health motives, the competent authority needs to liaise with the Ministry to decide on whether exceeding limit values is accepted while being aware of current case law on the issue. Finally, given the problems in the Netherlands to meet EU air quality limit values, a 'National Program for Cooperation on Air Quality' has been set up by the national government. The programme combines various approaches and measures and is the national government's attempt to experiment with improving air quality. Municipalities, provinces and specific government agencies are all involved in this programme.

In *odour policies*, the size of the installation causing the odour nuisance is decisive in clarifying whether provinces or municipalities are responsible for interpreting the national indicative limit values within a local context. While interpreting, the responsible authority needs to take the indicative limit values and the national requirement to prevent 'serious nuisance' into account. As 'serious nuisance' is not truly defined and hard to interpret, the Dutch Council of State (i.e. a judicial court) often has the final word in deciding whether provinces or municipalities have correctly made their case. Finally, making things even more complicated, many agricultural installations are subject to their own set of policies and regulations that address odour caused by manure.

Energy policies are still fairly ill developed in the Netherlands. Effectively, municipalities only need to implement national EPC standards, while all houses need to have an 'Energy Label' to show their energy efficiency. Other policies are based on voluntary participation, where the semi-government Agendschap.nl usually takes the lead.

Finally, *soil policies* are possibly the most complicated of all when it comes to the allocation of responsibilities. When attempting to uncover the detailed allocation of responsibilities, a true maze of competences, regulations, exceptions and agencies comes forward. There is, first of all, the question of *who takes the lead* in coping with polluted soil. The soil needs to be checked before any new activities are employed on or with this soil (e.g. urban renewal, Greenfield developments, use of soil). Then, the competent authority (i.e. which issues permits for this new activity) is responsible for checking the soil: the state for national projects, provinces for provincial projects and municipalities for local projects. If pollution is discovered, it is up to the competent authority to assess whether clean up is required, based on the indicative limit values and future use. Second, there is *the question of funding*. There are national funds, funds that are issued through the Investment Budget for Urban Renewal (ISV) and legal structures that indicate the liability of the owner of the site and the polluter of the site (e.g. the former owner) to pay for the clean up. Establishing who is liable to which share of the costs is often very complicated and often requires legal procedures and the use of case law.

The former Ministry of Housing, Spatial Planning and the Environment recognizes the extreme fragmentation of Dutch soil policies (VROM 2007a) and considers it not feasible to continue working with them. As of 2007, the new 'Besluit Bodemkwaliteit' (Soil Quality Ordinance) addresses these issues and relieves municipalities of some former strict requirements. In addition, there is the 2009 'Convenant Bodemontwikkelingsbeleid en aanpak spoedlocaties', based on a partnership between state, provinces and municipalities addressing the clean-up of the most pressing cases of soil pollution. Despite these initiatives, the complexity of the

234 *Beyond the minimum in the Netherlands*

policy field remains overwhelming and legal procedures and case law persists. This is also illustrated by a quick scan of relevant policy reports, decrees, advice and regulations on national soil policies in mid-2008. This quick scan produced about 30 documents that explain various relevant aspects of soil policies. As this quick scan was not meant to be complete and was done swiftly, it is safe to assume that there will be several more relevant documents. Understandably, it is far from easy to keep track of all these relevant soil policy documents and it will be a serious challenge for local professionals to read them, connect them and remember the details.

When overlooking the allocation of responsibilities for the five policy themes addressed, it becomes evident that it is not easy to keep track of 'who needs to do what'. This was also confirmed during interviews conducted with over 20 local experts that were included in later phases of the research (see next sections). The interviews revealed that *nobody* felt confident enough to claim they actually had a clear oversight of Dutch environmental policies. Still, producing integrated and coordinated approaches at a local level does require a (fairly) clear oversight of Dutch environmental policies. The question of who is supposed to give this oversight remains unanswered. It is at least doubtful that municipal staff will be able to do so in the buzz of everyday work . . .

What it all means

Around the time of the 1990 first NMP, Dutch environmental policies were considered well-coordinated, coherent and subject to limited fragmentation. The worries that during the last two decades environmental policies have witnessed a reduced coherence and increased fragmentation are, however, clearly supported by my study of five key policy themes. Each of the different environmental policy themes studied is organized in its own way. The allocation of responsibilities, the status of legal limit values and the status of ambition values are all different for each of these five themes. Consequently, Dutch environmental policy is no longer recognizable as a single coherent policy field. Instead, it has become a set of different and rather isolated policy fields; i.e. soil policy, noise policy, air policy, etc. In the midst of this fragmentation, it is hard if not impossible to recognize and define what the national government means with a 'base quality'. It is not clearly defined in energy policies, context specific in odour and soil policies, subject to a sliding scale in noise policies and possible to deviate from in air policies. In addition, identifying what 'beyond the minimum' means is even more difficult. After all, if the minimum (i.e. base quality) is not defined or only so at a sliding scale, what does 'beyond the minimum' mean?

The current content and organization of Dutch environmental policies is making the development and delivery of integrated and proactive approaches to the local environment difficult. The number of different policies and regulations, and the differentiation in the allocation of responsibilities and the status of objectives and standards make it very hard to implement Dutch environmental policies. This puts a serious stress on their resources (time, funding and staff). Faced with these stresses, there is a risk that these municipalities are limited in the resources they

Beyond the minimum in the Netherlands   235

can spend on the development and delivery of their own proactive and integrated policies. In addition, the fragmentation of environmental policies also makes it difficult to choose for a single and straightforward strategy that will present all relevant environmental interests and stakes in local governance. Rather, for each theme municipalities rely on the different rules, regulations and allocation of responsibilities applicable in the various environmental policy fields (see also VROM 2007a).

Our main conclusion in this section is that the current organization of Dutch environmental policies does not provide the desired robust foundation of coordinative policies upon which policies can be based for promoting proactive, integrated and tailor-made approaches. Before I come back to this conclusion, I will in the next section (6.5) address the consequences that the changes to Dutch environmental policies have had on local performance in environmental management. In the subsequent section (6.6), where I will explain these consequences, I will come back to the impact of the organization of Dutch environmental policies on local performance.

6.5 Local performance

This section addresses the performance of Dutch municipalities in environmental policy in developing proactive and integrated approaches so as to deliver an environmental quality that goes 'beyond the minimum'. The results presented in this section were informed by three separate inquiries. The first inquiry was a desk study of research reports on the performance of Dutch municipalities in environmental management. They were based on research conducted by provincial and national institutions that are responsible for supervising municipal performance. The second inquiry was a series of interviews with more than 40 experts from various governmental and non-governmental organizations, each involved in supervision of municipal performance or advising municipalities. The third inquiry consisted of surveys in 28 Dutch municipalities. The surveys focused both on levels of municipal performance and on explaining the causes for success or failure that will be addressed in the next section. More details on these inquiries are presented in the text below (see also Spreeuwers et al. 2008).

Inquiry 1: research reports

The reports studied are based on nationwide studies conducted during the 2000s to report on the compliance of *all* Dutch municipalities with Dutch environmental legislation. These studies were conducted and presented in reports by the 'Interprovincial Consultative Body' (IPO) in conjunction with the Ministry of Housing, Spatial Planning and the Environment and the Dutch Association of Municipalities (IPO 2003, 2005), the VROM inspection (2003a, 2005a, 2005b, 2006a, 2007a, 2007b) and the ECWM (2001). The studies were focussed on municipalities' efforts and success to comply with minimum legal requirements (standards); i.e. municipal efforts and success in going 'beyond the minimum'

236 *Beyond the minimum in the Netherlands*

were of secondary importance or even left outside the scope of these studies. With the assumption that municipalities that have trouble to meet their legal obligations will also have problems in going beyond 'the minimum', the conclusions of these studies are, however, considered indicative.

The reports unite in the conclusion that compliance to Dutch environmental regulations is *not evident in a significant number of Dutch municipalities*. Although most municipalities do to a large degree comply with legal obligations, the IPO (2003) finds that "none of the competent authorities complies with all requirements, whilst about half the authorities has problems to meet even half these requirements" (Huberts & Verberk 2005; p. 54). Based on this conclusion, the IPO gave advice to professionalize municipal organizations by IPO. Municipalities subsequently improved their performance (IPO, VROM, VNG, UvW en V&W 2005). Nevertheless, a follow-up report revealed that 2/3 of all municipalities still did not meet all legal obligations (IPO, VROM, VNG, UvW en V&W 2005). In addition, as IPO noted, several municipalities merely improved their performance due to the fact that there would be a follow up. Many municipalities thus only reacted to the follow up and made improvements ad hoc. The IPO report also addressed local efforts for pursuing integrated environmental quality. Unsurprisingly, IPO concluded that there was a general lack of interest to do so (IPO, VROM, VNG, UvW en V&W 2005; pp. 15–16). Despite this conclusion, IPO did note that local authorities became increasingly aware of the relevance to pursue an integrated environmental quality in the 2002–2005 period.

The VROM inspection conducted similar research as IPO in the period 2003–2007. The VROM inspection was slightly more optimistic. They translated the outcomes of their research in a score for municipalities, which is either 'adequate' or 'inadequate'. These scores also take into account how not fulfilling legal requirements can be considered tolerable if municipalities face difficult circumstances such as high background levels. In 2006a (p. 29) they, however, conclude that "one out of each four or five municipalities scores inadequate on enforcing environmental legislation", while "one in six scores inadequate in assigning permits".

The relatively old ECWM (2001) report focussed on the use of environmental management plans in Dutch municipalities. Although these plans are voluntary, they are considered important tools by the Ministry of VROM for developing and delivering proactive and integrated approaches. Thematic environmental plans or policies were developed in up to 70% of all municipalities, but only about 30% uses *integrated* environmental plans or policies (ECWM 2001; p. 11). A more recent follow-up report by the VROM inspection (2006a) suggests not much has changed. Rather, the inspection states (p. 18) that most municipalities choose to "limit themselves to carry out national and provincial environmental policies and refrain from developing their own environmental ambitions". Proactive and integrated policies were still far from common practice in Dutch municipalities.

Inquiry 2: expert interviews

The second inquiry was based on a total of 41 interviews with experts from various Dutch professional environmental organizations. The interviews were mainly conducted by phone due to the travel distances, while several face-to-face interviews were conducted. The interviews were semi structured. They were mainly designed to address the development of proactive and integrated policies in municipalities and the degree of municipal success in delivering higher levels of performance than the minimum requirements. In addition, questions were also asked regarding the main causes for the degrees of success and failure encountered, which will be discussed later in this chapter (section 6.6).

The choice of experts was based on their overview of Dutch municipal environmental management. They were selected from three groups of experts involved in supervision of municipal performance and in advising municipalities. The first group consisted of 15 experts from the 'VROM inspection', which is the national organization responsible for monitoring the implementation of national legislation in municipalities. The second group consisted of 13 experts working for 'Municipal Health Departments' (GGD's), which have to be formally involved in all municipal plans and projects where public health is at stake. The third group were a total of 13 experts from the Dutch 'Provincial Environmental Federations' and the 'Society for Nature and Environment' (SNM). They advise on environmental issues for anyone who hires their expertise; i.e. they have no formal role in Dutch environmental policy. They are critical followers of economic and spatial developments with a keen eye on the related ecological and environmental aspects. The complete list of experts is presented by Spreeuwers et al. (2008).

Expert interviews: supervising agencies

Fifteen interviews were conducted with representatives from the VROM inspection. All these representatives work with a multitude of municipalities. The VROM inspection's central task is to control for compliance of municipal plans to environmental legislation and checking whether municipal policies are developed in line with national spatial policies. Hence, the VROM inspection only checks for compliance with the National Spatial Strategy ('Nota Ruimte') and not whether these policies are in line with the NMP4 or other environmental plans. Consequently, there is no structural supervision on local environmental policies (such as environmental management plans or other thematic plans and policies). All interviewees thus stated that they were not involved in assessing local efforts to 'go beyond the minimum'. Rather, "if a municipality complies with legal requirements, this is sufficient for the VROM Inspection". Similarly, they also do not check for the use of the MILO instrument which is a central tool in moving 'beyond the minimum'. Still, their wide overview of local work put them in a good position to assess municipal performance in pursuing higher ambitions in environmental policy, which was also confirmed by the discussions with them.

Four of the 15 interviewees express they are somewhat positive about municipal ambitions to 'go beyond the minimum'. Six others are very explicit in stating that this ambition is not at all a priority in most of the municipalities they deal with. Rather, as one respondent states "environmental health and hygiene is only considered a limiting condition to most municipalities". Two others suggest that municipalities see environmental health and hygiene as 'inconvenient' or

238 *Beyond the minimum in the Netherlands*

'annoying' while aiming for local (re)development. In addition, most respondents state that environmental health and hygiene is not a strong interest in comparison to more development-oriented interests or financial interests.

Eleven of the 13 respondents of the Municipal Health Departments note that, despite the legal obligation to be involved, municipalities often do not invite them to advise on plans and projects with significant health consequences, or only invite them to advise on "spatial plans that are already largely finalized". Hence, advice from the Municipal Health Departments often has little or no impact. Furthermore, the Municipal Health Departments find it difficult to take initiative themselves; their limited capacities (time and staff) mean that they have to turn to ad hoc advice on the most urgent issues. Despite these limitations, Municipal Health Departments have much experience with environmental issues where public health is at stake.

The answers given by the experts of the Municipal Health Departments are largely similar to those given by experts of the VROM inspection. Ten out of the 13 respondents stated that municipalities assign little urgency to public health and environmental hygiene. In addition, all respondents confirmed that "the large majority of municipalities is only focused on meeting minimum legal requirements". In fact, most respondents stated that many municipalities search for what is legally possible so as to go as far as possible with bending the rules. Second, four respondents expressed concerns regarding the fact that the municipalities they deal with do not take the cumulative effect of various environmental stressors into account. Fulfilling limit values on each individual theme is the main criterion, which means that it is possible that there are areas faced with multiple forms of environmental stress close to the limit value which is far from advisable from a public health perspective.

Expert interviews: critical followers

In addition to respondents from the VROM inspection and the Municipal Health Departments that have a formal role concerning municipal environmental policies, also respondents of two non-government organizations were interviewed. These are the Dutch 'Provincial Environmental Federations' and the 'Society for Nature and Environment' (SNM), which are independent agencies who function as advisory bodies on environmental and possible ecological matters in the Netherlands. They offer expertise, advice and training, while they are also critical followers of environmental policies and developments in the Netherlands. Municipalities are among the prime actors that both ask for advice and are critically followed by the 'Provincial Environmental Federations' and the 'Society for Nature and Environment'. Therefore, staff of both organizations is well informed when it comes to local performance in environmental management.

Thirteen respondents were interviewed that worked for the 'Provincial Environmental Federations' and the 'Society for Nature and Environment'. There was much consensus among the respondents. All 13 respondents argued that they experience that "by far most municipalities limit themselves to what is strictly necessary". Hence, most conclude that "municipalities are only to a very limited degree pursuing an environmental quality that is higher than the minimum base quality". Three respondents state that "for most municipalities environmental aspects are still of secondary importance" and "more and more municipalities are searching for the absolute minimum of what is legally possible". In being critical followers, it was essential to also establish whether these claims could be backed up by the respondents. Most respondents did provide backup for their claims, often by pointing towards the outcomes of the research into municipal performance I discussed in the previous sub-section. Some important nuances are made by the respondents though. Eight respondents indicate that there are serious differences between the ambitions set for different environmental themes. Many municipalities have chosen to pursue higher ambitions for one or two priority themes. In addition, various respondents noted that the complexity of local environmental problems and possible high background levels of environmental stress can also prevent municipalities from achieving high ambitions.

Inquiry 3: the survey

The third inquiry was based on surveys sent to at least two representatives dealing with strategic environmental policies in 28 Dutch municipalities. It was filled in by 42 municipal officers, since the two representatives filled in the survey together in various municipalities. Similar to the expert interviews, the surveys also had two main objectives that guided their contents (i.e. questions). The main motive was now to uncover the main causes for success or failure of municipal performance. In the next section (6.6) I will address the survey outcomes in relation to the main causes for success and failure. Here, I limit myself to using the survey outcomes in relation to its second objective, which was to establish a more detailed understanding of how municipalities were responding to the national call to produce proactive and integrated policies and the related ambition to 'go beyond the minimum'. Hence, I now focus on 1) whether and how municipalities work with actively pursuing proactive and integrated approaches to the environment, and 2) whether they succeed in delivering higher levels of performance than the minimum requirements.

Importantly, the ambition to uncover the causes for success or failure of municipal performance did influence the selection of survey municipalities, which is not a random sample of municipalities. Instead, a choice was made for a large variation of municipalities. Criteria included differences in size for assessing the role of economies of scale and differences in geographical location to account for the impact of different degrees of background pollution on local environmental management. In addition, municipalities were selected based on variation in the degree of success and failure of meeting minimal requirements. Differences in performance were chosen to ensure that the survey could uncover factors that contribute to both success and failure. The research reports studied earlier were instrumental for this selection, as several of these reports contain explicit ranking of either all, or a selection of Dutch municipalities, measured by their performance

240   *Beyond the minimum in the Netherlands*

in environmental management (IPO 2003, 2005, SNM 2002, 2005, VROM inspection 2003a, 2005a, 2005b, 2006a, 2007a, 2007b) Based on these rankings, municipalities have been classified as 'frontrunners', (9) 'moderate performers' (9) and 'laggards' (10) in environmental management. More details on the contents of the survey, the list of interviewed local experts and the choice of municipalities are provided by Spreeuwers et al. (2008) and in English by Zuidema (2011).

The survey: levels of performance

The first part of the survey addressed the question of how environmental interests were introduced and integrated in local planning and policymaking. Not surprisingly, 32 respondents noted that legal requirements were the main drivers for introducing environmental interest. Still, 34 of the 42 respondents indicated that strategic plans were used, whilst 22 respondents indicated they used specifically designed instruments that helped them introduce environmental interests in spatial planning processes. These instruments were both national tools such as MILO, but also locally designed tools. In addition, 26 respondents indicated that frequent meetings with senior staff of various departments were common in their municipalities. These help with the introduction of environmental interests into local planning and policymaking. Another 24 respondents mentioned the development of informal working to help the introduction. Only two respondents (from one municipality) indicated that environmental interests were introduced purely on an ad hoc basis.

The second part of the survey addressed the question of whether integrated and proactive ambition levels were aimed for, and to what degree these ambitions were delivered. On the one hand, municipalities were asked how they coped with environmental themes that suffer from a lack of national regulations (e.g. energy, public green). Respondents from only four of the 28 municipalities indicated their municipality had no policies at all for addressing these themes, whilst even these four indicated that individual staff members did address such themes in debate. Despite the high number of municipalities that actively address these poorly regulated themes, respondents in only six of the 28 municipalities were optimistic about the delivery of results. Of the other 22 municipalities, respondents in nine noted that a degree of success did occur but was confined to specific projects or isolated themes. In 10 municipalities, respondents stated that success was rare. In three municipalities there were no clear answers.

On the other hand, questions were asked regarding the ambition to 'go beyond the minimum'. The difference between so-called 'streefwaarden' (ambition values) and 'grenswaarden' (limit values) is an indication. Both are part of Dutch environmental policy and express the difference between minimum requirements and a higher quality. Formally, municipalities are even obliged to aim for levels higher than the limit values, although there are no real legal incentives that push municipalities to do so. Despite the status of the difference between limit and ambition values, four of the 42 respondents noted that they did not know this difference. Others do know the difference, but only five of the respondents confirmed that ambition values are actual targets in their municipalities.

Third then, the survey showed that 'going beyond the minimum' is an ambition in 11 of the 28 municipalities, while in seven other municipalities ambitions values were used as an 'inspiration' to 'possibly do more than fulfilling minimum requirements'. To respondents of 11 other municipalities, there was no intent to do more than meet limit values; i.e. the minimum was considered good enough. As they survey suggests, a minority of the municipalities is structurally involved in aiming for a quality which 'goes beyond the minimum', while some others do so on a more ad hoc basis. In the meantime, actually fulfilling ambitions proves far from evident. Only about half of the respondents note that ambitions that are set will be implemented. In the meantime, many strategic plans and policies focus on how minimum requirements can best be met and dealt with, instead of focusing on higher ambitions: the use of instruments and plans for improving the integration of environmental interests into alternative policy domains, as just discussed, are largely focused on fulfilling minimum requirements not higher ambitions.

Finally, the group of surveyed and studied municipalities was deliberately selected to have a wide variety of municipalities in geographic location and population size, whilst also to represent more and less successful municipalities. The results from the survey also showed some differences between municipalities classified as 'frontrunners', 'moderate performers' and 'laggards'. Frontrunners were indeed found to be relatively advanced in their environmental management and were reasonably ambitious. But even within this group it was found that the delivery of higher ambitions remains problematic. In the meantime, differences between moderate performers and laggards could not be established. All municipalities assigned to these two groups have limited or no environmental ambitions apart from legal minima, while success is even more modest. To conclude, success in the *development* of proactive and integrated policies for achieving a higher local environmental quality is largely confined to the group of frontrunners. Most municipalities do not aim for much more than minimum requirements, except for some isolated themes. In addition, the *delivery* of a higher local environmental quality is much more limited, even in frontrunner municipalities. The results, although not drawn from a representative sample, indicate that Dutch municipalities consist of a group of frontrunners with a degree of success, and a big group consisting of 'the rest' without much success.

To conclude, the surveys confirm the results of the reports and expert interviews; i.e. going beyond the minimum is far from evident in Dutch municipalities. In interpreting the survey results, it is furthermore important to take into account that the survey sample is not representative. It was designed to generate a wide variety of municipalities in geographic location and population size, whilst also to represent more and less successful municipalities. While this is beneficial for establishing the main factors contributing to local success or failure to pursue higher ambitions in environmental policy, it does potentially cause a bias when interpreting the share of (un)successful municipalities. The mix chosen for the

242 *Beyond the minimum in the Netherlands*

surveys contained a quite higher proportion of municipalities that were classified as good performing (frontrunners) by the nationwide performance studies (see Spreeuwers et al. 2008). The set is thus biased towards on average better-performing municipalities. Despite this bias, the survey results still confirm that delivering an environmental quality that 'goes beyond the minimum' occurs in only a minority of Dutch municipalities.

First conclusions

All three inquiries point towards the conclusion that most municipalities do not structurally pursue an environmental quality that exceeds the legal minimum. Those municipalities that do, often do so to a limited degree or on few specific environmental themes. Consequently, legal standards are the main or sole reason for locally defending environmental interest in the majority of municipalities. Apart from legal standards, environmental ambitions and health issues are no criteria in spatial planning for the majority of municipalities, even if municipalities are aware of serious environmental stress (i.e. close to limit values) or of cumulative environmental stressors that are together producing high levels of stress. In addition to these findings, municipalities are only to a limited degree involved in the development of *integrated* environmental policies, although many did develop thematic environmental policies. The results indicate that a limited number of Dutch municipalities was involved in developing and delivering proactive and integrated approaches and of an environmental quality that goes 'beyond the minimum'. They also indicate that many municipalities face serious problems to even comply with minimum legal requirements. Furthermore, as the surveys show, those municipalities that do actively pursue an environmental quality that goes 'beyond the minimum' note that success often remains modest.

6.6 Explaining local performance

In this section I present the main factors contributing to success and failure to go 'beyond the minimum'. Important indications and suggestions of factors that contribute to local success and failure to go 'beyond the minimum' were already found in the reports analyzed, while the expert interviews further provided input for identifying these factors. The survey, as was discussed in the previous section (6.5), also allowed respondents to explain and report the main causes of success and failure they recognized in their own municipalities (see also Spreeuwers et al. 2008). To elaborate on these findings, eight in-depth case studies were conducted (Table 6.1). Central to these case studies was addressing the main causes for success and failure in more detail. Again, a large variation in the sample – drawn from the survey sample – was favoured above a representative sample so as to address factors that contribute to both success and failure. Differences in size, the level of success attributed to the environmental management of these municipalities and their geographical distribution were used to inspire a sample rich in variation (Table 6.1). The case studies were based on a desk study of relevant

Table 6.1 Selected case municipalities

Type ↓ Size →	Large (>100K)	Average (25–100K) inh.	Small (<25K)
Frontrunners	Maastricht	Rheden	*
Moderate performers	Zoetermeer	Meppel	Harlingen
Laggards	A'dam-Bos and Lommer	Westland Geldermalsen	*

local environmental policy documents, face-to-face interviews with local experts (a minimum of two in every case) and, where necessary, additional interviews with relevant stakeholders. More details on the selection of cases, interviewed persons, means of data collection and an extended presentation of the results can be found in Spreeuwers et al. (2008).

In this section I partly draw from the reports analysed and the expert interviews, while the survey and case studies are the main basis for my findings. To present my results in this section, I have structured the results into eight categories of explanatory factors that I identified based on the analysis of the data generated. Five of these factors focus on problems related to the *local ability* to pursue proactive and integrated approaches. They are 1) the organization of national policies, 2) local resources, 3) local competences, 4) economies of scale and 5) the local sphere of influence over environmental issues. The last three categories address *local willingness* and focus on 6) the weak profile of the environment, 7) national support and 8) *national willingness*.

1 Organization of national policies

The complexity and size of the existing set of environmental laws, regulations and policies makes implementing them difficult. The quick scan of five policy fields in section 6.4 revealed that each of the different environmental policy themes studied is organized in its own way, both when it comes to the allocation of responsibilities and the status of legal limit values and of ambition values. Consequently, it is hard to combine them and gain a clear picture of the distinction between a minimum 'base quality' and a quality that 'goes beyond the minimum'. In the meantime, each of these policy fields is itself also made up of a multitude of rules, regulations and policies. Hence, implementing sectoral regulations often requires up to date and advanced technical or judicial knowledge of these regulations. In response, the national government continuously issues policy manuals to explain how all these policies should be implemented. Nevertheless, none of the experts and municipal officers asked during my research claimed able to oversee Dutch environmental policies. And this does not come without consequences (see also VROM 2007a).

The cases and survey revealed that many professionals no longer recognize a single environmental policy field. In fact, all municipalities were to a degree critical of the current organization of national policies that, as the respondents argued, were fragmented, complicated, changing rapidly, rather theoretical and hard to oversee. First, respondents point towards the *fragmentation of environmental policies* in many sectoral and thematic policies on, for example, air, water, soil, noise, etc. This fragmentation undermines a coordinated implementation of environmental policies and, related to that, the development and delivery of more integrated approaches. To Maastricht, fragmentation meant that when locally aiming for integrated projects "you enter the process working integrated and are eventually forced to go out of the process working sectorally". The fragmentation of environmental policies was also considered to undermine the introduction of environmental interest in planning processes. Respondents in Rheden stated that this fragmentation forces the introduction of environmental quality into overall planning projects as a set of separate interests and targets. Doing so undermines the possibility to defend and introduce 'the environment' as an overall objective. In addition, the need to focus on isolated interests and targets, which are expressed as legal requirements, contributes to an inward looking attitude of environmental specialists.

Second, the kind of *technical expertise* needed for working with issues such as air pollution models, safety risk calculations or soil remediation was found to be quite challenging. To illustrate this, two respondents noted that expertise on dealing with safety risks was so specific that they considered only four or five people in the whole country able to deal with them. Due to the high technical complexity of policies, municipalities often need highly skilled experts. As this is often too demanding, many municipalities have to improvise or attract external expertise that is financially challenging. Complexity is further aggravated by what municipalities consider a bias towards theoretical regulations. Respondents in Maastricht, for example, stated that they "are simply unable to translate these policies to people around us". Examples of this theoretical nature are given by Spreeuwers et al. (2008) and in English by Zuidema (2011).

Third then, many respondents considered policies and regulations to *change too often* and *too suddenly*. Although it was widely acknowledged that changes are natural, the respondents were critical of the national government. As a respondent in Zoetermeer noted, "it can take a very long time before an alteration of a law actually passes, and then from one day to the following, it is there". This makes it very hard to swiftly respond. Rather, as respondents in Maastricht, Geldermalsen, Harlingen and Meppel note, municipal employees are often not ready to respond to these changes. They have no experience with the new policies and can hardly prepare for these policies as it is unclear when and which changes will become finalized. Closely related are the many changes municipalities see in models for the calculation of, for example, air quality or noise levels. These changes are not just causing additional workloads as recalculations have to be made; recalculations can also result in different outcomes during the course of projects and hence, serious changes to these plans.

Fourth, various respondents during the expert interviews, survey and cases stated that environmental regulations have now acquired a '*bad reputation*'. While environmental regulations are already often seen as 'annoying hindrances' to developments by market parties and development-oriented municipal departments, this is aggravated by the complexity and changing nature of these regulations. As a respondent noted, environmental officers have to deal with "the bad name which environmental regulations have, the large risks of failure in the face of changing regulations, and unclear procedures that are poorly coordinated with spatial planning procedures". To another respondent, the bad reputation also followed "a lack of flexibility and dogmatism" of environmental regulations.

Finally, complaints were also made about *conflicts between regulations and policies*. In addition to some minor complaints about conflicts between environmental regulations, the most pressing problems followed the housing quota assigned to Dutch municipalities. The national government issues, sometimes with the help of provinces, quota for the number of houses that ought to be built in municipalities. These quotas can be quite challenging, especially in densely populated areas (e.g. Amsterdam, Geldermalsen, Westland and Zoetermeer). Faced with high quotas and little space, municipalities need to build in high densities, while high land prices force them to build against low costs to keep houses affordable. Using space for buffers against noise, safety risks or poor air quality are therefore financially challenging, if not considered impossible. This was most evident in Geldermalsen, while respondents in Bos and Lommer noted the need for houses in the Amsterdam region forced them to build in an area that the respondents themselves found unfit to live in due to the rather pressing environmental issues (see Spreeuwers et al. 2008).

In the face of the complexities of Dutch national environmental policies, municipalities find it difficult to coordinate implementation, let alone combine separate regulations into an integrated and coherent strategy. According to the VROM inspection (2007a), the lack of coordination and occasional contradictions between regulations is one of the causes of severe local coordination problems. To the VROM inspection, these coordination problems are also seen as an important cause of municipalities' lack of compliance to legal requirements. Municipalities are no longer able to routinely implement minimum requirements, but rather, need to improvise and invest heavily in keeping up and dealing with these requirements. As a respondent in Meppel concluded, even after reading newsletters, advisories and picking up the phone to talk to colleagues in other municipalities, "you sometimes are simply not sure whether you are actually doing things in the proper way". In the meantime, decentralization and deregulation result in even more work. As two respondents in Geldermalsen noted, "we do not have the experience that deregulation actually produces less work" and "municipalities are now handed over additional tasks, with a minimum of additional budget".

2 Local resources

In the face of the complexities of Dutch environmental policies, it is hardly surprisingly that a lack of local resources was a returning item in almost all phases of the research. The research reports from the IPO (2003), IPO, VROM, VNG,

UvW en V&W (2005) and the VROM inspection (2006a, p. 88) explicitly refer to a shortage of people and resources. One of the conclusions of the IPO study (IPO, VROM, VNG, UvW en V&W 2005) is, for example, that municipalities often have problems to either employ or hire sufficient staff. As a consequence, the IPO found, many municipalities do not meet criteria such as 'sufficient expertise', 'financial means to hire specialists' and 'systems for controlling on responsibilities' (IPO, VROM, VNG, UvW en V&W 2005). In a somewhat earlier study, the ECWM states that "given the current efforts and results, municipalities are overall having about a 20–30% shortage of capacities" (ECWM 2001; p. 89). Hence, the VROM inspection (2007a, see also Huberts et al. 2005) notes that a relative ignorance of municipal staff of legal requirements in combination with complex and fast changing legislation is the main cause for municipal failure to meet legal requirements.

The conclusions drawn in the research reports studied were confirmed during the expert interviews. In all four groups of experts, problems with limited time, staff, resources and expertise were frequently mentioned. Surprisingly, lack of time was raised by just two municipalities as a problem during the surveys. During the cases, however, it was repeatedly addressed (six out of the eight cases). As the eight cases were all part of the survey as well, this at least suggests that more survey municipalities face problems with time and manpower. A possible response to a limited number of staff members and a lack of expertise is for municipalities to hire external experts. However, this is financially challenging, even for short-term projects. Given their limited financial means, municipalities often simply have to improvise.

As municipalities find it difficult to implement even minimum requirements, proactive and integrated approaches are not always prioritized. Respondents in Rheden, for example, noted that there was simply not enough time for developing proactive and integrated policies. Keeping pace with permits, licences and pressing issues already strains the staff. The same goes for Geldermalsen, where catching up with legal responsibilities is already seriously stressing environmental staff. Hence, Geldermalsen explicitly states that "ambitions and targets that go beyond the minimum legal requirements will only be pursued when, in addition to the resources and capacities invested in implementing legal regulations, any capacity remains" (Geldermalsen 2006; p. 5). In Zoetermeer respondents noted that their role in spatial planning processes had to remain limited. While spatial planners and urban designers do the bulk of the work, the ideal picture would be that environmental experts would stay close and would advise when needed. However, given their limited time, "in such a process it is simply impossible to always look around the corner". Respondents in Meppel confirmed this and noted that the lack of time to be continuously involved in discussions leads to missed opportunities and overall, a lower environmental quality. Therefore, as a respondent noted, "the thought that municipalities should produce so-called 'tailor-made' outcomes is very nice and all, but it remains the question whether municipalities are actually up to it". In asking this respondent about his municipality the reaction was simple: "we are not ready yet".

3 Quality and competences

A third key explanatory factor relates to the qualitative capacities of municipal organizations and their staff. This not only follows the difficulties to attract sufficient technical experts. Realizing environmental ambitions and positioning environmental interests within area-specific approaches also requires managerial skills which are not evident in most municipalities. Various (eight) survey respondents argued for improved working relations between especially spatial planning and environmental departments in pursuing proactive and integrated policies. However, a lack of competences such as visionary thinking, communicative skills and strategic planning of environmental officers was noted, just as a general absence of a 'pro-environmental' and 'integral' attitude.

The first main problem encountered with cross-sectoral working is the *late introduction* of environmental issues and ambitions into more integrated spatial planning processes (e.g. raised as key factor by 15 of the 42 respondents and in six of the eight case study municipalities). Instead of considering environmental aspects as quality criteria before a plan is made, these aspects are used to do a final check whether the plans fits in with legal requirements after most of the planning has been done. For many respondents, such a late check prevents environmental principles and ambitions to inspire spatial choices to deliver a higher environmental quality. As a respondent in Maastricht explained, "after a while, the puzzle is simply complete and only then can you add the piece with the environment".

Second, there are large *cultural differences* between more technical environmental specialists and more strategic-oriented spatial planning generalists. On the one hand, spatial planners are not always open to the influence of environmental specialists. Rather, as an interviewed expert noted, spatial planners "are very ambitious, which means that they sometimes simply forget certain environmental aspects, causing problems later in the planning process". Indeed, various respondents referred to the dominance of urban planners and designers in the development of spatial projects (e.g. cases Meppel, Zoetermeer). Communication between environmental staff and planners or designers is then often overlooked. On the other hand, environmental specialists also have difficulties presenting themselves in planning processes (e.g. Maastricht, Rheden and Zoetermeer). Environmental officers and policies are highly dependent on 'rules', 'data' and on 'what can *not* be done'. Hence, environmental departments are used to work with restrictive and often technical regulations. The bulk of the current staff of environmental departments therefore consists of technical specialists who are not trained or used to participate in more strategic, visionary and integrated thinking as is common for spatial planners and urban designers.

To improve cross-sectoral working, many respondents argue for protocols to ensure regular consultation takes place and for improving the communication between spatial planning and environmental departments. Many municipalities have already established working protocols, sometimes more informally (in smaller municipalities) or as formal procedures (larger municipalities). Many of these protocols still focus largely on fulfilling minimum legal requirements.

248 *Beyond the minimum in the Netherlands*

Exceptions were found though, such as the use of the so-called 'MIRUP' tool in Zoetermeer and Westland; i.e. a process management tool to incorporate environmental aspects in spatial planning processes (see De Roo et al. 2012).

To improve communication, respondents often refer to enthusiastic and inspirational environmental officers (noted during the expert interviews, five survey respondents and by respondents from four of the municipalities). These officers can bridge differences and generate a more proactive and ambitious municipal 'culture' towards environmental quality. They can also generate support from politicians, the public and the private sector. Various municipalities have recognized the relevance of attracting environmental officers that have adequate managerial skills for improving communication with spatial planners especially. Hence, these municipalities have split their environmental department into a division of specialists working on permits and control and a division of generalists working on more strategic and integrated policies. Maastricht and Zoetermeer already worked with such a division for a long time, whilst Meppel, Rheden and Westland have more recently organized their department as such. The idea is that strategic staff members can function as a bridge between environmental specialists on one side and spatial planners and urban designers on the other. In Maastricht, the idea of creating a bridge has become the job description of some of its staff members, whom they call 'environmental coordinators'. Consequently, the environmental department in Maastricht is now often involved immediately in spatial planning processes. The early introduction of environmental expertise helps harvesting opportunities that produce a higher environmental quality, which, as Maastricht concludes, need not imply additional financial costs. Success mostly follows creative solutions.

4 Economies of scale

In the face of limited resources, qualities and competences, various respondents pointed at the economies of scale involved in dealing with environmental policies. To begin with, the difference between smaller and larger municipalities is raised in each of the research reports studied (ECWM 2001, IPO 2003, 2005, VI 2006a). The VROM inspection, for example, notes that financial and personnel capacities mean that larger municipalities are "more able to pursue changes and improvements" (2006a; p. 32). In addition, the ECWM recognizes that small municipalities face difficulties due to "the organisational and financial implications of coping with environmental issues and the relation with other policies" (2001; p. 11). Still, respondents in the fairly large municipality of Maastricht indicated that even there it can be quite problematic to fill vacancies for specialists.

The difference in size of municipalities is also a prominent issue for many experts. Large municipalities are, for example, found more able to navigate bureaucracy in searching for subsidies and funds. Such funds are not just important for ensuring the implementation of higher environmental ambitions, but can also allow municipalities to pursue innovative example projects. In addition, large municipalities were considered in a better position to hire expert staff, acquire subsidies and develop more integrated environmental policies. Smaller municipalities on the other hand, often rely on one or two environmental employees that cover the entire width of environmental management.

To cope with their problems to attract specialists, many municipalities in the Netherlands have joined forces and created *regional environmental agencies,* which are financed from the municipal budgets of the participating municipalities. These agencies can create the desired economies of scale and hence, employ specialists on, for example, safety risks, air quality and noise nuisance that can do the work for the various municipalities covered by the agencies. Experiences in Harlingen en Geldermalsen with such agencies were considered positive, whilst Rheden and Westland were also using expertise from similar agencies. Warnings and doubts on the use and misuse of environmental agencies were also given. Regional cooperation proved especially suited for attracting technical expertise and improving the implementation of legal responsibilities. When it comes to strategic environmental tasks, municipalities want to retain their responsibilities.

5 Local sphere of influence

The research shows that municipalities face several environmental challenges that they can hardly influence. First, much environmental stress in the Netherlands is caused by sources that are largely outside of the sphere of influence of individual municipalities. On the one hand, this is caused by jurisdictional fragmentation in the face of issues with *external effects.* In a country as urbanized and densely populated as the Netherlands, there are often high background levels of noise and air pollution. Such issues cannot be solved by a single municipality, and sometimes not even by the national government. Maastricht, for example, faces serious air pollution issues that are influenced by the nearby Belgian city of Liège. On the other hand, many motorways, railways or factories are subject to state or provincial regulations. Good examples were found in Amsterdam – Bos and Lommer where the A10 motorway, the Amsterdam ring railway and two other major transport routes were all regulated by other agencies than the city borough. Again, municipalities have little or no influence in coping with the environmental stress caused by these sources.

Second, the municipal sphere of influence is constrained as several environmental issues are considered *too complicated* to be dealt with within reasonable financial and temporal frames. The VROM inspection noted that it is sometimes impossible to reduce environmental stress below legal thresholds without making extreme and radical sacrifices (VROM inspection 2006a). The expert interviews confirmed this conclusion. Even the most critical group of experts interviewed (of the Provincial Environmental Federations and the Society for Nature and Environment) expressed that several municipalities are simply unable to do more than focus on minimum requirements, while even these requirements are sometimes too big a challenge. High background levels of stress are among the causes of the complicatedness that municipalities face, but highly mixed urban areas can also cause severe challenges. After all, it is no exception that removing factories,

250 *Beyond the minimum in the Netherlands*

closing roads or demolishing houses are the only real solutions to meet standards. Such solutions are usually considered socially unacceptable, which means that a more flexible and longer term strategy is needed. In the face of such complex problems, efforts to 'go beyond the minimum' are barely if at all realistic, at least when it comes to certain environmental stressors. To illustrate this, in Bos and Lommer and Maastricht, air quality standards are highly challenging, whilst noise issues in, for example, Rheden and again Bos and Lommer can hardly be prevented.

Third, municipalities are also limited in their sphere of influence due to the *dispersal of power* in local governance. Societal actors such as landowners, project developers and housing corporations can play key roles in the development and delivery of environmental plans and policies. Project developers and housing corporations often need to co-invest in sustainable construction, energy savings or, for example, sound remediation measures. This forces municipalities to enter into processes of bargaining, negotiations and the development of partnerships and contracts. For many municipalities, this is still fairly new and challenging. Although spatial planning departments often have experience, it is especially the traditionally technical environmental specialists that are not used or competent to enter into bargaining and negotiations. As we have seen, some larger municipalities nowadays work with more strategic staff members (e.g. 'environmental coordinators'). These are not just crucial for cross-sectoral working within municipal administrations. They can also be important people for crossing the public–private divide. Such officers, however, are not common in most municipalities, while especially smaller municipalities can also hardly be assumed able to hire them.

Even if municipalities have the right competences and resources to enter into bargaining and negotiations with private parties, they are hindered by the 'weak profile' of the environment in combination with their financial resources. On the one hand, as a survey respondent noted, there is often a general "lack of financial resources to realise ambitions and targets". A total of eight survey respondents confirm this notion. Similarly, half of the case study municipalities note that financial constraints undermine reaching additional environmental ambitions. The costs of, for example, creating buffer zones for noise or safety risks (e.g. Geldermalsen, Meppel), sustainable buildings (e.g. Geldermalsen) and the actual implementation of mostly sustainable ambitions (e.g. Meppel, Zoetermeer) can be problematic in the face of financial constraints.

On the other hand, municipalities also face serious problems if they have limited or no landownership. Geldermalsen, for example, noted that it only owned about 10% of the land for the new 'Garstkampen' project. Consequently, Geldermalsen relies on bargaining to persuade (market) parties to agree to more ambitious policies such as energy savings and sustainable construction. Respondents in Geldermalsen noted that such bargaining was not easy, as market parties were found reluctant to invest in ambitious environmental measures (see also Westland). The opinion that market parties are not interested in more ambitious environmental measures was not confirmed by all. Respondents in Zoetermeer noted that market parties do recognize the potential added value of more ambitious environmental policies. In addition, respondents in Maastricht explained, many pro-environmental measures need not have a negative cost–benefit ratio, even in the short term.

6 A 'weak profile'

The research also showed that the 'weak profile' of environmental interests contributed to the lack of a 'pro-environmental' attitude and problems to pursue integrated and cross-sectoral policies. First of all, a large majority of experts stated that "the environment is not an important issue in most municipalities". Rather, as one respondent illustrated "in many municipalities the environment is seen as a limiting condition to desired developments". Consequently, as two other experts noted, environmental objectives are often seen as 'awkward' or 'annoying'. Second, indications of limited political and administrative will in municipalities were also found during the surveys and cases.

The survey was most explicit in revealing this 'weak profile'. To the Dutch national government, environmental quality should be one of the underlying criteria in working towards an 'optimal living quality'. Only 10 of the 42 respondents, however, considered environmental values to have a 'strong' or 'relatively strong' position in comparison to policies related to mobility, spatial development, social wellbeing and development, economy, etc. (Spreeuwers et al. 2008). To 16 others, the environment was considered as just one of the many interests that needed to be balanced to such policies with no particular added importance. In addition, 11 respondents considered the environment to have a 'weak' position, leaving five respondents without an answer. Interestingly, various respondents (six in total) explicitly noted that the position of the environment depends largely on legal requirements; i.e. strong with standards, weak without. Apart from some frontrunners, most municipalities therefore focus on fulfilling minimum legal requirements (see also section 6.4). Environmental interests are rarely underlying criteria to inspire plans and projects, apart from legal requirements. Rather, financial or spatial design principles are respected as leading criteria for plans and policies. During the survey, a further five respondents noted a lack political and administrative support, while 12 noted a limited sense of urgency and a limited awareness of environmental issues that mostly undermined support for environmental ambitions other than within the legal framework.

During the case studies, good examples of the relative weak position of the environment were also found. In Bos and Lommer, pressing socio-economic problems and seriously outdated residential buildings dominated the policy agenda. A similar story was told in Rheden, where ambitious policies for building a train tunnel in a town centre were triggered not by environmental issues, but by issues such as mobility and the current physical barrier between two town parts. Hence, as a respondent in Rheden concluded "environmental criteria are not decisive for the spatial layout of the area". Finally, respondents in Harlingen noted that "Harlingen is a city of workers and a lot of unemployment, so the 'soft side' [non regulatory] of the environment does not really count".

252 Beyond the minimum in the Netherlands

According to the respondents, a lack of willingness can be found in both the political arena and the administrative organization. To some, the role of politicians is considered most important. After all, as respondents in Rheden noted, it is within the *political arena* were budgetary choices are made. Such choices are essential in either expanding the environmental staff to draft strategic and integrated policies or in paying for ambitious environmental programmes. Similarly, respondents in Zoetermeer noted that it was hard to achieve any ambitions that go beyond minimum requirements without political commitment. This was especially hard on themes such as sustainability, as benefits are neither typically local, nor short term. On the flipside, if the municipal council is ambitious, this is also reflected in municipal policies (e.g. Maastricht, Meppel, Rheden and Zoetermeer) and partly in delivery (mostly Maastricht and Zoetermeer).

Administrative enthusiasm and, more generally, *a pro-environmental municipal culture* were also seen as important triggers for good performance in environmental management (e.g. mentioned by six survey respondents and during most cases). In Maastricht respondents noted that in their city the municipal culture was 'to go for quality'. This culture is also reflected in their environmental policies, expressed in much tolerance for additional or innovative measures. As a respondent put it, "you have the opportunity to do things". Two other case municipalities also indicated that the administrative organization and a 'pro-environmental' and 'integral' attitude contribute to success. However, such an attitude was not found to be common in most municipalities. While this was aggravated by the weak profile of environmental interests and the bad reputation of environmental regulations, it is also a general lack of knowledge that *minimum* requirements are meant to safeguard against excessive environmental pollution and stress, and not equal to a high environmental quality. In Geldermalsen even environmental officers believed that "if you as a municipality meet the legal limit values, this also means that you deliver a good public health situation". In fact, only the respondents in Maastricht were convincing in showing a 'pro-environmental' and 'integral' attitude, which leaves the other seven municipalities without such a culture.

Finally, the weak profile of the environment also came forward in response to the delivery of environmental ambitions. In the face of budgetary challenges, environmental ambitions are often the first to be dropped. For example, high ambitions for noise were not met in the development of the 'Oosterheem' neighbourhood in Zoetermeer. During the urban design of 'Oosterheem', reducing the number of houses in order to create buffer zones for traffic noise was not considered acceptable despite the high noise ambitions. Instead, this was considered 'a waste of space'. Another good example of a problematic implementation came from the town of Meppel (Spreeuwers et al. 2008). Hence, it was repeatedly raised during the survey that "it is often a mistake to think that finishing a plan is enough to deliver results". In response, the operationalization, delivery and monitoring of ambitions were all mentioned by various respondents as critical factors in ensuring implementation would take place in environmental management. Finally, four respondents noted that targets needed to be clear to deliver them.

7 *National support*

Faced with limits to local willingness and abilities to pursue proactive and integrated policies, or to 'go beyond the minimum', national support is crucial. The Dutch national government, however, uses only policy recommendations and minor financial support as stimuli. There are no 'checks and balances' such as the need to account for local results, legal or contractual obligations or financial sanctions, which urge municipalities to do so. Those 'checks and balances' that are in place have to do with fulfilling minimum legal requirements (e.g. air quality, safety risks) or with deciding limit values locally (e.g. noise, odour). The result is that minimum legal requirements continue to dominate municipal environmental management. After all, as a respondent in Rheden stated, "it is these criteria that the Ministry of VROM uses to judge you".

Therefore, as a respondent of the VROM inspection noted, municipalities are in need of "a trigger to aim for a higher environmental quality". Four other VROM inspection respondents suggested that the national government should "give more attention to informing local authorities about the national government's desire that municipalities aim for such a higher environmental quality". In addition, respondents of the Municipal Health Departments suggested that more could be done by the national government. To one of the respondents, the key was to ensure additional funding would become available to do more at a local level. Several others noted that more national pressure to ensure that the legal requirement for municipalities to involve the Municipal Health Departments was needed.

Instead of 'checks and balances', the national government relies on persuasive tools as presented to municipalities through policy recommendations and some voluntary financial provisions to stimulate municipalities to develop and deliver proactive and integrated policies and to 'go beyond the minimum'. The national government funds some example projects to fuel the development of knowledge and gain information about 'best practices'. These funds are, however, used rather ad hoc, in isolated projects and are limited in the amount of municipalities that can use them. The most important stimuli the national government uses therefore relies on knowledge dissemination, policy advice and persuasive policies. Of these, the MILO [Milieu in de Leefomgeving; 'Environment in the Living Environment', VROM 2004b] approach stands out. It was initially intended to be based on a coalition of the national government, the provinces and the municipalities, but is now only a voluntary instrument. Hence it has become only one of the many policy advisories, papers and folders that are disseminated or are accessible on websites.

The tools the national government uses to stimulate and persuade municipalities to 'go beyond the minimum' make little impact in a local realm. During the survey, questions were asked regarding the familiarity of respondents with some key concepts and instruments the national government uses. Extensively used in national policies and advice are the concepts of an 'optimal living quality' and 'quality of the living environment'. However, more than one third (16) of the 42

254   *Beyond the minimum in the Netherlands*

respondents indicated they did not know or recognize these concepts. In addition, these concepts proved hard to make operational by those who did know them. In asking respondents to explain what these concepts meant, a wide variety of answers was generated. For some respondents, environmental aspects were at the heart of these concepts, while others considered them not part of the concepts, apart from legal requirements. It shows that these concepts are not convincing in showing what 'going beyond the minimum' implies.

As MILO was designed as the main instrument to persuade and help municipalities to achieve the desired 'optimal living quality', questions regarding the familiarity with the approach were part of the survey. Surprisingly, only half of the respondents (21) were familiar with the MILO approach. Similarly, only nine of the 28 municipalities actually used any of the national instruments that meant to help them achieve the desired 'optimal living quality', including MILO. It shows that the national policies are not reaching their target population well, let alone that they have a strong impact in encouraging municipalities to 'go beyond the minimum'. The limited familiarity of the MILO approach is even more pressing when taking the differences between 'frontrunner' municipalities and 'moderate performers' and 'laggards' into account. To push local environmental management forward, it is mostly the large group of moderate performers and laggards that need to be reached as they currently mostly fail to 'go beyond the minimum'. Of the 16 respondents that did not know the concepts of an 'optimal living quality' and 'quality of the living environment', however, 14 came from non-frontrunner municipalities. Furthermore, of the 15 who did not know the MILO approach, 12 came from the 27 non-frontrunner respondents. If the Dutch national government hopes to push especially the bulk of (non-frontrunner) municipalities forward, these numbers are not promising.

8 National political climate

The disappointing results of the recent decentralization efforts should also not be isolated from the political climate in the Netherlands during the first one-and-a-half decades of this century. During this time governmental policies were dominated by a conservative and right-wing government that prioritized deregulation and seemed little interested in environmental policies. The fourth and last National Environmental Policy Plan was published in 2001. It was supposed to be drafted every four years, at least according to the law as set in the Environmental Management Act of 1993. Since 2001, the national government has also been dominated by a stronger neoliberal tradition. Where in the 1980s the rise of a neoliberal tradition translated itself in expanding environmental policies beyond government controlled interventions, the doubts during the 1990s meant that more conservative and right wing–oriented political parties began to increasingly confide in market-based instruments, decentralization, deregulation and privatization (see also De Roo 2004). With these parties dominating governments from 2001 onwards, environmental policy making is no longer a clearly visible part of the state's activities. Implementing existing regulations and new EU policies are what remain.

A key illustration of the current governmental attitude can be seen when reflecting on the problems with meeting the 'new' European Union air pollution standards on fine particles and nitrogen oxides during the mid-2000s. Although other countries also had difficulties, the Dutch situation was amongst the worst (e.g. Backes et al. 2005). Although caused partly by its geographical location and density, the response has been peculiar and highly illustrative of the current environmental discourse. After the air pollution standards came into force in 2004, the Dutch saw some serious court cases that literally stopped existing urban development and infrastructure projects. A very strict interpretation of the European Union standards meant that all projects that add air pollution in areas not yet meeting the required standards would not get a permit. It was a peculiar interpretation, different than what happened in most other European countries. The response was to swiftly ask for postponement of meeting standards from the European Union. This was granted, only under the condition that the Dutch would develop a national action plan for improving air pollution. The Dutch did so in 2009 with the National Collaboration Programme on Air pollution, NSL. The NSL was a collaborative effort of various national ministries, provinces and municipalities and required all the different levels of government to take their own measures. These measures would include changes in the modal shift, reduced speeds, the banning of cars in city centres, altered locations for roads and urban developments, etc. Interestingly, within two years, the national government withdrew one of its main contributions: the installation of road pricing. According to calculations, it would have been one of the main measures for the NSL to succeed. The state did not replace it with anything comparable (e.g. Busscher et al. 2014). Instead, it even chose to increase the maximum speed on highways from 120 km/h to 130 km/h, further exacerbating the air pollution problem. The NSL stopped running in 2015 but is planned to be continued for another two years, in part because the problems it was supposed to solve remain. That is, the national governments showed quite modest support for solving air pollution issues. Instead, the NSL seemed mostly meant to simply continue building and get the desired postponement; exactly what Busscher et al. (2014) also conclude.

In recent years some changes are happening. To begin with, the Dutch are developing a new integration on the physical environment ('omgevingswet') for all regulations related to the physical environment: housing, infrastructure, water, environment, nature, etc. It is an ambitious attempt that has been running since 2011. Initially, it seemed a promising attempt to at least reduce the current fragmentation of policies and to regain focus in where environmental policies should focus. As of summer 2015 the first results will also be subject to debate in the parliament. What is clear are at least four points. The first is the most promising. Regulations will be more streamlined and the national government clearly recognizes the limited ability of municipalities to cope with many of the tasks they face. As a result, they have installed so-called 'Regional Environmental Agencies' ('Regionale Milieudiensten') that operate as the administrative centres for implementing regulations. Instead of each municipality having to deal with these regulations, economies of scale are created by uniting work in these agencies to perform administrative tasks for smaller municipalities. The three other results are less promising. There was and is no substantive vision or policy underlying the new law. Rather, it is purely a rearranging of existing regulations. Second, the law does hint at vision building at a regional and local level, but does not urge these to be developed. Rather, regional and local governments are merely meant to implement existing regulations. The development of area-based strategies, it seems, is left fully up to local units. Finally, when it comes to these environmental issues, the plan is to simply follow the European Union. Clearly, the new integrated Act for the Physical Environment is not meant as a context for innovation in environmental policies on regional and local levels to take place.

The Ministry of Environment and Infrastructure has, however, also recently (I&M 2014) published a new attempt at setting a policy agenda. It is the first attempt since 2005 and, as that one failed, the possibility to develop real policies for the first time since 2001. Under the heading of 'Modernisation Environmental Policy', the government does express some new ambitions. Apart from improving and streamlining regulations and their implementation, two main elements are interesting. One: the national government puts some new issues on the agenda that require extra efforts, most notably related to water quality and air quality. In other words, it argues that new measures are needed. Second: the national government realizes it needs to form new coalitions with other societal partners, both in an international and national arena. It recognizes that its role in stimulating a more sustainable society needs a government that sets the right conditions for innovation and supports green economic activities. Clearly, then, there seem to be some first signs of a national government that returns to a focus on pushing for environmental ambitions and policies. It is, however, a very modest response consisting of a lengthy letter to parliament, with many ideas and little clarity and that, most of all, at this point still misses budget and a vision. Environmental policies thus remain marginalized with merely some hopeful signs, where area-based policies remain simply a possibility that only willing and able local authorities might pursue.

Next steps

Faced with the results of this section, I am now able to confront the approach as the Dutch have taken in pursuing proactive and integrated approaches to the environment with the arguments that help to inform an increase in the role of the local level in environmental policy as discussed in Chapter 4. I will do so in the following section (6.7). Based on that, section 6.8 contains the final conclusions of this chapter.

6.7 Bringing it together

In Chapter 4 I presented four arguments that can help inform attempts to increase the role of the local level in environmental policy, and also when it comes to the pursuit of proactive and integrated environmental policies. In this section I will return to all four arguments. On the one hand, my aim is to reflect on these arguments based on the data I collected on the approach taken by the Dutch national government to advance the role of the local level in environmental policy. In doing so, I hope to establish whether these indeed help to understand the consequences of the approach taken by the Dutch national government. On the other hand, I also hope to better explain the consequences of the approach taken by the Dutch national government to advance the role of the local level in environmental policy as discussed in this chapter. Hence, in this chapter I aim to explain the levels of performance encountered by discussing the main factors contributing to these levels of performance, including the national policies used to support and stimulate local willingness and ability.

Argument 1: complexity as a criterion

The first argument I discussed in Chapter 4 was that the degree of complexity of the issues faced can be a key argument in support of an increased role of the local level. Based on the experiences of the early 1990s, the Dutch national government also embraced the idea that complex local environmental issues require an area-based and tailor-made approach. The experiences in the early 1990s thus convinced the national government that common governance formats are limited in their capacity to cope with complex and unique local circumstances. In response, decentralization and deregulation were pursued to severely reduce the impact of sectoral regulations for various environmental themes. In addition, the City and Environment approach and the Investment Budget for Urban Renewal (ISV) further enabled local authorities to make their own decisions in environmental policies. The Dutch national government now sees itself responsible for protecting a so-called 'base quality' dictated by minimum requirements. Local authorities are subsequently asked to aim for higher levels of performance than these minimum requirements, based on specific and unique local circumstances. It is a change which is compatible with the claim made in Chapter 4 that the complexity of issues faced is an argument in support of a greater role of the local level.

Complexity has clearly been an inspiration for the alterations to Dutch environmental policies during the last decade and a half. The degree to which complexity is currently embedded as a criterion in the organization of environmental policies to deal differently with issues of different degrees of complexity is, however, less clear. In Chapter 4 I argued that common policy formats that can be routinely implemented are better suited for dealing with uniform issues that show limited degrees of complexity than leaving this to local authorities. In a Dutch context the previous strong reliance on coordinative policies and 'strict and ambitious' national targets has now been replaced with a strong reliance on local decisions regarding many local environmental issues. It certainly accommodates municipalities to cope with more complex issues based on local circumstances. Decentralization and deregulation have, however, also reduced reliance on common policy formats that can be routinely implemented, also when it comes to uniform issues of limited complexity. This has not been without problems.

258 *Beyond the minimum in the Netherlands*

It is true that common policy formats remain in place and are meant for dealing with rather uniform issues, for example, when coping with air quality and noise and odour issues through indicative limit values. Especially for noise and odour policies, municipalities are also invited to make their own policies within a context of checks and balances that allow for deviations from indicative limit values. These checks and balances remain fairly weak, however, leaving much room for municipalities to also choose for deviations and the development of area-specific approaches even if this is not strictly needed due to the complexity of the circumstances faced. Motives such as room for additional urban development or financial gain are often triggers, rather than intervening local circumstances that defy meeting indicative standards. In the meantime, common formats for soil remediation and energy policies hardly exist. Municipalities often have to develop their own approach for dealing with these issues, based on a multitude of national regulations and policies (soil policies) or with only limited national guidance (energy). Especially for soil remediation, the complexity of the issue is in itself often not the motive for a shift to specific approaches, but the complexity of the regulations makes it necessary.

In the wake of the many changes of the past 15 years, municipalities often have to choose their own strategy with regards to how they interpret national policies and regulations and with regards to the many possibilities to deviate from the targets set by these regulations. They not only have to interpret these policies, but also have to ensure that while implementing them, they comply with the many rules that exist regarding whether they are allowed to choose their own interpretations and the motivations needed for possible deviations from indicative limit values. Many municipalities struggle to deal with these newly acquired responsibilities and up to about one in every four municipalities is not meeting procedural requirements and limit values. Many municipalities are no longer even aware of all the legal requirements they have to comply with. Before decentralization operations were instigated, most local environmental issues could be dealt with by municipalities based on a routine implementation of the common policy formats (see also Borst et al. 1995, De Roo 2003). This has been replaced by a constant struggle by municipalities to understand and implement national policies and regulations.

Argument 2: a focus on protection

In adopting a post-contingency approach, I have argued that decentralization can conflict with the desire to pursue single fixed objectives, such minimum levels of protection against environmental stress. At first sight, the Dutch national government seems well aware of this consequence of decentralization. After all, it considers itself responsible to safeguard a minimum 'base quality' that is inspired by the protection of humans and ecosystems to intolerable levels of environmental stress. Local authorities are only responsible for pursuing higher levels of performance than these minimum requirements and for possible deviations from these minimum requirements under complex circumstances. I however uncovered two problems with this Dutch approach.

First, it is unclear what is meant by this 'base quality'. It is not clearly defined in energy policies, context-specific in odour and soil policies, subject to a gliding scale in noise policies and possible to deviate from in air policies. As a consequence, it is now often unclear which levels correspond with tolerable or intolerable levels of environmental stress in the face of the protection of human health and ecosystems. What is clear is that municipalities can focus on environmental ambitions that are below levels required by motives of human health and wellbeing as represented by the standards of the early 1990s. I argued that introducing minimum levels of protection means that decentralization will have to take place within a context of generic policies that guarantee that these minimum levels are maintained. This brings us to a second problem.

Second, then, we can witness that decentralization and deregulation allow for municipalities to deviate from levels of protection that were common in the early 1990s. As it is not uncommon for municipalities to focus on the absolute minimum, there are many cases where environmental ambitions are indeed below this level. The consequence in these cases is that municipalities focus on lower quality levels as compared to the situation before decentralization, with possible health effects. As I argued in Chapter 4, highly complex issues might defy generic policy approaches even if health issues are at stake and make such deviations acceptable. However, even in 'green field' sites where it would be fairly easy to meet indicative limit values for noise or odour, municipalities have chosen to deviate from these indicative limit values. These minimum legal requirements (limit values) are based on public health criteria. They are now being deviated from, not intended as a last resort to deal with highly complex cases, but as a possibility for additional development and financial gain. It provides backup to the claim that local governance should take place in a context of generic policies and hierarchical control if guarantees should be maintained regarding minimum levels of protection.

Argument 3: responding to constraints

While protection provides decision makers with an argument to retain central governance formats, I argued that the constraints on local willingness and ability to pursue environmental ambitions can also be an argument to retain a degree of central guidance. To begin with *economies of scale*, the last section has clearly indicated that municipalities face problems in attracting sufficient numbers of skilled specialists (i.e. technical expertise) and attracting the required managerial skills. Rather, a widespread lack of local capacities to respond to current environmental policies and regulations was encountered, which means that many municipalities even fail to keep up their legal responsibilities. While this relates to the complexity of Dutch environmental policies (see below), it also relates to limited resources, staff and expertise. Particularly, smaller municipalities have difficulties acquiring sufficient managerial and technical expertise. It suggests

260 *Beyond the minimum in the Netherlands*

that economies of scale are indeed a constraint on local ability to pursue proactive and integrated approaches.

In Chapter 4, I argued that central policies can be important means of support to local ability in the face of limited resources, funding and knowledge to pursue proactive and integrated approaches to the environment. However, the Dutch national government has for a long time not responded to these constraints by providing additional resources or staff. In the absence of such central support, a post-contingency approach suggests we would see limited follow up at a local level. This can be supported by my conclusion (section 6.5) that most municipalities do not structurally pursue an environmental quality that exceeds the legal minimum. Furthermore, many municipalities have even sought to develop economies of scale on their own initiative, for example, through the installation of regional environmental agencies. This also provides backup for the claim that limited economies undermine local authorities' capacities to deal with environmental issues. It is therefore also a hopeful sign that as of 2013 these regional environmental agencies have become obligatory by law. It is a sign suggesting that the national government realizes some of the mistakes made in the past and is also actively correcting these.

In Chapter 4 I claimed that *external effects* make up a second constraining factor for local performance in environmental policy. The data also provides support for this claim. The main problems regarding external effects relate to the high background levels of environmental stress in the densely populated Netherlands and the fragmentation of competences needed for dealing with local environmental issues. High background levels not only require regional, national or even international cooperation, but also make it hard for many municipalities to even meet minimum requirements; i.e. environmental issues become too complicated to be dealt with within reasonable financial and temporal frames. Fragmentation of competences means that many sources of environmental pollution are controlled by agencies outside of municipal control. Examples include sources of pollution such as traffic on motorways and railways and large (industrial) installations. The national response to these issues is largely confined to efforts to reduce overall environmental stress through, for example, reducing pollution at the source (e.g. cleaner cars, lower speed on highways, less emissions from agriculture or households, etc.). The other main response has been to increase the flexibility of environmental targets. This has not corrected for the external effects which municipalities face, but rather implies that some degree of failure to pursue more ambitious environmental targets is accepted in some places, at least in the relatively short term. Possible alternative responses such as support for municipalities to pursue more ambitious policies when it comes to issues such as energy, noise or air policies are mostly limited to persuasive policies. This also brings us to the response of the national government to the weak profile of the environment.

Our data also suggests that the third main constraining factor on local performance in proactive and integrated approaches towards the environment is relevant: the *weak profile* of the environment. This 'weak profile' is on the one hand manifesting itself in a lack of political and administrative willingness to deliver more ambitious environmental policies. About half of the municipalities

Beyond the minimum in the Netherlands 261

have no ambition at all to go 'beyond the minimum', while implementation of well-meant ambitions in other municipalities also proved problematic. This also relates to wide cultural differences between spatial planning and environmental departments, which undermines a pro-environmental municipal culture. On the other hand, the weak profile is related to the dispersal of power in local governance, which undermines more ambitious policies especially if municipalities face a limited landownership. It calls for bargaining, negotiations and the development of partnerships and contracts. These are activities that most municipalities find difficult, especially because of limited financial resources and the required managerial skills. In the face of these constraints, national stimuli were argued for in Chapter 4, such as rewards, flexible regulations or possibly contracts such as (voluntary) agreements to 'go beyond the minimum' between the national government, lower levels of authority or stakeholders. Several respondents of the national government and NGOs also asked for such stimuli (see also section 6.6). In addition, I argued that the absence of such stimuli could easily result in disappointing results in practice, which we also find in practice (section 6.5).

The national government initially argued for a legally robust stimulus as proposed in the NMP4. The MILO approach was to become this legally robust stimulus. MILO, likely influenced by the existing political climate, eventually became a voluntary policy recommendation, similar to other persuasive policies. Consequently, there are no 'checks and balances' that urge municipalities to 'go beyond the minimum'. Although there are provisions in noise and odour policies that urge municipalities to motivate decisions to exceed indicative limit values, it remains unclear whether we should consider these indicative limit values to be 'beyond the minimum' or, in contrast, should see these indicative values as minimum requirements that can be deviated from. Rather, the conclusion seems justified that the only 'checks and balances' the national government relies on are a few financial provisions and persuasive tools. Stronger incentives, such as contracts, covenants, subsidies or more flexible regulations are not used. Hence, the coordinative model of governance and its associated central guidance are not used to help set the conditions required for decentralization and the associated use of proactive, area-based and integrated approaches to result in their envisioned outcomes. Rather, the national government opted for the dismantling of the coordinative model of governance so as to allow localities to develop proactive, area-based and integrated approaches. The consequences are clear. Persuasive tools have met little follow up and local success is limited. Most policy advisories are not known by the majority of Dutch municipalities. When advisories are read by municipalities, concepts such as an 'optimal living quality' and 'quality of the living environment' prove hard to make operational. Consequently, their use is confined to those municipalities where proactive, enthusiastic and motivated environmental officers take the lead.

Argument 4: robust and dynamic

The last and final argument I discussed in Chapter 4 is that adding value to minimum requirements in a dynamic 'area-specific' setting should be preceded by

262 *Beyond the minimum in the Netherlands*

the capacity to routinely implement sectoral environmental regulations. In the Dutch context, however, evidence suggests that instead of routine implementation, municipalities face conditions of management overload. The organization of Dutch national environmental policies causes serious implementation problems for municipalities. Changes to sectoral policies following the decentralization and deregulation operations during the last 15 years have caused an 'implosion' of Dutch environmental policies. Therefore, Dutch municipalities currently face environmental policies and regulations that are considered fragmented, theoretical, complex and too dynamic and large in number to keep track of. It seems utterly impossible to acquire a clear overview of Dutch environmental policies, and implementing minimum quality levels still proves to be a serious challenge in the Netherlands. Many municipalities are no longer aware of all the legal requirements they have to comply with. The conclusion can be no other than that the current organization of national environmental policies in the Netherlands makes it quite hard to be 'able' to implement them. And while municipalities find it difficult to coordinate implementation of minimum legal requirements, proactive and integrated approaches are not prioritized (section 6.6). It provides support for the claim made in Chapter 4, which is that it is hardly realistic for local authorities to increase levels of performance to exceed minimum requirements when they face too fragmented, complex and voluminous central policies and regulations focused on this minimum. It is a conclusion that also the national government has accepted and that has resulted in its attempts to streamline its policies. This is also confirmed with the work currently taking place on the new law for the physical environment ('omgevingswet') that started in 2011 (see section 6.6). As this law is scheduled for finalization in 2018, it is currently still too early to tell if it will have the desired effect.

6.8 Conclusions

The aim of this chapter was to investigate the practical consequences of the approach the Dutch national government has taken to increase the role of the local level in environmental policy and to explain why these consequences have occurred. In doing so, I also aimed to reflect on the theoretical arguments I put forward in Chapter 4, which meant to help inform attempts to increase the role of the local level in environmental policy. In this concluding section, I return to both ambitions.

Consequences

During the last two decades, confidence in the traditional functioning of the coordinative model – with the central governance on top – has decreased significantly in Dutch environmental policy. The faith in environmental standards was greatly nuanced whilst more integrated and interactive approaches gained in popularity. Fuelled by these changes, the Dutch chose to decentralize important parts of their environmental policies whilst deregulation meant that standards became less ambitious or more flexible. The national government seriously loosened its grip on the local environment. Supported only by tools associated with the argumentative model (e.g. policy advice and manuals), municipalities are left to take the lead in filling the gap that was left by the retreat of the national government.

The national government also motivated the choice to let municipalities take the lead in pursuing higher environmental ambitions. In the NMP4 it states that such higher environmental ambitions should be based on the local circumstances, as "the quality of the living environment differs from place to place, from person to person and also from period to period" (VROM 2001; p. 328, translation CZ). Given their proximity to local circumstances, municipalities are considered in the best place to develop policies that are tailored to the local situation. Municipalities are therefore supposed to choose their own priorities and strategies in aiming for a higher environmental ambition pursued with proactive and integrated local policies. Only the protection of the absolute minimum, albeit poorly defined, remains based on the coordinative capacity of the national government. The outcome in terms of the environment should be a quality that, if possible, 'goes beyond the minimum'. With this approach, the Dutch national government aimed for a transition in environmental management, as shown in Figure 6.1, from position 1 to 2a.

Our data show that that the envisioned transition in Dutch environmental policies from position 1 to position 2a (Figure 6.1) has not taken place, except for some good practice examples in a minority of municipalities. In fact, only a limited number of municipalities succeed in at least partly coping with their new responsibilities to 'add' quality to the so-called 'base quality'. Most municipalities are not equipped to do so or don't have the ambition. Consequently, by far most Dutch municipalities have not succeeded in making a transition from position 1 to position 2a.

There is however more to conclude on. After all, the majority of Dutch municipalities are still largely focusing on environmental standards and other regulations. Local authorities have not filled in the gap left by the retreat of the national

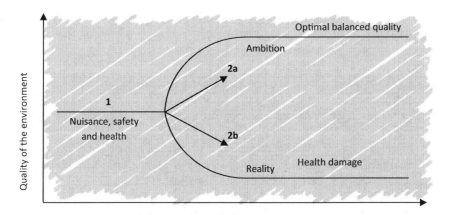

Figure 6.1 Transitions in Dutch environmental policy between 1990 and 2009

264   *Beyond the minimum in the Netherlands*

government. Dutch environmental policy, in practice, thus still largely functions as an example of the coordinative model of governance. However, the coordinative instruments have undergone important changes during the last years. On the one hand, minimum legal requirements have become less ambitious during the last years (i.e. in noise, soil and odour policies). As municipalities are still predominantly using these minimum requirements as a benchmark, they are therefore aiming for a lower level of quality than before. Examples of this were indeed encountered during the case studies (Spreeuwers et al. 2008). Instead of a transition from position 1 to 2a, a transition from position 1 to 2b took place; i.e. the average environmental quality delivered is devaluating with potential adverse health effects.

Explaining the consequences: a post-contingency approach

The second aim of the chapter was to better understand why the consequences just noted have occurred and, while doing so, to reflect on the theoretical arguments put forward in Chapter 4, which meant to help inform attempts to increase the role of the local level in environmental policy. A key motivation for decentralization and deregulation in the Netherlands changes was to increase the local ability to cope with complex and unique local circumstances; i.e. complexity was a motivation for decentralization. In Chapter 4 I argued that decentralization has important consequences and that in responding to these consequences, central governments can play an important role in supporting and stimulating local willingness and ability to perform decentralized tasks and responsibilities. We can indeed find good examples of such consequences in my research and also can relate this to the role of central policies and influence.

First, we have seen that *the Dutch national government lost focus on the need to maintain and protect the coherency of sectoral environmental regulations.* Consequently, decentralization and deregulation have caused an 'implosion' of Dutch environmental policies and regulations as compared to the situation before decentralization. Faced with serious problems to oversee and cope with these policies, regulations and rules, implementing minimum requirements is not a routine operation and many municipalities fail to do so. As fulfilling minimum requirements is no longer a routine operation in most Dutch municipalities, aiming for higher environmental ambitions is at least constrained, if not simply considered irrelevant by municipalities. The remaining body of national coordinative policies and instruments are not resulting in a robust and stable foundation for municipalities to build upon when pursuing more proactive, integrated and thus dynamic policy approaches in a local setting. Rather, it prevents them from doing so. It is a conclusion that also led the national government to readdress the coherency of its existing policies with its focus on a new law on the physical environment ('omgevingswet'). Nevertheless, despite the importance and relevance of doing so, it is a focus that came too late to prevent the problems with implementing regulations when decentralization processes were in full swing.

Beyond the minimum in the Netherlands 265

A second conclusion is that *the Dutch national government has been extremely optimistic, if not to say naïve, about municipalities' willingness and ability to pursue ambitious environmental policies.* The national government reduced its grip on local environmental governance. Dismantling of the coordinative model of governance, however, has gone so far that there are hardly any stimuli for local authorities to pursue proactive, area-based and integrated approaches. Decentralization and deregulation have left Dutch environmental policies with limited checks and balances on local performance, leaving much to be decided upon by municipalities themselves. The national government hoped and expected municipalities to fill in the gap left and furthermore, take on the responsibility to actively pursue environmental qualities that 'go beyond the minimum' as far as they could given the locally existing conditions.

Given the optimism of the national government, it has largely refrained from providing support for municipalities in the face common constraints on local performance, such as willingness, time resources, skills and a limited sphere of influence. Central influence over local environmental governance is on the one hand confined to maintaining minimum requirements. These are, however, partly incompatible with providing safeguards against health consequences previously (early 1990s) considered unacceptable. The limits of how far decentralization should go based on health motives are therefore arguably exceeded. The degree of local autonomy is such that even common and fairly simple environmental issues in noise and odour policies are sometimes being addressed through specific local approaches. The incentive to do so is not the high complexity of the situations encountered, but rather the possibility of economic and financial gain. In other words, the functional choice for protection of humans and ecosystems through national generic policies as explained in the NMP4 has not been translated into the necessary (centralized) structural configuration.

On the other hand, the Dutch national government tries to encourage local environmental governance to increase performance to levels above fulfilling minimum requirements. It is however not always clear to municipalities what is expected of them, nor are there clear checks and balances or procedural formats that guide the transition to 'go beyond the minimum'. The national call to 'go beyond the minimum' is only communicated with persuasive instruments, such as policy recommendations and manuals. Some prime concepts and instruments the national government uses to this end are little known (MILO) or considered difficult to use ('optimal living quality' and 'quality of the living environment'). In addition, in the midst of the large pile of advice and manuals coming their way, many municipalities don't hear this call. Those who do are merely 'asked' to take on new responsibilities. Faced with no rewards, and while struggling to comply with existing environmental regulations, most municipalities see going beyond the minimum as a merely additional and hardly interesting pursuit for which they have limited time, competences and resources. In practice, only about four in 10 proved willing, whilst many find themselves at least partly unable to cope with the new tasks responsibilities.

266   *Beyond the minimum in the Netherlands*

A response

If the Dutch hope to alter the current lack of success, at least three complementary activities seem advisable. First and foremost, fulfilling minimum legal requirements should again become a routine operation. It would require not only a clarification and simplification of what these minimum requirements are, it would also require a serious reduction of the time and resources needed for municipalities to implement them. Ideally, the minimum 'base quality' would be explained for all relevant environmental themes in one single policy document, which would also explain under which conditions deviations are and are not possible. In the face of the voluminous, fragmented and complex organization of current Dutch environmental policies, writing such a document might seem too ambitious. It is illustrative of the problem Dutch environmental policies face. After all, municipalities need to do exactly the kind of research that would allow for such a document to be written. The current work on the new environmental policy act means to tackle problems of incoherencies and fragmentation. It, however, seems motivated more by attempts to deregulate and to keep close to existing EU legislation that it is motivated by an attempt to explain something that could be called a 'base quality'. Nevertheless, it does seem to be a key step in the right direction to at least accommodate an easier implementation of regulations. It might be too early to tell if the new law will have its desired impact, but it is clear that it will not be a law meant to allow for a combination of routinely implemented generic regulations and tailor-made local strategies. In not actively promoting local policy development, it seems to only focus on improving the implementation of generic regulations.

Second, after the base quality would again become a routine operation, guidance is needed that explains the responsibilities and ambitions of various levels of authority when aiming for a transition to 'go beyond the minimum'. It would not only require an explanation of what going beyond the minimum means (contents), but also how such a transition can be achieved (process). In addition, the limited local willingness is an argument for introducing some incentives and national guidance. These can involve much more than persuasive policies, as is illustrated by, for example, the use of moderate coercive tools, subsidies, awards, contracts and (voluntary) agreements or eco-labels. That is, a partial return to coordinative instruments and central guidance and linking these with financial or argumentative instruments can help to set the conditions for local authorities to perform the tasks envisioned by the national government. It might be that the recent new attempts at developing environmental policies become a context in which such a focus on 'going beyond the minimum' would mean a partial return to coordinative instruments and central guidance. But as of yet, this is not part of the policy brief written (I&M 2014), nor does it seem likely given its focus on other issues.

Third and finally, the required strategic competences and working routines are not highly developed in most municipalities, leaving many municipalities not yet ready to cope with what the national government expects of them. Given the economies of scale involved in attracting expertise, some effort to create economies of scale to deal with tasks that require high demands of technical expertise seems appropriate. One step to respond to these problems has been taken. The installation of regional environmental agencies is a clear sign of the national government's acknowledgement of the problem. This might also support municipalities in focussing more on developing their own area-based environmental policies, as they can leave the traditional administrative tasks to these regional environmental agencies. Nevertheless, it would also be helpful to support them in gaining managerial expertise to integrate various policy interests, and to cross the public–private domain also demands attention. The national government might help by providing municipal experts with courses and training. Either way, a focus on stimulating local willingness and ability is needed in addition to a well-explained vision of what is expected of municipalities. If not, success in going 'beyond the minimum' will continue to depend on highly skilled, enthusiastic and hardworking professionals in combination with a high degree of local political will. And as political will proves to be limited and 'super employees' are an exception, success is also likely to remain such an exception.

7 The relevance of a post-contingency approach

Most of the Western governments that are currently involved in governance renewal operations are attempting to move away from a reliance on the coordinative model of governance and its associated central government control. The coordinative model is increasingly seen to be incompatible with the challenges of our complex and plural societies. Governance renewal operations have therefore been initiated with the hope and ambition that they will expand the societal capacity to govern. Traditional approaches to governance are now being replaced, sometimes swiftly, by new approaches. Common changes involve strategies such as decentralization, deregulation and privatization in the context of increasingly popular collaborative or communicative rational approaches, whilst they also mean to increase room for approaches that are dynamically responding to local circumstances. As a result, regional and local governments are gaining importance in developing and delivering on policies, whilst power and responsibility are partially transferred to the private sector and the community.

In recent decades, governance renewal operations have also influenced environmental health and hygiene policies. In my research, I addressed the renewal of these policies, while focusing on two decentralization operations in the EU and the Netherlands. Decentralization, among other things, is seen as a means for improving the responsiveness of governance to more complex issues. These issues are often characterized by conflicting objectives and mutually dependent interests and stakeholders, and often have a unique local manifestation. In response, decentralization operations are supposed to allow policies to be dynamically adapted to local circumstances and interests by stimulating local integrated and communicative approaches.

In my research, I accept that decentralization can help governance to respond to the challenges of our complex and plural societies. Based on my research, I nevertheless suggest *reconsidering the process of the diminishing importance of the coordinative model of governance*. This is partly because I conclude that the coordinative model still has some important benefits. These are, for example, benefits in coping with less complex and common policy problems and for guaranteeing that minimum levels of protection against environmental stress are met. In addition, I also conclude that the robust framework of policies provided by the coordinative model and its technical/instrumental rational approach makes the coordinative model *a crucial foundation upon which to build new dynamic policy approaches*. In other words, technical/instrumental rational approaches can function in addition to and as a basis for more dynamic, flexible and communicative rational approaches. In this final chapter I return to these two key conclusions.

I begin by reflecting and concluding on my theoretical inquiry. This inquiry focussed on developing theoretical arguments for providing us with clarity regarding how to choose between various approaches to governance, also when it comes to policies on environmental health and hygiene. Section 7.1 returns to the theoretical and philosophical difficulties that exist while searching for such arguments. In section 7.2 I will continue with my response to these difficulties, in the form of my proposal for a post-contingency approach. Section 7.3 addresses the practical inquiry conducted. It will show how a post-contingency approach can provide advice to the practical setting by indicating the consequences of such an increase. In section 7.4 I come back to one of the crucial assumptions I made in this book, which is to largely address 'complexity' from a so-called 'static perspective'. As I will explain, this is simplifying matters and hence calls for reflection so as to assess how post-contingency can also relate better to the rise of more 'dynamic perspectives' on complexity as are increasingly common in planning and governance debates. As I will explain, post-contingency does offer a starting point to help combine such dynamic perspective in planning and governance debates, but it is an offer that does require additional study.

Finally, then, in section 7.5 I will come back to the central issue of this book and conclude on the added value of adopting a post-contingency approach, specifically concerning governance renewal operations in environmental health and hygiene policies. Most notably, I will show why the coordinative model of governance is not outdated. Instead, I will argue that it continues to play a key role in providing guarantees for meeting minimum levels of protection against environmental stress and, in addition to such minimum levels, for stimulating local authorities to pursue higher environmental ambitions. Hence I will revisit the conclusion that the robust framework of policies provided by the coordinative model makes its continued use a *precondition* for innovative and dynamic policy approaches to be used in our quest for a 'liveable' future for our spatial environment.

7.1 Navigating the plural

Environmental policy has long depended on central government control and the instruments and institutions associated with the coordinative model. Despite important successes, many of these traditional environmental policies have been criticized for their 'react and mend' nature and for causing fragmentation of policies into various poorly coordinated thematic policy agendas. One of the arguments is that these reactive and fragmented policies are not well adapted to unique local circumstances and it is difficult to implement them alongside non-environmental policies, such as those related to spatial planning, economic growth strategies or social development policies. In the meantime, new policy ambitions such as sustainable development are highlighting the desire to focus on more proactive and

270 *Relevance of a post-contingency approach*

integrated policies. It has only further illuminated the drawbacks of fragmented and 'react and mend' approaches and, hence, has fuelled the desire for renewing environmental policy.

In response to the drawbacks associated with the coordinative model, various Western governments such as Sweden, Norway, the United Kingdom and the Netherlands have already started by partly dismantling their coordinative policies and instruments through, for example, decentralization and deregulation. Such renewal operations have important risks. This is certainly true when it comes to environmental policy, as the protection of humans and ecosystems against environmental stress is among its prime objectives. It calls for a keen awareness of the direction of governance renewal and clarity regarding its consequences. But while attempts at renewing governance unite in their relative shift away from a reliance on the coordinative model, their destinations are less clear. Rather, they are dominated by extensive administrative experimentation and seem characterized by what Offe (1977) predicted would be a 'restless search' for adequate governance approaches. It therefore leaves open the question of how and to what degree decentralization operations should be pursued and coordinative policies and instruments can be replaced by alternative policy approaches (e.g. De Vries 2000, Fleurke & Hulst 2006, Lemos & Agrawal 2006 and Prud'homme 1994). Ideally, then, theoretical debates on governance, particularly regarding environmental health and hygiene policies, can help us to answer these questions.

Theoretical debates on governance share the sceptical attitude of practitioners towards the applicability of the coordinative model of governance. This follows from the current post-positivist attitude that characterizes most theoretical debates on governance. Post-positivism holds a sceptical attitude towards the positivist foundations of our 20th century social sciences and the related idea that governments, supported by empirical inquiry and rationally designed bureaucracies, can exert a high degree of control over the social and physical world (e.g. Allmendinger 2002b, Mumby 1997). In response, alternative theoretical proposals regarding the organization of governance have emerged. The result is what Allmendinger (2002b) calls a post-positivist planning landscape, subject to a range of different and sometimes incommensurable theoretical proposals regarding the organization of governance, i.e. what Healey et al. (1979) refer to as 'theoretical plurality' (see also Baert 1998, Bernstein 1983, Fisher and Forester 1993, Harrison 2002, Stoker 1998). Consequently, theory is no longer providing practitioners with much clarity as to how to choose between various governance approaches. In aiming to provide practice with guidance for choosing between various approaches to governance in the realm of environmental health and hygiene, my ambition was to respond to this theoretical plurality in this book. It resulted in an inquiry into the philosophical roots of this theoretical plurality, so as to ground my response.

Coping with a theoretical duality

At the foundations of the theoretical plurality faced are different philosophical positions upon which academics have drawn in proposing their theoretical

Relevance of a post-contingency approach 271

approaches. In being characterized by a post-positivist attitude, most share what is called a relativist epistemology. This implies that facts and values are increasingly considered as intertwined, while knowledge and rationality are considered to be socially constructed (e.g. Baert 1998, Mumby 1997). Therefore, claims to knowledge and rationality are considered at least partly relative to the different frames of reference from which people perceive, interpret and judge what they experience. This not only contributes to a tendency for people to hold on to different claims to what is 'real' and 'rational', but also implies that we have problems with formulating criteria for choosing between these claims. After all, whose perspective should we rely on if we are to develop criteria for choosing between these various claims? As Allmendinger (2002b) rightly notes, post-positivism thus hints at a form of relativism.

Most post-positivists consider a strong relativist position as a too-radical perspective, possibly resulting in an 'anything goes' perspective on both rationality and governance. In response, various post-positivists have proposed strategies that can help us distinguish between claims as to what is real and rational. Although it is simplifying matters somewhat, a duality can be found within these post-positivist strategies between those who stay close to a *realist* perspective and those who stay close to a *relativist* perspective. Some theorists largely draw upon a (critical) realist perspective. While disagreeing with the positivists' reliance on full objectivity, establishing what is real and rational is, for them, still largely based on an *object-oriented focus*, i.e. evidence based. In still confiding in the assumption of a realist ontology, they give priority to trying to distinguishing between claims as to what is real and rational should be based first and foremost on our experiences in the world in which we act and on logically explaining the causal relations in this world that can explain the events experienced. Others stay closer to a relativist position and the associated idea that what is real and rational is established through *intersubjective-oriented action*, i.e. argument based. Rather than assuming a realist ontology, these scholars focus on arguing for the validity or invalidity of certain claims through communicative interaction, i.e. what is real and rational can be established on the basis of agreeing upon the rules and concepts that make most sense for those involved.

The duality we face within post-positivist philosophical debates between (critical) realism and relativism has also manifested itself in debates on governance. Many newer theoretical approaches to governance draw upon a more relativist position. Planning action is now mostly about agreeing upon which descriptions of the phenomena faced, which outcomes and which courses of action make most sense to those involved. In this context, some largely rely on collaboration between various stakeholders (e.g. Healey 1997, Innes 1996), while others refer to the (power) struggles and competitions needed to come to agreements (e.g. Booher & Innes 2001, Flyvbjerg 1998, Susskind & Cruikshank 1987). Either way, they are theoretical positions accepting that what is rational is mostly a human construction, decided upon by those involved. This has resulted in theoretical positions regarding planning and governance that stay close to the notion of communicative rationality. Other proposals are more modest in their acceptance of

272   *Relevance of a post-contingency approach*

communicative rationality (see also Allmendinger 2002a, Fainstein 2000 and Harvey 1997). Instead, these proposals suggest that planning action begins with developing knowledge about the structures and mechanisms of the world in which we act. They are proposals that thus largely draw upon a (critical) realist position. In doing so, planning is dominated by translating the knowledge we gain about the world into plans and action for delivering our objects. Although often with important nuances, these approaches do stay closer to a technical/instrumental rational perspective. It is also this perspective that for a long time dominated environmental health and hygiene policies. Knowledge about the impact of environmental stress was a basis for setting policy objectives such as environmental standards, which were implemented based on central government control and administrative tools such as licences, permits, taxes and government sanction. From a communicative rational perspective, it should also be highlighted that environmental health and hygiene policies are debated in a societal and often local realm, where other (local) interests and priorities determine what makes sense.

The duality between realist and relativist positions can now be considered as on the one hand contributing to the theoretical plurality we face in our governance landscape. On the other hand, this same duality is also manifesting itself in two conflicting philosophical positions regarding how we might distinguish between these theoretical approaches for combining them. As a consequence, we seem to be left without a defined common ground that serves as a starting point when choosing between different approaches to planning and decision making, also when it comes to decentralization in environmental policy. In this book I have argued against this conclusion. Instead, I posited that, despite the underlying philosophical dilemma, we can, and in being proactive also do, develop arguments that help in choosing between various approaches to governance.

The response

On the verge of the 21st century, philosophical debates continue to attempt to bridge the duality between realism and relativism (e.g. Archer et al. 1998, Billig & Simons 1994, Edwards, Ashmore & Potter 1995 and Lõhkivi 2001). I acknowledge that there is much to be gained by advancing these philosophical debates. Nevertheless, I also argued that we already have the means available for responding to this dilemma in a pragmatic way. We can do so, I argued, by taking the pragmatic stance that *these dual positions can be considered as complementary*. It is based on the argument that it is hard to ignore the fact that people draw upon their experiences (an object-oriented focus) as well as upon intersubjective debates in distinguishing between the objects, cases and situations they face and in attributing meaning to them. Therefore, choices between various approaches to governance also depend on both people's experiences, the logical explanations of these experiences, the knowledge they hence acquire of the issues and circumstances they face and how they interpret and evaluate this. Hence, I draw upon both (critical) realism and relativism by considering the complementarity of the solutions they offer for distinguishing between claims to what is real, rational and considered desirable and valued.

Relevance of a post-contingency approach 273

In my response I have further taken inspiration from *contingency theory* and its claim that the performance of different organizational structures and strategies (i.e. approaches to governance) is contingent on the circumstances encountered. Contingency theory itself is founded on a realist perspective and the idea that we can choose between different approaches to governance through an object-oriented focus, i.e. based on our knowledge of the circumstances and how these influence the performance of our approaches. I go beyond the singular perspective of an object-oriented focus and also take an intersubjective-oriented focus. This resulted in my proposal for expanding contingency theory by incorporating the duality between a realist and relativist position into what I have called a post-contingency approach. It is in adopting a post-contingency approach that I argue to have found a 'common ground' that serves as a starting point for navigating the plural governance landscape.

7.2 A Post-contingency approach

Contingency theory emerged during the 1960s and 1970s as a response to the 'one size fits all' notion that dominated planning and policy sciences. Such a 'one size fits all' notion implies that there is, in principle, a single best approach to governance. In practice, this was equated with the coordinative model of governance. Contingency theorists considered this perspective too simplistic. Instead, they suggest that the right way to organize management and governance should be *contingent on* the circumstances encountered (e.g. Bryson & Delbecq 1979, Fiedler 1994 and Lawrence & Lörsch 1967). A contingent perspective on reality, therefore, does not accept one singularly defined reality but considers a reality that is context related and situation dependent. This proved to be an important step forward in organizational and policy sciences and gained strong empirical support. However, contingency theory also received some important critiques.

First, classical writings on contingency tend to overlook the difference between the function and structure of governance approaches; i.e. between what decision makers intend to achieve and how they organize governance. This difference, however, can help us navigate the plural governance landscape. Before I return to this difference, I will address the second point of critique I focussed on: the lack of attention paid in classical writings to contingency upon the societal context in which planning and policy-making take place.

Contingency theory states that the performance of different organizational structures and strategies depends on the circumstances encountered. The process of coming to such a selection, however, proved largely coherent with the ruling doctrines of the day: an object orientation and an instrumental/technical approach. An object orientation comes forward trying to develop uniform knowledge and standards upon the performance of different structures and strategies under different circumstances. This knowledge would subsequently inform the choice for a structure or strategy, where given the circumstances encountered and objectives set, this choice is a mere logical consequence of having such knowledge; i.e. where a deterministic logic consistent with a technical/instrumental rationale comes forward. Although more recent contingency studies often take a more

274   *Relevance of a post-contingency approach*

nuanced approach, it was hardly a surprise that the rise of post-positivist notions in planning and administrative sciences also inspired critique to emerge upon these 'classic approaches' to contingency theory.

Dominant among the critiques was the notion that contingency theory largely ignores the societal context in which knowledge is interpreted and contingency choices are made. It led to an acceptance within contingency approaches that people face limitations in observing and interpreting contextual circumstances and in observing and interpreting which organizational structures and strategies perform best given these circumstances. Although this nuances the deterministic logic of classic approaches, it hardly challenges its object orientation and rationale. Furthermore, these nuances also do not explicitly acknowledge that what is considered as performing 'best' is also a possible matter of interpretation and debate. Classical approaches to contingency, I therefore suggest, tend to overlook the fact that planning and policy-making take place in a societal realm where values often matter more than facts. It is in accepting the importance of this social context, in which issues and objectives are interpreted and carry different meanings, that a post-contingency approach adds value to classical approaches to contingency. Post-contingency implies we should further redefine and interpret the idea of contingency choices. Post-contingency does not deny the value of taking an object-oriented focus, but understands that its limitations force us to look beyond only taking such an object-oriented focus. Hence, it moves beyond the object-oriented and deterministic logic of these classic approaches by adding an intersubjective focus to the process of making contingency choices and adjustments.

Post-contingency

With a post-contingency approach I first accept that *defining* issues and the circumstances surrounding them is at least partly relative to the different frames of reference from which people perceive, interpret and judge what they experience. Indeed, practice often forces us to accept that issues are defined largely based on dominant institutional beliefs, political ideologies and social interpretations. A post-contingency approach accepts that an object-oriented approach will not offer definite closure regarding the validity of such definitions. After all, as objective knowledge is considered impossible, uncertainties will remain upon the different and conflicting interpretations that people continue to adhere to. Similarly, we will not be able to predict in detail how our possible approaches or actions will perform under such circumstances, as we are, again, facing the same lack of objective knowledge. Uncertainty and discussion on what is 'real', therefore, simply will remain, forcing us to accept different interpretations and beliefs upon the situations and issues we face and how our responses might impact on them.

Second, a post-contingency approach also accepts that, when people *value or judge* the issues and circumstances they face, they do so while adhering to different preferences and objectives. Therefore, apart from possibly disagreeing upon what they 'know' about these issues and circumstances, people can also disagree upon what this knowledge *means*, i.e. they value and judge it differently. Some

Relevance of a post-contingency approach 275

consider global warming as an opportunity for increased agricultural outputs or tourism; others see it as the biggest human catastrophe that has ever occurred. It is now no longer simply a matter of who is wrong or right and, hence, not just a matter of improving our knowledge of the issues and circumstances faced (object oriented). It is now, instead, more a matter of values and choices, i.e. which preferences and values people adhere to (intersubjective oriented).

With a post-contingency approach I thus accept both the socially mediated nature of knowledge and the different preferences and values that people adhere to when judging what the issues and circumstances they face mean to them. Therefore, a post-contingency approach *does not* aim to provide decision makers with definite answers as to which interpretations they should rely on or how they should subsequently respond. Rather, the approach considers that both these interpretations and responses are, at least to a degree, socially mediated *matters of choice*. Nevertheless, we have also seen how a post-contingency *does allow us to reflect* on these interpretations and choices by investigating their likely consequences. It is here where both an object-oriented and intersubjective-oriented approach are combined (see also Figure 3.2).

Starting with an object-oriented approach: degree of complexity

Adopting an object-oriented approach follows up on classical contingency theory's assertion that the performance of different approaches to governance is contingent on the circumstances encountered. Therefore, I investigated how different approaches to governance relate to different circumstances. In doing so, I refrained from picking up on studies that address the numerous linkages between individual environmental variables and organizational variables; i.e. an approach that can easily be seen as reductionist. Such approaches run the risk of overlooking the dynamic ways in which rich varieties of environmental circumstances and governance approaches truly interact. Instead, I searched for how we understand how more commonly accepted stereotype governance approaches tend to link to alternative environmental circumstances. Regarding the various governance formats available, I categorized approaches to governance by their relative reliance on either a technical/instrumental rational or a communicative rational approach. In categorizing issues, I followed one of the most dominant contingency arguments developed in the past; i.e. the argument that different conditions of complexity would logically coincide with different planning approaches and organizational formats (section 3.6). But while using the degree of complexity as an argument for making planning choices, post-contingency also acknowledges that what is considered as more or less complex is not self-evident.

Many contingency studies chose to define complexity without paying explicit attention to the different actors involved. They merely equated complexity with uncertainty regarding the factors or causes that contribute to the phenomena we intend to address, how these factors relate and change over time (their stability) and the possible intervening circumstances that could influence these cause-and-effect relations. Here I also explicitly relate complexity to the actors involved

276   *Relevance of a post-contingency approach*

and the different perspectives, interpretations and behaviours they have. There-fore, complexity is also influenced by how circumstances are interpreted, valued and translated into opinions and preferences. To me then, an increased degree of complexity involves, among other things, diminishing certainties, increased contextual instability, diminishing direct causal relations and increasingly diverse and possibly conflicting interpretations and judgements of the issues, their con-textual circumstances and the preferred approaches to them. The result is that it becomes increasingly difficult to 'objectify' the issue and its circumstances and, instead, that there is an increased need to intersubjectively establish how the issue and its circumstances are seen and which approaches might make sense to those involved.

Complexity implies that we deal with 'fuzzy' entities that are strongly embed-ded in their changing and unstable contexts that behave highly unpredictable. While we are able to gain knowledge of their behaviour, as we aim to describe, understand and deal with these issues, we will continuously face many (contex-tual) factors and mutually dependent actors that interact in unpredictable ways, influencing how issues behave, change, are interpreted and evaluated. Relying on a technical/instrumental rational approach and its confidence in predictability will now be fundamentally limited. Rather, as I explained, such complexity provides an argument for shifting the focus towards a communicative rational approach. In operating from an object-oriented focus, therefore, it is possible to connect the degree of complexity with choices between various approaches to governance.

The value and limitations of an object-oriented approach

Adopting an object-oriented approach involves connecting different circum-stances to the performance of various approaches to governance. I started with conditions of limited complexity where cause-and-effect relations can safely be assumed as well known and stable, and where there is also agreement regarding which objectives to pursue. In other words, we know what we want and know how to get there. A technical/instrumental rational approach is now likely to have both predictable and widely supported outcomes. In contrast, a communicative rational approach will make little sense, as there is simply little to discuss. Rather, such an approach could even end in time-consuming processes that might well frustrate progress.

Under conditions of increased complexity, cause-and-effect relations are typi-cally less clear and are mutually dependent (and non-linear), while it is also often difficult to define issues as strictly separated entities. Rather, context and entity seem to overlap, circumstances are changing and unstable, while many interven-ing variables constantly alter the issues we face. In the meantime, we might also face conflicting interests and objectives that should be pursued, not infrequently represented by various mutually dependent actors. Hence, the objectives to pursue and the means of achieving them are both subject to debate. It will be problematic to rely on the kind of straightforward action associated with a focus on fulfill-ing predefined objectives that is propagated by a technical/instrumental rational

Relevance of a post-contingency approach   277

approach. Not only is it uncertain whether it will indeed result in its envisioned outcomes due to the many uncertainties we face regarding the issues and circumstances we deal with, but it is also likely to lead to strong resistance from those who interpret the issues differently or adhere to different objectives that conflict with those pursued. Especially if such resistance exists in a context of power dispersal, there is also the potential to block progress by use of resources and access to decision making. In response, a communicative rational approach allows stakeholders to discuss the knowledge that is available and the various interpretations and objectives to which they adhere. It thus allows them to cope with complex circumstances by agreeing upon *what is considered to be known* and *which objectives* to pursue.

Adopting an object-oriented approach shows us how the degree of complexity helps illuminate the likely consequences of choosing between different governance approaches. More specifically, through taking an object-oriented approach we are able to gain knowledge of the degree of complexity encountered and express expectations upon the consequences of relying on different governance approaches when applied under these perceived conditions of complexity. We see these consequences by contingently relating a shift along one spectrum (the degree of complexity) to a shift along another spectrum (technical/instrumental rational to communicative rational). Whilst this is not a contingency to be treated as a mathematical certainty, it is a contingency showing us why combining two contingently related categorizations can help us act. Categorizing issues in terms of their relative degree of complexity simply helps us assess the consequences we can expect when we shift from a technical/instrumental to a communicative rational approach. In more practical terms: we can now see how the degree of complexity influences the consequences of relying on different governance approaches.

Gaining a greater understanding of the degree of complexity involved, post-contingency proposes, is thus a helpful strategy for informing choices between governance approaches. Post-contingency, however, goes beyond a singular reliance on an object-oriented approach, where the social context is reduced to only a characteristic of the circumstances faced instead of influencing decision making itself. Instead, post-contingency also accepts that how decision makers want to *respond* to these consequences is still an intersubjectively mediated choice. People will typically adhere to different preferences and objectives and, hence, will evaluate and judge the consequences differently. Investigating the degree of complexity merely helps them to reflect on these choices, by getting a better grip on the likely consequences of their actions. Thus, the perceived or deduced degree of complexity is an *argument* for choosing between various approaches to governance; how decision makers *respond* to this argument remains a matter of choice.

The role of the intersubjective emerges not only in accepting that responding to the perceived or deduced degree of complexity is a matter of choice. I also noted that relying on an object-oriented focus should not be conflated with developing fully objective knowledge of the degree of complexity faced and, hence, of the kind of consequences we can expect when relying on different governance approaches. A contingency should not be considered in any way as a mathematical

278 *Relevance of a post-contingency approach*

certainty: it is a construct that helps us structure the linkage between environmental circumstances and the governance approaches we might choose to rely on. It remains crucial to see that knowledge upon such a contingency is also influenced by the frames of reference from which people perceive and interpret what they experience, i.e. intersubjectively mediated. People can come to different interpretations of the degree of complexity faced and, hence, of the consequences they expect when relying on different governance approaches. This highlights the limitations of relying on an object-oriented focus for reflecting on choices between governance approaches by predicting their likely consequences. While using contingency as a construct, we are thus able to produce knowledge upon consequences to expect, but such knowledge might also remain highly contested.

Proposing that defining issues in terms of their degree of complexity and subsequently choosing how to respond to these definitions are, by and large, socially mediated matters of choice does not mean these choices are 'free'; or to better phrase it, without argument. One key argument lies in simply accepting that an object-oriented approach can still produce valuable knowledge that helps us support our decisions; i.e. it does constrain what we can plausibly believe and expect. Furthermore, existing political or ideological arguments might prevail, whilst existing power relations also constrain the freedom of making choices. Finally, arguments can also be found without a reliance on an object-oriented approach. That is, arguments can be found without trying to understand and interpret the situation we face, *but from the consequences that we aim to achieve*. The argument follows the idea that different approaches are known to have alternative benefits (consequences) that we might value regardless of the exact situation we face. It is a search for arguments that brought me back to the difference between function and structure, which is often largely overlooked in classical approaches to contingency (e.g. Gresov & Drazin 1997, Miller 1986).

Beyond an object-oriented approach: structure and function

The contingency between function and structure is based on arguing that some configurations of function and structure perform better than others (e.g. Gresov & Drazin 1997, Merton 1968, Miller 1986, 1994 and Payne 2001). I categorized the functional focus by a reliance on predefined single fixed objectives on the one hand, and a reliance on multiple composite or integrated objectives on the other hand. Structure has been categorized by a relative reliance on a centralized or decentralized organization. I subsequently showed that a focus on predefined single fixed objectives matches a centralized structure, as is common in a technical/instrumental rational approach. In addition, I noted that a focus on multiple composite or integrated objectives matches a decentralized structure, as is common in a communicative rational approach. Hence, I showed that a shift from a focus on predefined single fixed objectives towards multiple composite or integrated objectives *contingently relates* to a shift from a centralized towards a decentralized structure. I showed that ignoring this contingency will result in reduced performance (i.e. non-matching) of configurations of structure and function.

Relevance of a post-contingency approach 279

The contingency between function and structure allows for reflection on the choices of decision makers between various approaches to governance. In searching for matching configurations of structure and function, decision makers know that their choices regarding function have consequences for their choice of structure and vice versa. To illustrate, they know that relying on a centralized structure has benefits, especially if they decide to prioritize a predefined single fixed objective (function). They benefit from the effectiveness of hierarchical control and the efficiency of a routine implementation of centrally issued common policy formats. Clearly, if there are motives to choose for single fixed objectives, the consequence is that a centralized structure becomes logical and vice versa. Alternatively, once decision makers shift focus towards pursuing multiple composite or integrated goals, the consequence is that it becomes more problematic to benefit from centralized policy formats. The pursuit of multiple composite or integrated goals is easily outside of the 'span of control' of central governments. Instead, relying on a decentralized structure has benefits because it helps governance to adapt to the local circumstances, interests and stakeholders. In doing so, however, decision makers also have to accept that policy outcomes will be dependent on local expertise, willingness and resources. Finally, it might also be that stakeholder and citizen involvement are highly valued, which makes it problematic to rely on setting single fixed objectives but will instead beg for more flexibility in allowing multiple composite or integrated solution strategies. To conclude, then, the contingency between function and structure helps us to identify consequences to expect when choosing between various approaches to governance, without having full a priori knowledge of the issues and circumstances involved.

7.3 An empirical inquiry

In my empirical inquiry I applied a post-contingency approach for analysing and explaining the consequences of decentralization operations in practice. I focussed on attempts by the EU and the Dutch government to decentralize power and authority to the local level in the realm of environmental health and hygiene policies. While doing so, I used the theoretical arguments put forward in Chapter 4 that help to inform attempts to increase the role of the local level in environmental policy, while allowing myself to reflect on these arguments.

The objective of both the decentralization operations assessed was to improve the capacity of governance to cope with urban environmental issues. Both the EU and Dutch government noted that existing central policies fell short in dealing with urban environmental issues due to the complexities involved in dealing with them. The EU argued that existing central regulations are too (sectorally) fragmented, reactive and lacking in vision to deal with urban environmental issues. The Dutch government added that common policy formats were often too rigid, inflexible and ambitious for implementation in the face of complex urban environmental issues. In response, the EU opted for active and integrated management of urban environmental issues, whereby local authorities should take the lead 'since no two urban areas are the same'. The Dutch went even further and gave local

280   *Relevance of a post-contingency approach*

governments 'as much integrated responsibility for the local living environment as possible'. In doing so, local governments could adapt to unique local circumstances through integrated, flexible and communicative local policy approaches.

In line with adopting a contingent and a post-contingency approach, both the EU and Dutch government considered decentralization as a response to conditions of complexity. The aim of decentralization is for local authorities to develop dynamic policy approaches that are tailored to complex and unique local circumstances. From such a perspective, central guidance and the associated coordinative model of governance would only be used for dealing with issues that have a common manifestation in various localities and are thus characterized by conditions of limited complexity. With our post-contingency approach, however, I argued that, although complexity is indeed an argument for decentralization, decentralization is primarily a matter of choice that can also be informed by other arguments. It is here where I move beyond the reliance of contingency theories on an object-oriented approach, but also consider choosing between various organizational structures and strategies as intersubjectively mediated. In this context, the contingency between function and structure has also helped to better understand and predict the consequences of decentralization and the conditions under which to expect them. It is in responding to these possible consequences that I have also formulated various arguments for retaining a degree of central policy formats and control, even when we are confronted with complex issues and when decentralization seems in itself a sensible strategy.

First, I argued through a post-contingency approach that decentralization involves a reduced governance capacity for meeting single fixed goals. Decentralization (a change in the structure) is thus linked to greater uncertainty regarding the ultimate outcome of the policy (the functional focus). The desire to install minimum levels of protection against environmental stress in a local realm is, however, an important motive for setting single fixed goals. I argued that introducing minimum levels of protection means that decentralization will have to take place within a context of generic policies that guarantee that these minimum levels are maintained. In the Dutch case I have also discussed what happens if this is not done. The Dutch government did decide to retain a set of generic policies for protecting a minimum 'base quality'. However, rather than providing guarantees, many existing generic policies were made less ambitious and more flexible. This was to allow localities to deviate from national generic targets in the face of highly complex local circumstances, even if this might involve adverse health effects. However, as a result of the degree of flexibility offered in allowing for deviations, local authorities can also deviate from national targets even if they face conditions of limited complexity. Consequently, deviations were also inspired by motives such as scope for additional urban development or financial benefit. Many Dutch municipalities currently focus on environmental ambitions that are below the targets set by the national government as minimum levels of protection for human health.

Second, with a post-contingency approach I argued that a key consequence of decentralization is that policy outcomes become increasingly dependent on local expertise, willingness and resources. Decentralization therefore has its

Relevance of a post-contingency approach   281

limits once we encounter localities that are no longer willing or able to perform decentralized tasks and responsibilities. My empirical inquiry also showed that many local authorities are limited in their willingness and ability to pursue proactive and integrated approaches to the environment, and this has contributed to disappointing levels of performance. Common constraints encountered are a lack of managerial expertise for organizing cross-sectoral working, political bargaining, process management, partnerships, etc. In addition, limited resources were frequently mentioned. In the Dutch context I found municipalities failing to attract even basic managerial skills and technical expertise due to their small environmental departments. Finally, political and administrative leadership were often limited.

I argued in Chapter 4 that central policies can support, stimulate and guarantee local authorities to cope with newly decentralized tasks and responsibilities. This is the case not only if there is an ambition to meet single fixed goals such as minimum environmental ambitions, but also to help local authorities respond to the aforementioned constraints. Central policies and the associated reliance on the coordinative model of governance are then used to set the conditions required for decentralization to result in its envisioned outcomes. However, neither the EU nor the Dutch government has chosen to provide much support or stimuli for local authorities. Instead, they largely rely on voluntary policy recommendations and have not chosen to use support based on regulations, funding, contracts or covenants, etc. The empirical inquiry shows that, without much support and stimuli from the EU and Dutch government, many towns and cities continue to be at least partly unable to respond to the call by the EU and Dutch government to pursue proactive and integrated approaches to the environment.

Finally, I argued in Chapter 4 that local willingness and ability also depend on a well-organized set of central policies that can be routinely implemented. The empirical inquiry also illustrates what happens if this is not the case. Within the EU there were many complaints about the fragmentation and bureaucracy of EU regulations. Many stakeholders argued that this undermined their capacity to pursue proactive and integrated approaches. In the Dutch context, the consequences were even more pronounced. Decentralization and deregulation have caused an 'implosion' of Dutch environmental policies. This has resulted in a complex, fragmented, voluminous, rapidly changing and sometimes conflicting body of policies. None of the respondents interviewed even claims to have a clear overview of these policies, but they recognize a mosaic of policy themes, organizational routines, standards and objectives. My study did not find a routine implementation of policies and regulations, but instead shows that localities find it difficult to get national policies and regulations implemented. Most municipalities are in a constant situation of crisis management to try to navigate the complexity of the policies and regulations they face. While municipal employees try to navigate the complexities of environmental regulations, pursuing more dynamic proactive and integrated approaches is not a priority. Central policies are not helping local authorities to develop dynamic policy approaches that are tailored to local circumstances. Instead, they frustrate these attempts and undermine the local ability to implement even minimum requirements.

282 *Relevance of a post-contingency approach*

7.4 Complexity revisited

My proposal for a post-contingency approach draws from debates on complexity theory, albeit in a modest fashion. I have chosen to use it while staying close to a static complexity perspective; i.e. without directly including the element of time. Relying on a static perspective to complexity allows us to categorize different issues and contingently relate planning approaches to them. However, despite this advantage, a static perspective seems to fall short of grasping the richness of what complexity sciences might teach us planners; something that exactly a dynamic perspective means to explore (e.g. Batty 2005, Byrne 2003, De Roo & Silva 2010, De Roo et al. 2012, Hillier 2008 and Portugali 2000).

The relevance of dynamic perspectives

Dynamic complexity perspectives explicitly incorporate the element of time and are focused on the non-linear and adaptive behaviours of social and physical phenomena. It stands for a world view where reality is seen as constantly changing in often unpredictable ways due to its non-linear character. As a result, a dynamic conception of change and complexity challenges the argument that shifting towards a communicative rationale is sufficient for coping with conditions of complexity. Relying on a communicative rationale has helped planners cope with uncertainties being the result of various interacting actors and their different interpretations, values and interests. It does, however, not necessarily help them address uncertainties that are the result of a non-linear world of discontinuous change. A communicative rationale is about creating a shared perspective upon planning issues and how to address it now and in the future. It does not explicitly address the changeable nature of this issue, nor of possibly intervening circumstances or altering perceptions or opinions. If it does, processes of change are assumed to continue in accordance to past trends and current observations, much similar to a technical rationale. Hence, decision making is typically conceptualized as making choices to influence possible futures based on knowledge and predictions of the current circumstances faced, and how these circumstances are currently interpreted and valued by decision makers. Put differently, we focus on decisions made based on the 'here and now'. The dynamic, changeable and unpredictable nature of the processes and issues planners deal with, therefore, is hardly being addressed (Ahern 2011, Hillier 2007).

In recent years planners and policy scientists have proposed alternative approaches that do aim to respond to conditions of change and dynamics, which we also touched upon in Chapter 3. Most notably are those debating the creation of more adaptive institutions and modes of governing (e.g. Duit 2012, Folke et al. 2005, Gupta et al. 2010 and March & Olsen 2006). Adaptive governance approaches allow for flexibility in altering the chosen policy paths, while they also see planning more as a process of experimentation. Through monitoring and learning from our actions and ongoing trends, we can increase not just our understanding of the dynamics of the issues we deal with, but also develop new

Relevance of a post-contingency approach   283

understandings and ideas upon how to deal with them (e.g. Duit et al. 2010, Folke et al. 2010, Foxon et al. 2008, Hillier 2007, Loorbach 2010, Pahl-Wostl 2009 and Van der Brugge & Van Raak 2007). It is also here where the inter-subjective oriented perspective of a communicative rationale comes forward, as such processes of learning involve communication and debate to help 'make sense' of what such new information means with regards to defining the problem, the policy objectives to pursue and the strategies to adopt (see also Busscher et al. 2014).

More adaptive forms of governance rely on institutional arrangements that have "the capacity to deal with surprise, to learn, and to support flexible institutions" (Van der Brugge & Van Raak 2007; p. 2). Adaptive governance typically relies on "polycentric institutional arrangements that are nested, quasiautonomous decision-making units operating at multiple scales" (Olsson et al. 2006; p. 19). These arrangements try to balance centralized and decentralized control and include horizontal linkages between organizations and societal groups to increase the diversity of response options (Folke et al. 2005; p. 449). Adaptive governance approaches, therefore, do not replace technical and communicative rationales, but merely position them in an 'evolving' rationale where new information, changed interpretations and institutional responses co-evolve so as to produce a continuous capacity to adapt. This also offers an interesting avenue for investigation with regards to how we might reconcile a post-contingency approach with a more dynamic perspective on complexity.

Post-contingency and complexity

To begin with, the notion of adaptiveness can inspire improvements with regards to using a post-contingency approach. If circumstances change, also conditions of complexity change. Planners are quite familiar with many of those changes, such as the availability of new knowledge and technologies, swiftly changing societal opinions or media attention, unexpected policy failures, or sudden events such as the recent economic crisis, etc. Such changes suggest a reconsideration of how to both interpret and approach policy issues. Finding governance approaches and organizational formats that 'fit in' with the contextual conditions encountered, therefore, need not be seen as a static one-time event. Instead, as Donaldson (2001) suggests, for example, there is a need for a more dynamic understanding of 'fit'. Such a 'fit', according to Donaldson does not emerge as a static equilibrium between two sets of conditions. Rather, he proposes a 'disequilibrium theory' of organizations, implying an ongoing and cyclical process where changed contextual conditions, based on how they are interpreted, urge for organizations to adapt their strategies and structures so as to improve their 'fit' with these changed conditions. Labelled as 'structural adaptation to regain fit' (SARFIT), contingency is then expressed as a search for 'fit' within an adaptive policy process. The SARFIT approach might offer us an interesting first pathway for further exploring the value of a contingency argument within an evolving world, where our planning approaches and institutions might well be seen as co-evolving with the societal challenges and developments they aim to address.

284   *Relevance of a post-contingency approach*

Alternatively, a post-contingency approach might also help us reflect on the use of more adaptive approaches to governance. Although relying on a static perspective, post-contingency accepts that more adaptive approaches to governance can also be among the different approaches we might contingently relate to the circumstances we face. Post-contingency simply suggests that this involves an element of choice. This has three key repercussions that offer pathways for research to reconcile a post-contingency approach with a more dynamic perspective on complexity.

First, the conditions of complexity we encounter might well suggest that it is advisable to choose for a more adaptive and flexible approach to governance. It could then be a preferred choice to rely on approaches that mean for planning and governance to become more adaptive to change and that propose planning as an ongoing process of experimentation and learning. Nevertheless, post-contingency suggests that there might also be other arguments that could warn against such a choice. This suggestion resonates with, for example, the work of Ulrich Beck (2006). Beck suggests we are embedded in a so-called 'risk society'. The increased complexity of many societal processes makes it hard to oversee the risks associated with them, whilst these risks are potentially extremely high. It is a an argument also illustrated by, for example, Duit and Galaz (2008), who explain that radical changes can occur in complex adaptive systems once a certain 'threshold' or 'tipping point' is reached (see also Folke 2006, Gunderson & Holling 2002 and Walker & Meyers 2004). Under such conditions we are not only facing risks that we can oversee, we also face risks we are unable to even identify. Climate change is a key example of such a risk, as Beck (2006) notes. Under conditions of high uncertainty, Beck argues, it might be tempting not to take radical action and to avoid irreversible choices. Instead, it might be smart to choose to rely on adaptive modes of governance where we can remain flexible and alter our approaches in response to possible surprises. However, the irony is that especially under such conditions of high uncertainty, we might also be highly induced to take radical action because if we do not, the risks might well increase manifold. Essentially, then, we face a trade-off between different sets of consequences and risks we cannot oversee where the 'precautionary principle' might be a decisive factor. It is then mostly a matter of 'making sense' of what we do know and make choices regarding how we proceed to act upon this knowledge; i.e. similar to what a post-contingency approach proposes.

Second, and in line with the first point just made, adopting a post-contingency approach implies that responding to highly uncertain, complex and changing circumstances is not only about making governance more flexible and adaptive. We do not *just* need dynamic policy approaches that are open to adaptive change and tailored to complex circumstances. We also need a robust foundation of policies that enable, stimulate and support the development of such dynamic and adaptive approaches. It is a lesson which also resonates with arguments made within the realm of complexity theory. Bertolini (2010), for example, refers to what he calls "evolutionary planning" as a combination of "robust measures" and flexible policies that allow for "options that can and should be left open" (p. 93). In a similar

Relevance of a post-contingency approach   285

stance, Lempert et al. (2003) argue that coping with uncertain and complex situations should be based on "robust . . . strategies that perform 'well enough' by meeting or exceeding selected criteria across a broad range of plausible features *and* alternative ways of ranking the desirability of alternative scenarios" (p. 45, emphasis added). With these proposals, we might be able to alter organizations, plans or strategies in order to make them more adaptable and resilient, so as to co-evolve with their changing contexts (e.g. Dietz et al. 2003, Stacey 1996, see also Senge 1990). These conclusions suggest that it is worthwhile to further assess these debates in complexity theory to possibly support or improve our post-contingency approach.

Third, and finally, proposals towards more adaptive, evolutionary and flexible approaches to governance are also showing variation. Not every proposal relies on the same premises, nor is it intended to cope with the same circumstances. To illustrate, some focus on understanding or guiding processes of self-organization (e.g. Boonstra & Boelens 2011, Portugali 2000), others on adapting to changing circumstances (e.g. Berkes et al. 2000, Duit & Galaz 2008, Folke et al. 2005). It might well make sense to come to categorizations of alternative circumstances of complexity and dynamics and contingently relate alternative approaches to governance to them. It is a challenging attempt that has inspired some pioneering work (e.g. de Roo 2012). Most notably, however, it is again a possible pathway in further understanding the relevance of a post-contingency approach while taking a dynamic perspective to complexity.

7.5 Final reflections

The empirical inquiry allows for reflection on adopting a post-contingency approach. In doing so, I first discussed that conditions of increased complexity provide an argument for decentralization. Governments are, among other things, also using this argument to partly dismantle the coordinative policies in response to the challenges of our 21st-century world. This was most evident in Dutch environmental policy (Chapter 6), while also EU attempts at stimulating more dynamic and integrated policy approaches appreciated this argument (Chapter 5). But while conditions of increased complexity provide an argument for decentralization, doing so is still a matter of choice that can be informed by alternative motives. This brings me to the second key conclusion of adopting a post-contingency approach.

Contingently shifting focus between various approaches to governance is not just a response to the matter of degree of complexity identified. It is also a matter of choice that can be informed by arguments other than the perceived degree of complexity. I discussed various key arguments for informing such choices, which were informed by recognizing the contingency between the function and structure of governance approaches. This allows us to see how each choice regarding either function or structure has consequences, which are at least partly predictable. First, a consequence of decentralization is a reduced governance capacity to meet single fixed goals such as minimum health requirements. The empirical inquiry also

286 *Relevance of a post-contingency approach*

confirmed that decentralization can easily result in problems in meeting minimum levels of protection. If this is considered unacceptable, decentralization should take place within a context of generic policies that guarantee that these minimum levels are maintained. Second, decentralization implies that policy outcomes become increasingly dependent on local expertise, willingness and resources. In response, central policies can be important for enabling, stimulating and supporting local authorities in dealing with decentralized tasks and responsibilities. Local performance can easily be undermined without such central policies, even causing performance to fall below minimum levels. Therefore, I conclude, *the coordinative model of governance and its associated central guidance are important in providing a robust foundation of policies for the development of more dynamic policy approaches that are tailored to the circumstances.*

I fully support the argument that decentralization *can be* a means of renewing governance so as to cope more effectively with the challenges with which our complex 21st-century societies are confronting us. In addition, however, I argue that when we do so, we also need a robust basis upon which to build the kind of dynamic and tailor-made approaches that are meant to respond to these challenges. Therefore, I argue, the coordinative model need not be considered as a relic of the past, even if we shift towards more dynamic and tailor-made approaches for dealing with more complex issues. Rather, it can be maintained as a complementary model in addition to the new and flexible dynamic and tailor-made approaches. In doing so, the coordinative model can provide the necessary *guarantees* for meeting minimum outcomes and be a *stimulus* for local parties in developing and delivering these flexible dynamic and tailor-made approaches.

Governance renewal is often a reaction to the limitations and disadvantages of the coordinative model. Nevertheless, I argue, this book shows us not to overlook the merits of relying on the coordinative model of governance, even when pursuing governance renewal (see also Ashford 2002, Ashford & Hall 2011). I base this argument on the idea that adapting to complex circumstances is not just about developing policy approaches tailored to these complex circumstances. It is also about developing a robust foundation of policies that enable, stimulate and support the development of such dynamic approaches. Instead of adopting a perspective whereby we rely *either* on robust coordinative policies *or* on dynamic tailor-made approaches, my view is that we need robust coordinative policies *and* dynamic tailor-made approaches. This conclusion, as we have just seen, is also not uncommon in debates within complexity theory.

Final conclusions

The aim here was to navigate the plural governance landscape by emphasizing the complementarity between an object-oriented and an intersubjective-oriented approach. It was in doing so that we could also meet a second ambition, which was to reflect on recent governance renewal operations. Illustrated by my empirical inquiry, we have seen that a post-contingency approach can help us navigate the plural governance landscape and allow for the desired reflection on governance renewal operations. This does not imply that a post-contingency approach tells us which approach to adopt. Rather, as we have seen in Chapters 3 and 4, it provides us with a framework that can help to illuminate the consequences of choices between various approaches to governance, also in relation to the circumstances involved. It is subsequently up to decision makers, such as elected officials, to choose how they wish to respond to these consequences.

A post-contingency approach has also helped us to reflect on two recent attempts to renew governance through decentralization operations and helped us to understand when certain consequences can be expected when pursuing decentralization. It also showed us how we can subsequently provide practitioners with advice for responding to such consequences. Mostly, as I argued, it is important to be prudent when dismantling the coordinative model of governance, also when we accept the benefits of developing more dynamic and tailor-made approaches for dealing with the complex policy issues with which 21st-century societies are confronting us. The coordinative model is neither dated nor outdated. Instead, it is relevant in fulfilling a complementary role in terms of more flexible and dynamic policy approaches. It can provide the robust foundation of policies that guarantees that minimum policy outcomes are met, and that supports and stimulates the development of flexible, dynamic and even adaptive policy approaches. It is in doing so, I also conclude, that these policies can help environmental policies in their pursuit of the desired 'liveable' future for our spatial environment.

Bibliography

Adger, W.N. (2010) Social Capital, Collective Action, and Adaptation to Climate Change, in: M. Voss (ed.) *Der Klimawandel*, Springer, Berlin.

Ahern, J. (2011) From Fail-Safe to Safe-to-Fail: Sustainability and Resilience in the New Urban World, in: *Landscape and Urban Planning*, Vol. 100, No. 4, pp. 341–343.

Alawattage, C., D. Wickramasinghe (2007) *Management Accounting Change: Approaches and Perspectives*, Routledge, London.

Alexander, E.R. (1992) *Approaches to Planning: Introducing Current Planning Theories, Concepts and Issues*, Second Edition, Gordon & Breach, Amsterdam.

Allender, E., M.C. Loui, K.W. Regan (1999) Chapter 27: Complexity Classes, Chapter 28: Reducibility and Completeness, Chapter 29: Other Complexity Classes and Measures, in: M.J. Atallah (ed.) *Algorithms and Theory of Computation Handbook*, CRC Press, Boca Raton, FL.

Allmendinger, P. (2001) *Planning in Postmodern Times*, Routledge, London.

Allmendinger, P. (2002a) *Planning Theory*, Palgrave, Basingstoke.

Allmendinger, P. (2002b) Towards a Post-Positivistic Typology of Planning Theory, in: *Planning Theory*, Vol. 1, No. 1, pp. 77–99.

Allmendinger, P. (2009) *Planning Theory, 2nd Edition*, Palgrave & Macmillan, Basingstoke.

Alonso, S., J. Keane, W. Merkel (eds.) (2011) *The Future of Representative Democracy*, Cambridge University Press, Cambridge.

Amburgey, T.L., T. Dacin (1994) As the Left Foot Follows the Right? The Dynamics of Strategic and Structural Change, in: *Academy of Management Journal*, Vol. 37, No. 6, pp. 1427–1452.

Andersen, M.S. (1997) Denmark, The Shadow of the Green Minority, in: M.S. Andersen, D. Liefferink (eds.) *The Innovation of EU Environmental Policy*, Scandinavian University Press, Oslo.

Andersen, M.S., D. Liefferink (1997) *The Innovation of EU Environmental Policy*, Scandinavian University Press, Oslo.

Andersen, M.S., R.V. van Kempen (eds.) (2001) *Social Fragmentation, Social Exclusion and Urban Governance*, Ashgate, Aldershot.

Archer, M., R. Bhaskar, A. Collier, T. Lawson, A. Norrie (eds.) (1998) *Critical Realism: Essential Readings*, Routledge, London.

Arimura, T.H., A. Hibiki, H. Katayama (2008) Is a Voluntary Approach an Effective Environmental Policy Instrument?: A Case for Environmental Management Systems, in: *Journal of Environmental Economics and Management*, Vol. 55, No. 3, pp. 281–295.

Arts, B., J. van Tatenhove (2005) Policy and Power – A Conceptual Framework Between the 'Old' and 'New' Policy Idioms, in: *Policy Sciences*, Vol. 37, No. 3–4, pp. 339–356.

Arts, E.J.M.M. (1998) *EIA Follow-up: On the Role of Ex Post Evaluation in Environmental Impact Assessment*, Geo Press, Groningen.

Ashford, N.A. (2002) Government and Environmental Innovation in Europe and North-America, in: *American Behavioral Scientist*, Vol. 45, No. 9, pp. 1417–1434.

Ashford, N.A., R.P. Hall (2011) The Importance of Regulation-Induced Innovation for Sustainable Development, in: *Sustainability*, Vol. 3, No. 1, pp. 270–292.

Association of London Government (2004) Communication from the Commission "Towards a Thematic Strategy on the Urban Environment", *COM (2004) 60 Final, A Response from the Association of London Government*, Association of London Government, London.

Axelrod, R.S., S.D. VanDeveer (2014) Introduction: Governing the Global Environment, in: R.S. Axelrod, S.D. VanDeveer (eds.) *The Global Environment; Institutions, Law and Policy*, Fourth Edition, Sage, Thousand Oaks, CA.

Baartmans, R., B. Van Geleuken (2004) *Decentralisatie in het milieubeleid, een verkenning van wensen, ervaringen en aanbevelingen*, TNO-MEP, Apeldoorn.

Bache, I., M. Flinders (2004) Multi-Level Governance and the Study of British Politics and Government, in: *Public Policy and Administration*, Vol. 19, No. 1, pp. 31–52.

Backes, Ch.W., T. Van Nieuwerburgh, R.B.A. Koelemeijer (2005) Transformation of the First Daughter Directive on Air Quality in Several EU Member States and its Application in Practice, in: *European Environmental Law Review*, Vol. 14, No. 6, pp. 157–164.

Baert, P. (1998) *Social Theory in the Twentieth Century*, Polity Press, Cambridge.

Baker, S., M. Kousis, D. Richardson, S. Young (1997) *The Politics of Sustainable Development: Theory, Policy and Practice within the European Union*, Routledge, London.

Bakker, R., R. Bergh, G. Beukema, F.Giskes, J.J. de Graeff, J. Rijsdijk, G. Rodewijk (2005) *Ontwikkel kracht! Eindrapport van de adviescommissie gebiedsontwikkeling*, Lysias Consulting Group B.V., Amersfoort.

Baldwin, R. (2005) Is Better Regulation Smarter Regulation?, in: *Public Law*, Autumn Volume, pp. 485–511.

Bardach, E. (1977) *The Implementation Game: What Happens After a Bill Becomes a Law*, MIT Press, Cambridge, MA.

Barnes, P.M., I.G. Barnes (1999) *Environmental Policy in the European Union*, Edward Elgar, Cheltenham.

Barrett, S., C. Fudge (eds.) (1981) *Policy and Action*, Methuen, London.

Barrow, C.J. (1995) *Developing the Environment: Problems and Management*, Longman, Harlow.

Batty, M. (2005) *Cities and Complexity; Understanding Cities with Cellular Automata, Agent-Based Models, and Fractals*, The MIT Press, Cambridge, MA.

Bauman, Z. (1992) *Intimations of Postmodernity*, Routledge, London.

Beauregard, R. (1989) Between Modernity and Postmodernity: The Ambiguous Position of U.S. Planning, in: *Environment and Planning D: Society and Space*, Vol. 7, No. 4, pp. 381–395.

Beck, U. (1992) *Risk Society: Towards a New Modernity*, Sage, New Delhi.

Beck, U. (2006) Living in the World Risk Society, in: *Economy and Society*, Vol. 35, No. 3, pp. 329–345.

Beck, U., E. Beck-Gernsheim (2002) *Individualization: Institutionalized Individualism and Its Social and Political Consequences*, Sage, London.

Beder, Sharon (2006) *Environmental Principles and Policies: An Interdisciplinary Approach*, UNSW Press, Sydney and Earthscan, London.

290 *Bibliography*

Benz, A., B. Eberlein (1999) The Europeanization of Regional Policies: Patterns of Multi-level Governance, in: *Journal of European Public Policy*, Vol. 6, No. 2, pp. 329–348.

Berger, P.L., T. Luckmann (1967) *The Social Construction of Reality: A Treatise in the Sociology of Knowledge*, Anchor, New York.

Bergström, O., P. Dobers (2000) Organising Sustainble Development: From Diffusion to Translation, in: *Sustainable Development*, Vol. 8, pp. 167–179.

Berkes, F., J. Colding, C. Folke (2000) Rediscovery of Traditional Ecological Knowledge as Adaptive Management, in: *Ecological Applications*, Vol. 10, pp. 1251–1262.

Berkes, F., J. Colding, C. Folke (eds.) (2003) *Navigating Social-Ecological Systems: Building Resilience for Complexity and Change*, Cambridge University Press, Cambridge.

Berkes, F., and C. Folke (eds.) (1998) *Linking Social and Ecological Systems: Management Practices and Social Mechanisms for Building Resilience*, Cambridge University Press, New York.

Berkhout, F., A. Smith, A. Stirling (2004) Socio-technological Regimes and Transition Contexts, in: B. Elzen, F.W. Geels, K. Green (eds.) *System Innovation and the Transition to Sustainability: Theory, Evidence and Policy*, Edward Elgar, Cheltenham.

Bernard, N. (2002) *Multi-Level Governance in the European Union*, Kluwer Law International, The Hague.

Bernstein, R. (1983) *Beyond Objectivism and Relativism: Science, Hermeneutics and Praxis*, University of Pennsylvania Press, Philadelphia.

Berting, J. (1996) Over Rationaliteit en Complexiteit, in: P. Nijkamp, W. Begeer, J. Berting (eds.) *Denken Over Complexe Besluitvorming: een panorama*, Sdu Uitgevers, Den Haag.

Bertolini, L. (2010) Complex Systems, Evolutionary Planning? in: E. Silva, G. de Roo (eds.) *A Planners Meeting with Complexity*, Ashgate, Aldershot.

Bettis, R.A., C.K. Prahalad (1995) The Dominant Logic: Retrospective and Extension, in: *Strategic Management Journal*, Vol. 16, pp. 5–14.

Bevir, M. (2004) Governance and Interpretation: What Are the Implications of Postfoundationalism, in: *Public Administration*, Vol. 82, No. 3, pp. 605–625.

Bevir, M. (2010) *Democratic Governance*, Princeton University Press, Princeton.

Bevir, M., R.A.W. Rhodes (2001) *A Decentered Theory of Governance: Rational Choice, Institutionalism, and Interpretation*, Institute of Governmental Studies, UC Berkeley.

Bevir, M., R.A.W. Rhodes (2003) Searching for Civil Society: Changing Patterns of Governance in Britain, in: *Public Administration*, Vol. 1, pp. 41–62.

Biermann, F. (2007) 'Earth System Governance' as a Cross Cutting Theme of Global Change Research, in: *Global Environmental Change*, Vol. 17, pp. 326–337.

Biersteker, T.J. (1989) Critical Reflections on Post-Positivism in International Relations, in: *International Studies Quarterly*, Vol. 33, No. 3, pp. 263–267.

Billig, M., H.W. Simons (eds.) (1994) *After Postmodernism: Reconstructing Ideology Critique*, Sage, London.

Blanco, I., V. Lowndes, L. Pratchett (2011) Policy Networks and Governance Networks, in: *Political Studies Review*, Vol. 9, No. 3, pp. 297–308.

Boelens (2010) Theorizing Practice and Practicing Theory: Outlines for an Actor-Relational-Approach in Planning, in: *Planning Theory*, Vol. 9, No. 1, pp. 28–62.

Bogason, P., J.A. Musso (2006) Democratic Prospects of Network Governance, in: *The American Review of Public Administration*, Vol. 36, No. 1, pp. 13–18.

Booher, D.E., J.E. Innes (2001) *Network Power in Collaborative Planning*, Working Paper, University of California, Institute of Urban and Regional Development, Berkeley.

Bibliography 291

Boonstra, B., L. Boelens (2011) Self-organization in Urban Development: Towards a New Perspective on Spatial Planning, in: *Urban Research and Practice*, Vol. 4, No. 2, pp. 99–122.

Borst, H., G. de Roo, H. Voogd, H. van der Werf (1995) *Milieuzones in Beweging*, Samsom H.D. Tjeenk Willink, Alphen aan den Rijn.

Börzel, T.A. (1997) What's So Special About Policy Networks? An Exploration of the Concept and Its Usefulness in Studying European Governance, in: *European Integration Online Papers (EIoP), European Community Studies Association Austria (ECSA-A)*, Vol. 1, No. 8, pp. 1–28.

Börzel, T.A. (1998) Shifting or Sharing the Burden. The Europeanisation of Environmental Policy in Spain and Germany, in: *European Planning Studies*, Vol. 6, No. 5, pp. 537–553.

Bouwman, M.E., E. Hermans, G. de Roo en J. Visser (2005) *Nieuwbouw naast de snelweg. Laveren tussen procesoptimalisatie*, norm en verantwoordelijkheid, Groningen.

Brandsen, T., M. Holzer (eds.) (2009) *The Future of Governance, Selected Papers from the Fifth Transatlantic Dialogue on Public Administration, National Center for Public Performance*, Rutgers, Newark, NJ.

Breheny, M., R. Rockwood (1993) Planning the Sustainable City Region, in: A. Blowers (ed.) *Planning for a Sustainable Environment*, Earthscan, London.

Bressers, H., S.M. Kuks (2003). What Does Governance Mean? From Concept to Elaboration, in: H. Bressers, W.A. Rosenbaum (eds.) *Achieving Sustainable Development: The Challenge of Governance Across Social Scales*, Preager, New York-Westpoint-London.

Briassoulis, H. (1999) Policy and Practice, Who Plans Whose Sustainability? Alternative Roles for Planners, in: *Journal of Environmental Planning and Management*, Vol. 42, No. 6, pp. 889–902.

Briassoulis, H. (2005) Policy Integration: Realistic Expectation or Elusive Goal?, in: H. Briassoulis (ed.) *Policy Integration for Complex Environmental Problems, The Example of Mediterranean Desertification*, Ashgate, Aldershot.

Brown, K. (2011) *Rethinking Progress in a Warming World: Interrogating Climate Resilient Development*, paper presented at the EADI/DSA Annual Conference, University of York, 19–22 September.

Brown, R.H. (1994) Rhetoric, Textuality, and the Postmodern Turn in Sociological Theory, in: S. Seidman (ed.) *The Postmodern Turn, New Perspectives on Social Theory*, Cambridge University Press, Cambridge.

Brugge, R. van der, R. van Raak (2007) Facing the Adaptive Management Challenge: Insights from Transition Management, in: *Ecology and Society*, Vol. 12, No. 2, p. 33.

Bryson, J.M., A.L. Delbecq (1979) A Contingent Approach to Strategy and Tactics in Project Planning, in: *American Planning Association Journal*, Vol. 45, pp. 167–179.

Bulkeley, H. (2005) Reconfiguring Environmental Governance: Towards a Politics of Scales and Networks, in: *Political Geography*, Vol. 24, pp. 875–902.

Bulkeley, H., V.C. Broto (2012) Government by Experiment? Global Cities and the Governing of Climate Change, in: *Transactions of the Institute of British Geographers*, Vol. 38, No. 3, pp. 361–375.

Burns, T., G.M. Stalker (1961) *The Management of Innovation*, Tavistock, London.

Burr, V. (2003) *Social Constructionism*, Routledge, Hove.

Burström, F., I. Korhonen (2001) Municipalities and Industrial Ecology: Reconsidering Municipal Environmental Management in: *Sustainable Development*, Vol. 9, pp. 36–46.

Busch, P., H. Jörgens (2005) International Patterns of Environmental Policy Change and Convergence, in: *European Environment*, Vol. 15, pp. 80–101.

292   *Bibliography*

Busscher, T., C. Zuidema, T. Tillema, E.J.M.M. Arts (2014) Bringing Gaps: Governing Conflicts between Transport and Environmental Policies, in: *Environment and Planning A*, Vol. 46, No. 3, pp. 666–681.

Butler, C.D., W. Oluoch-Kosura (2006) Linking Future Ecosystem Services and Future Human Well-being, in: *Ecology and Society*, Vol. 11, No. 1, pp. 30–46.

Buttel, F.H. (2000) Ecological Modernization as Social Theory, in: *Geoforum*, Vol. 31, pp. 57–65.

Byrne, D.S. (2003) Complexity Theory and Planning Theory: A Necessary Encounter, in: *Planning Theory*, Vol. 2, No. 3, pp. 171–178.

Cameron, J., N. Odendaal, A. Todes (2004) Integrated Area Development Projects: Working towards Innovation and Sustainability, in: *Urban Forum*, Vol. 15, No. 1, pp. 311–331.

Campbell, S. (1996) Green Cities, Growing Cities, Just Cities: Urban Planning and the Contradictions of Sustainable Development, in: *Journal of the American Planning Association*, Vol. 62, No. 3, pp. 296–312.

Carlsson, L. (2000) Policy Networks as Collective Action, in: *Policy Studies Journal*, Vol. 28, pp. 502–520.

Carson, R. (1962) *Silent Spring*, New York, Houghton Mifflin.

Castells, M. (1996) *The Rise of the Network Society, The Information Age: Economy, Society and Culture* Vol. I., Blackwell, Oxford.

Castells, M. (2000) *The Rise of the Network Society*, Blackwell, Oxford.

Casti, J. (1979) *Connectivity, Complexity and Catastrophe in Large-Scale Systems*, John Wiley & Sons, Toronto, Canada.

Castree, N. (2008a) Neoliberalising Nature: The Logics of Deregulation and Reregulation, in: *Environment and Planning A*, Vol. 40, pp. 131–152.

Castree, N. (2008b) Neoliberalising Nature: Processes, Effects, and Evaluations, in: *Environment and Planning A*, Vol. 40, pp. 153–173.

CEC, Commission of the European Communities (1990) *Green Paper on the Urban Environment*, Commission of the European Communities, Brussels.

CEC, Commission of the European Communities (1992) *The Fifth Environment Action Programme of the European Community Environment: Towards Sustainability. A European Community Programme of Policy and Action in Relation to the Environment and Sustainable Development*, Office for Official Publications of the European Communities, Luxembourg.

CEC, Commission of the European Communities (1996) *European Sustainable Cities*, Office for Official Publications of the European Communities, Luxembourg.

CEC, Commission of the European Communities (1997) *Communication from the Commission, 'Towards an Urban Agenda in the European Union'*, CEC, Brussels.

CEC, Commission of the European Communities (2001a) *The Sixth Environment Action Programme of the European Community Environment 2010: Our Future, Our Choice*, Office for Official Publications of the European Communities, Luxembourg.

CEC, Commission of the European Communities (2001b) *European Governance, a White Paper*, Office for Official Publications of the European Communities, Luxembourg.

CEC, Commission of the European Communities (2004) *Towards a Thematic Strategy on the Urban Environment*, Office for Official Publications of the European Communities, Luxembourg.

CEC, Commission of the European Communities (2005) *Commission Staff Working Document, Annex to the Communication on Thematic Strategy on the Urban Environment Impact Assessment*, Commission of the European Communities, Brussels.

Bibliography 293

CEC, Commission of the European Communities (2006) *Thematic Strategy on the Urban Environment, COM (2006) 718 Final*, Brussels.

CEC, Commission of the European Communities (2007a) *Integrated Environmental Management, Guidance in Relation to the Thematic Strategy on the Urban Environment*, Technical Report 2007–013, CEC, Brussels.

CEC, Commission of the European Communities (2007b) *Sustainable Urban Transport Plans Preparatory Document in Relation to the Follow-up of the Thematic Strategy on the Urban Environment*, Technical Report 2007–018, CEC, Brussels.

CEC, Commission of the European Communities (2009) *Promoting Sustainable Urban Development in Europe; Achievements and Opportunities, European Commission*, Directorate-General for Regional Policy, Brussels.

CEC, Commission of the European Communities (2010a) *Making Our Cities Attractive and Sustainable; How the EU Contributes to Improving the Urban Environment*, Office for Official Publications of the European Communities, Luxembourg.

CEC, Commission of the European Communities (2010b) *EUROPE 2020: A European Strategy for Smart, Sustainable and Inclusive Growth, COM (2010) 2020*, Commission of the European Communities, Brussels.

CEC, Commission of the European Communities (2012) *Smart Cities and Communities, European Innovation Partnership, COM (2012) 4701*, Commission of the European Communities, Brussels.

CEMR (2004) *CEMR Response to the Communication 'Towards a Thematic Strategy on the Urban Environment'*, Council of European Municipalities and Regions, Brussels.

Chandler, A.D. (1962) *Strategy and Structure: Chapters in the History of the Industrial Enterprise*, MIT Press, Cambridge.

Christensen, K.S. (1985) Coping with Uncertainty in Planning, in: *Journal of the American Planning Association*, Vol. 51, No. 1, pp. 63–73.

Coaffee, J., N. Headlam (2007) Pragmatic Localism Uncovered: Experiences from English Urban Policy Reform, in: *Geoforum*, Vol. 39, pp. 1585–1599.

Cohen, M.D., J.G. March, J.P. Olsen (1972) A Garbage Can Model of Organizational Choice, in: *Administrative Science Quarterly*, Vol. 17, pp. 1–25.

Committee of the Regions (2012) *Delivering on the Europe 2020 Strategy: Handbook for Local and Regional Authorities*, European Union Committee of the Regions, Brussels.

Connelly, S. (2004) *What Is Sustainable Development? Mapping a Contested Concept*, paper presented at the Association of the European Schools of Planning Conference, 1–4 July, Grenoble, Department of Town & Regional Planning, University of Sheffield, Sheffield.

Connelly, S. (2007) Mapping Sustainable Development as a Contested Concept, in: *Local Environment: The International Journal of Justice and Sustainability*, Vol. 12, No. 3, pp. 258–278.

Connerly, E., K. Eaton, P. Smoke (2010) *Making Decentralization Work: Democracy, Development, and Security*, Lynne Rienner, Boulder, CO.

COSLA, Convention of Scottish Local Authorities (2005) *Response to the Communication 'Towards a Thematic Strategy on the Urban Environment'*, 25 January 2005.

Council of Europe (2007) *Leipzig Charter on Sustainable European Cities, Council of Europe*, available online, http://www.eu2007.de.

Council of the European Union (2004) *Commission Communication 'Towards a Thematic Strategy on the Urban Environment', Council Conclusions*, Luxembourg, 14 October 2004.

294   Bibliography

Creedy, A., G. Porter, G. de Roo, C. Zuidema (2007) *Towards Liveable Cities and Towns, Guidance for Sustainable Urban Management*, EUROCITIES, Brussels.

Cromby, J., D.J. Nightingale (1999) What's Wrong with Social Constructionism? In: D.J. Nightingale, J. Cromby (eds.) *Social Constructionist Psychology: A Critical Analysis of Theory and Practice*, Open University Press, Buckingham, pp. 1–21.

Davidoff, P. (1965) Advocacy and Pluralism in Planning, in: *Journal of the American Institute of Planners*, Vol. 31, No. 4, pp. 331–338.

Davoudi, S. (2012) Resilience: A Bridging Concept or a Dead End? in: *Planning Theory & Practice*, Vol. 13, No. 2, pp. 299–307.

Dawes, R.M. (1980) Social Dilemmas, in: *Annual Review of Psychology*, Vol. 31, pp. 169–193.

De Boer, J., C. Zuidema (2015) Towards an Integrated Energy Landscape, in: *Urban Design & Planning*, Vol. 9, Nos. 3–4, pp. 315–331. DOI: 10.1680/udap.14.00041.

De Lange, M. (1995) *Besluitvorming rond strategisch ruimtelijk beleid*, Thesis Publishers, Amsterdam.

De Leeuw, A.C.J. (1984) *De wet van de bestuurlijke drukte*, Van Gorcum, Assen/Maastricht.

De Pater, B., H. van der Wusten (1996) *Het geografisch huis; de opbouw van de wetenschap*, Coutinho, Bussum.

De Roo, G. (1993) *De LAT-relatie tussen ruimte en milieu*; Gebiedsgericht beleid moet niet meer nummer drie zijn, in: *ROM Magazine*, No. 12, pp. 16–19.

De Roo, G. (2002) *De Nederlandse planologie in weelde gevangen: van ruimtelijk paradijs naar een leefomgeving in voortdurende staat van verandering . . .*, Oratie, Faculteit der Ruimtelijke Wetenschappen, Rijksuniversiteit Groningen, Groningen.

De Roo, G. (2003) *Environmental Planning in the Netherlands: To Good to Be True: From Command and Control Planning Towards Shared Governance*, Ashgate, Aldershot.

De Roo, G. (2004) *Toekomst van het Milieubeleid: Over de regels en het spel van decentralisatie – een bestuurskundige beschouwing*, Koninklijke Van Gorcum BV, Assen.

De Roo, G. (2007) Shifts in Planning Practice and Theory: From a Functional Towards a Communicative Rationale, in: G. de Roo, G. Porter (eds.) *Fuzzy Planning: The Role of Actors in a Fuzzy Governance Environment*, Ashgate, Aldershot.

De Roo, G. (2012) Spatial Planning, Complexity and a World 'Out of Equilibrium': Outline of a Non-linear Approach to Planning, in: G. De Roo, J. Hillier, J. Van Wezemael (eds.) *Complexity and Planning: Systems, Assemblages and Simulations*, Ashgate, Farnham – Surrey.

De Roo, G., D. Miller (eds.) (2000) *Compact Cities and Sustainable Urban Development; A Critical Assessment of Policies and Plans from an International Perspective*, Ashgate, Aldershot.

De Roo, G., E.A. Silva (2010) *A Planners' Meeting with Complexity*, Ashgate, Farnham – Surrey.

De Roo, G., G. Porter (eds.) (2007) *Fuzzy Planning: The Role of Actors in a Fuzzy Governance Environment*, Ashgate, Aldershot.

De Roo, G., H. Voogd (2004) *Methodologie van Planning; Over processen ter beïnvloeding van de fysieke leefomgeving*, Countinho, Bussum.

De Roo, G., J. Hillier, J. Van Wezemael (2012) *Complexity and Planning: Systems, Assemblages and Simulations*, Ashgate, Farnham – Surrey.

De Roo, G., J. Visser, C. Zuidema (2011) *Smart Methods for Environmental Externalities; Towards Liveable Cities, Enhancing the Interaction Between Urban Planning and Environmental Health and Hygiene*, Ashgate, Farnham – Surrey.

Bibliography 295

De Roo, G., M. Schwartz (eds.) (2001) *Omgevingsplanning, een innovatief proces; Over integratie, participatie en de gebiedsgerichte aanpak*, Sdu Uitgevers, Den Haag.

De Roo, G., W.S. Rauws (2011) Positioning Planning in the World of Order and Chaos . . . On Perspectives, Behaviour and Interventions in a Non-linear Environment, in: J. Portugali (ed.) *Complexity Theories of Cities*, Springer, Berlin.

De Vries, M.S. (2000) The Rise and Fall of Decentralization: A Comparative Analysis of Arguments and Practices in European Countries, in: *European Journal of Political Research*, Vol. 38, No. 2, pp. 193–224.

De Zeeuw, F.H. Puylaert, H. Werksma (2009) *Doorbreek de Impasse tussen Milieu en Gebiedsontwikkeling*, TU Delft, Delft.

Dedeurwaedere, T. (2005) The Contribution of Network Governance to Sustainable Development, in: *Les séminaires de l'Iddri*, no 13, Paris, Iddri.

deLeon, P., L. deLeon (2002) Whatever Happened to Policy Implementation? An Alternative Approach, in: *Journal of Public Administration Research and Theory*, Vol. 12. No. 4, pp. 467–488.

Deleuze, G., F. Gautari (1994) *What Is Philosophy?* Verso, London.

Derksen, J.W. (2001) *Lokaal bestuur*, Elsevier bedrijfsinformatie's, Gravenhage.

Derthick, M. (1972) *New Towns in Town, Why a Federal Program Failed*, Urban Institute, Washington, DC.

Deshmukh, A.V., J.J. Talavage, M.M. Barash (1998) Complexity in Manufacturing Systems: Part 1 – Analysis of Static Complexity, in: *IIE Transactions*, Vol. 30, No. 7, pp. 645–655.

Dewey, J. (1931) *Philosophy and Civilization*, G.P. Putnam's Sons, New York.

DG Environment (2004a) *EU Expert Group on the Urban Environment*, Meeting of 7 April 2004, Minutes, DG Environment, Brussels, available online, http://ec.europa.eu/environment/urban/expert_group_urban_env.htm.

DG Environment (2004b) *EU Expert Group on the Urban Environment*, Meeting of 24 September 2004, Minutes, DG Environment, Brussels, available online, http://ec.europa.eu/environment/urban/expert_group_urban_env.htm.

DG Environment (2005) *EU Expert Group on the Urban Environment*, Meeting of 17 May 2005, Minutes, DG Environment, Brussels (see http://ec.europa.eu/environment/urban/expert_group_urban_env.htm)

Dhakal, S., H. Imura (2003) Policy-based Indicator Systems: Emerging Debates and Lessons, in: *Local Environment: The International Journal of Justice and Sustainability*, Vol. 8, No. 1, pp. 113–119.

Dietz, T., E. Ostrom, P.C. Stern (2003) The Struggle to Govern the Commons, in: *Science*, Vol. 302, No. 5652, pp. 1907–1912.

Dixon, D., J.P. Jones (1998) My Dinner with Derrida, or Spatial Analysis and Post-structuralism Do Lunch, in: *Environment and Planning A*, Vol. 30, pp. 247–260.

Dóci, G., E. Vasileiadou, A.H. Petersen (2014) *Exploring the Transition Potential of Renewable Energy Communities* (Internal Report, Ecis Working Papers, No. 1406), Technische Universiteit Eindhoven, Eindhoven.

Dodder, R., R. Dare (2000) *Complex Adaptive Systems and Complexity Theory: Interrelated Knowledge Domains, ESD.83: Research Seminar in Engineering Systems*, Massachusetts Institute of Technology, Boston.

Donaldson, L. (1996) The Normal Science of Structural Contingency Theory, in: S.R. Clegg, C. Hardy, W.R. Nord (eds.) *Handbook of Organization Studies*, Sage, Thousand Oaks, CA.

296   *Bibliography*

Donaldson, L. (2001) *The Contingency Theory of Organisations*, Sage, Thousand Oaks, CA.

Donaldson, L. (2006) The Contingency Theory of Organizational Design: Challenges and Opportunities, in: R.M. Burton, B. Eriksen, D.D. Hakonsson, C.C. Snow (eds.) *Organization Design: The Evolving State-of-the-Art*, Springer, New York.

Douglas, M., A. Wildavsky (1983) *Risk and Culture*, University of California Press, Berkeley and Los Angles.

Dovers, S. (2003) Processes and Institutions for Resource and Environmental Management: Why and How to Analyse, in S. Dovers, S. Wild River (eds.) *Managing Australia's Environment*, The Federation Press, Sydney, pp. 3–14.

Drazin, R., A.H. Van de Ven (1985) Alternative Forms of Fit in Contingency Theory, in: *Administrative Science Quarterly*, Vol. 30, pp. 514–539.

Dryzek, J. (1987) *Rational Ecology*, Basil Blackwell, Oxford.

Dryzek, J.S. (1990) *Discursive Democracy: Politics, Policy, and Political Science*, Cambridge University Press, New York.

Dubbeldam, M., W. Goedmakers (2003) *Integraal Management: Instrument Van Verandering?* Van Gorcum, Assen.

Dublin City Council (2007) *Dublin Bay, An Integrated Economic, Cultural and Social Vision for Sustainable Development*, Dublin City Council, Dublin.

Duit, A. (2012) Adaptive Capacity and the Ecostate, in E. Boyd, C. Folke (eds.) *Adapting Institutions: Governance, Complexity and Social-Ecological Resilience*, Cambridge University Press, Cambridge.

Duit, A., V. Galaz (2008) Governance and Complexity – Emerging Issues for Governance Theory, in: *Governance: An International Journal of Policy, Administration, and Institutions*, Vol. 21, No. 3, pp. 331–335.

Duit, A., V. Galaz, K. Eckerberg, J. Ebberson, (2010) Introduction: Governance, Complexity and Resilience, in: *Global Environmental Change*, Vol. 20, pp. 363–368.

Eckersley, R. (1992) *Environmentalism and Political Theory*, UCL Press, London.

ECWM (2001) *Het gemeentelijk Milieubeleidsplan, van plan naar proces*, Evaluatiecommissie Wet Milieubeheer, Den Haag.

Edelman, M.J. (1977) *Political Language: Words that Succeed and Policies that Fail*, Academic Press, New York.

Edwards, D., M. Ashmore, J. Potter (1995) Death and Furniture: The Rhetoric, Politics and Theology of Bottom-Line Arguments against Relativism, in: *History of the Human Sciences*, Vol. 8, No. 2, pp. 25–49.

EEB, European Environmental Bureau (2003) *Thematic Strategy on the Urban Environment, Stakeholder Consultation Report for the NGO Sector*, EEB, Brussels.

EEB, European Environmental Bureau (2004) *NGO Stakeholder Recommendations on the Communication 'Towards a Thematic Strategy on the Urban Environment'*, EEB, Brussels.

EEB, European Environmental Bureau (2006) *Commission Adopts Hands-off Approach on Urban Environment, European Environmental Bureau*, available online, www. eeb.org

EGUE, EC Expert Group on the Urban Environment (1998) *Response of the EC Expert Group on the Urban Environment on the Communication 'Towards an Urban Agenda in the European Union'*. Brussels.

EGUE, EC Expert Group on the Urban Environment (2004) *Draft Final Report of the Working Group on Urban Environmental Management Plans and Systems*, Brussels.

Bibliography 297

EGUE, EC Expert Group on the Urban Environment (2005) *Final Report of the Working Group on Urban Environmental Management Plans and Systems, European Union Expert Group on the Urban Environment*, Working Group on Urban Environmental Management Plans and Systems, Brussels.

Ekins, P. (1999) European Environmental Taxes and Charges: Recent Experiences, Issues and Trends, in: *Ecological Economics*, Vol. 31, pp. 39–62.

Elmore, R.F. (1979) Backward Mapping: Implementation Research and Policy Decisions, in: *Political Science Quarterly*, Vol. 94, pp. 601–616.

Emery, F., E. Trist (1967) The Causal Texture of Organizational Environments, in: *Human Relations*, Vol. 18, pp. 21–32.

Enderlein, H., S. Wätli, M. Zürn (eds.) (2010) *Handbook on Multi-level Governance*, Edward Elgar, Cheltenham.

Enviplans (2006) *Enviplans Guidelines, Integrated and Sustainable Planning and Management of the Urban Environment*, Enviplans, Brussels.

Ericksen, N.J., P.R. Berke, J.L. Crawford, J.E. Dixon (2004) *Plan-making for Sustainability: The New Zealand Experience*, Ashgate, Aldershot.

Eurocities (2004a) *Commissioner Margot Wallström Meets with Representatives of European Cities to Discuss Proposals for Urban Environment Strategy, News Release*, 15 March 2004.

Eurocities (2004b) *Towards a Strategy on the Urban Environment, a Eurocities Statement*, Eurocities, Brussels.

Eurocities (2005) *2nd Eurocities Statement on the Proposed Thematic Strategy on the Urban Environment Towards Urban Sustainability in Europe*, Eurocities, Brussels.

Eurocities (2006) *3rd Eurocities Statement on the Proposed Thematic Strategy on the Urban Environment Towards Urban Sustainability in Europe*, Eurocities, Brussels.

European Environmental Advisory Councils (EEAC) (2003) *Environmental Governance in Europe*, Lemma, Utrecht.

Evans, B., M. Joas, S. Sundback, K. Theobald (2005) *Governing Sustainable Cities*, Earthscan, London.

Evans, P.B. (ed.) (1995) *Livable Cities? Urban Struggles for Livelihood and Sustainability*, University of California Press, Berkeley.

Fainstein, S.S. (2000) New Directions in Planning Theory, in: *Urban Affairs Review*, Vol. 35, p. 451.

Faludi, A., H. Mastop (1997) Evaluation of Strategic Plans: The Performance Principle, in: *Environment and Planning B: Planning and Design*, Vol. 24, pp. 815–832.

Feitelson, E. (ed.) (2004) *Advancing Sustainability at the Sub-National Level: The Potential and Limitations of Planning*, Ashgate, Aldershot.

Fiedler, F.E. (1967) *A Theory of Leadership Effectiveness*, McGraw-Hill, New York.

Fiedler, F.E. (1994) *Leadership Experience and Leadership Performance*, US Army Research Institute for the Behavioral and Social Sciences, Alexandria, VA.

Fiorino, D.J. (2006) *The New Environmental Regulation*, MIT Press, Cambridge, MA.

Fisher, F., J. Forester (eds.) (1993) *The Argumentative Turn in Policy Analysis and Planning*, Duke University Press, Durham, NC.

Flameling, A. (2010) *Verantwoording van het Groepsrisico, een excuusparagraaf?!, Masterthesis, Master Environmental and Infrastructureplanning*, University of Groningen, Groningen.

Fleurke, F., R. Hulst (2006) A Contingency Approach to Decentralization, in: *Public Organization Review: A Global Journal*, Vol. 6, No. 1, pp. 39–58.

298   *Bibliography*

Fleurke, F., R. Hulst, P.J. de Vries (1997) *Decentraliseren met beleid*, Sdu Uitgevers, Den Haag.

Flynn, B. (2000) Is Local Truly Better? Some Reflections on Sharing Environmental Policy between Local Governments and the EU, in: *European Environment*, Vol. 10, pp. 75–84.

Flyvbjerg, B. (1998) *Rationality and Power*, University of Chicago Press, Chicago.

Folke, C. (2006) Resilience: The Emergence of a Perspective for Social–ecological Systems Analyses, in: *Global Environmental Change*, Vol. 16, pp. 253–267.

Folke, C., S.R. Carpenter, B. Walker, M. Scheffer, T. Chapin, J. Rockström (2010) Resilience Thinking: Integrating Resilience, Adaptability and Transformability, in: *Ecology and Society*, Vol. 15, No. 4, p. 20.

Folke, C., T. Hahn, P. Olsson, J. Norberg (2005) Adaptive Governance of Social-ecological Systems, in: *Annual Review of Environment and Resources*, Vol. 30, pp. 441–473.

Forester, J. (1989) *Planning in the Face of Power*, University of California Press, Berkeley.

Foxon, T. J. (2010) A Coevolutionary Framework for Analysing a Transition to a Sustainable Low Carbon Economy, in: *Ecological Economics*, Vol. 70, No. 12, pp. 2258–2267.

Foxon, T.J., L.C. Stringer, M.S. Reed, (2008) *Governing Long-term Social-ecological Change: What Can the Resilience and Transitions Approaches Learn from Each Other?*, paper prepared for presentation at the 2008 Berlin Conference, "Long-Term Policies: Governing Social-Ecological Change", Berlin, 22–23 February 2008.

Franke, T., W. Strauss, B. Reimann, K. J. Beckman (2007) *Integrated Urban Development – a Prerequisite for Urban Sustainability in Europe*, Deutches Institut für Urbanistik, Berlin.

Friedmann, J.R.P. (1973) *Retracking America: A Theory of Transactive Planning*, Anchor Press/Doubleday, Garden City, NY.

Friend, J.K., N. Jessop (1969) *Local Government and Strategic Choice*, Pergamon, Oxford.

Frizelle, G., E. Woodcock (1995) Measuring Complexity as an Aid to Developing Operational Strategy, in: *International Journal of Operations & Production Management*, Vol. 15, No. 5, pp. 26–39.

Galaz, V. (2005) Social-Ecological Resilience and Social Conflict – Institutions and Strategic Adaptation in Swedish Water Management, in: *Ambio*, Vol. 34, No. 7, pp. 567–572.

Geelhoed, L.A. (1984) *Eindbericht Commissie vermindering en vereenvoudiging van overheidsregelingen*, TK, 17931, nr. 9.

Geels, F., R. Kemp (2000) *Transities vanuit socio-technisch perspectief, rapport voor de studie "Transities en Transitiemanagement" van ICIS en MERIT in opdracht van VROM tbv van NMP-4*, Okt 2000, UT, Enschede en MERIT, Maastricht.

Geels, F. W. (2011) The Multi-Level Perspective on Sustainability Transitions: Responses to Seven Criticisms, in: *Environmental Innovation and Societal Transitions*, Vol. 1, No. 1, pp. 24–40.

Geels, F.W., J.W. Schot (2007) Typology of Sociotechnical Transition Pathways, in: *Research Policy*, Vol. 36, pp. 399–417.

Gehring, T., S. Oberthür (2006) Introduction, in: S. Oberthür, T. Gehring (eds.) *Institutional Interaction in Global Environmental Governance; Synergy and Conflict Among International and EU Policies*, MIT Press, Cambridge.

Geldermalsen, Gemeente (2006) *Milieuprogramma 2007–2010*, Geldermalsen.

Gell-Man, M. (1994) *The Quark and the Jaguar*, W.H. Freeman, New York.

Gemeente Meppel (2004) *Wonen tussen de weilanden – Masterplan Nieuwveense Landen*, KRA landschapsarchitecten in opdracht van Gemeente Meppel, Meppel/Utrecht.

Gergen, K.J. (1999) *An Invitation to Social Construction*, Sage, London.

Bibliography 299

Gershberg, A.I. (1998) Decentralisation, Recentralisation and Performance Accountability: Building an Operationally Useful Framework for Analysis, in: *Development Policy Review*, Vol. 16, pp. 405–431.

Geurtsen, A. (1996) *Situatie- afhankelijke informatievoorziening, een onderzoek naar het verduidelijken van de realtie tussen situatie en informatieverzorging*, Drukkerij Elinkwijk, Utrecht.

Gibbs, D., A.E.G. Jonas (1999) Governance and Regulation in Local Environmental Policy: The Utility of a Regime Approach, in: *Geoforum*, Vol. 31, pp. 299–313.

Giddens, A. (1998) *The Third Way: The Renewal of Social Democracy*, Polity, Cambridge.

Gillespie, S. (2004). *Scaling up Community-Driven Development: A Synthesis of Experience*, FCND Discussion Paper No. 891, Food Consumption and Nutrition Division, International Food Policy Research Institute. Washington DC.

Giroux, H.A. (1988) *Schooling and the Struggle for Public Life: Critical Pedagogy in the Modern Age*, University of Minneapolis Press, Minneapolis.

Glachant, M. (ed.) (2003) *Implementing European Environmental Policy, the Impacts of Directives in the Member States*, Edward Elgar Publishing Ltd., Cheltenham.

Glasbergen, P., C. Dieperink (1989) Het Nationaal Milieubeleidsplan, de weg naar duurzaamheid? Over de noodzaak van bestuurskundige consequenties-analyses, in: *Milieu en recht*, No. 7–8, pp. 298–307.

Glasson, J., R. Therivel, A. Chadwick (2005) *Introduction to Environmental Impact Assessment*, Third Edition, Routledge, London.

Global Footprint Network (2010) *Climate Change Is Not the Problem*, Annual Report, Global Footprint Network, Oakland, available online, http://www.footprint network.org.

Godschalk, D.R. (2004) Land Use Planning Challenges: Coping with Conflicts in Visions of Sustainable Development and Livable Communities, in: *Journal of American Planning Association*, Vol. 70, No. 1, pp. 5–13.

Golub, J. (ed.) (1998) *New Instruments of Environmental Policy*, Routledge, London.

Gouldson, A., P. Roberts (2000) *Integrating Environment and Economy: Strategies for Local and Regional Government*, Earthscan, London.

Gresov, C. (1989) Exploring Fit and Misfit with Multiple Contingencies, in: *Administrative Science Quarterly*, Vol. 34, pp. 431–453.

Gresov, C., R. Drazin (1997) Functional Equivalence in Organization Design, in: *The Academy of Management Review*, Vol. 22, No. 2, pp. 403–428.

Grin, J., J. Rotmans, J., J. Schot (eds.) (2009) *Transitions to Sustainable Development: New Directions in the Study of Long Term Transformative Change, Routledge Studies in Sustainability Transitions*, Routledge, London.

Größler, A., A. Grübner, P.M. Milling (2006) Organisational Adaptation Processes to External Complexity, in: *International Journal of Operations & Production Management*, Vol. 26, No. 3/4, pp. 254–280.

Groff, R. (2004) *Critical Realism, Post-positivism, and the Possibility of Knowledge*, Routledge, New York.

Gunderson, L.H., C.S. Holling (2002) *Panarchy: Understanding Transformations in Human and Natural Systems*, Island Press, Washington, DC.

Gupta, J., K. Termeer, J. Klostermann, S. Meijerink, M. Van der Brink, P. Jong, S. Nooteboom (2010) Institutions for Climate Change: A Method to Assess the Inherent Characteristics of Institutions to Enable the Adaptive Capacity of Society, in: *Environmental Science & Policy*, Vol. 13, No. 6, pp. 459–471.

300   Bibliography

Guy Peters, B. (2002) Governance: A Garbage Can Perspective, in: *IHS Political Science Series*, No. 84.

Guy Peters, B. (2011) Governance as Political Theory, in: *Critical Policy Studies*, Vol. 5, No. 1, pp. 63–72.

Habermas, J. (1981) *The Theory of Communicative Action*, Beacon Press, London.

Hadfield, L., R. A. F. Seaton (1999). A Co-evolutionary Model of Change in Environmental Management, in: *Futures*, Vol. 31, No. 6, pp. 577–592.

Hajer, M., J.P.M. van Tatenhoven, C. Laurent (2004) *Nieuwe vormen van Governance, een Essay Over nieuwe vormen van bestuur*, RIVM rapport 500013004/2004, RIVM, Bilthoven.

Hall, P. (1980) *Great Planning Disasters*, Weidenfeld, London.

Hanf, K., F. Scharpf (eds.) (1978) *Interorganisational Policy Making: Limits to Coordination and Central Control*, Sage, London.

Hannan, M.T., J. Freeman (1977) The Population Ecology of Organizations, in: *American Journal of Sociology*, Vol. 83, pp. 929–984.

Hardin, G.J. (1968) The Tragedy of the Commons, in: *Science*, Vol. 162, pp. 1243–1248.

Harrison, P. (2002) A Pragmatic Attitude to Planning, in: P. Allmendinger, M. Tewdwr-Jones (eds.) *Planning Futures: New Directions for Planning Theory*, Routledge, London.

Harvey, D. (1989) *The Condition of Postmodernity: An Enquiry into the Origins of Cultural Change*, Blackwell, Cambridge, MA.

Harvey, D. (1997) The New Urbanism and the Communitarian Trap, in: *Harvard Design Magazine*, Winter/Spring, available online, www.gsd.harvard.edu/hdm/harvey.htm.

Harvey, D. (2005) *A Brief History of Neoliberalism*, Oxford University Press, Oxford.

Haughton, G., C. Hunter (1994) *Sustainable Cities*, Jessica Kingsley Publisher, London.

Healey, P. (1992) Planning through Debate—The Communicative Turn in Planning Theory, in: *Town Planning Review*, Vol. 63, No. 2, pp. 143–162.

Healey, P. (1997) *Collaborative Planning; Shaping Places in Fragmented Societies*, Macmillan Press Ltd., London.

Healey, P. (2007) Re-thinking Key Dimensions of Strategic Spatial Planning: Sustainability and Complexity, in: G. de Roo, G. Porter (eds.) *Fuzzy Planning: The Role of Actors in a Fuzzy Governance Environment*, Ashgate, Aldershot.

Healey, P. (2008) The Pragmatic Tradition in Planning Thought, in: *Journal of Planning Education and Research*, Vol. 28, pp. 277–292.

Healey, P., G. McDougall, M.J. Thomas (eds.) (1979) *Planning Theory, Prospects for the 1980s*, Pergamon Press, Oxford.

Heclo, H. (1978) Issue Networks and the Executive Establishment, in: A. King (ed.), *The New American Political System*, American Enterprise, Washington, DC, pp. 87–124.

Hekkert, M.P., M.E. Ossebaard (2010) *De innovatiemotor: Het versnellen van baanbrekende innovaties*, Koninklijke Van Gorcum, Assen.

Heller, J. (1961) *Catch-22*, Simon & Schuster, New York.

Héritier, A. (2010) Intergovernmental Decisions and Multi-level Governance: Producing Patchwork Policies, in: E. Ongaro, A. Massey, M. Holzer, E. Wayenberg (eds.) *Governance and Intergovernmental Relations in the European and the United States, Theoretical Perspectives*, Edward Elgar, Cheltenham.

Heylighen, F. (2008) Complexity and Self-Organization, in: M.J. Bates, M.N. Maack (eds.) *Encyclopedia of Library and Information Sciences*, CRC Press, Boca Raton, FL.

Heywood, A. (2002) *Politics*, Palgrave, Basingstoke.

Bibliography 301

Hildebrand, P.M. (2002) The European Community's Environmental Policy, 1957 to '1992': From Incidental Measures to an International Regime? in: A. Jordan (ed.) *Environmental Policy in the European Union: Actors, Institutions and Processes*, Earthscan, London.

Hillier, J. (2005) Straddling the Post-structuralist Abyss: Between Transcendence and Immanence?, in: *Planning Theory*, Vol. 4, No. 3, pp. 271–299.

Hillier, J. (2007) *Stretching Beyond the Horizon: A Multiplanar Theory of Spatial Planning and Governance*, Ashgate, Aldershot.

Hillier, J. (2008) Plan(e) Speaking: A Multiplanar Theory of Spatial Planning, in: *Planning Theory*, Vol. 7, No. 1, pp. 24–50.

Hjern, B., D.O. Porter (1981) Implementation Structures: A New Unit of Administrative Analysis, in: *Organisational Studies*, Vol. 2, No. 3, pp. 211–227.

Hoch, C. (1996) A Pragmatic Inquiry About Planning and Power, in: J. Seymour, L. Mandelbaum, R. Burchell (eds.) *Explorations in Planning Theory*, Center for Urban Policy Research, New Brunswick, NJ.

Hoch, C. (2006) What Can Rorty Teach an Old Pragmatist Doing Public Administration or Planning?, in: *Administration & Society*, Vol. 38, No. 3, pp. 389–398.

Hoch, C.J. (2007) Pragmatic Communicative Action Theory, in: *Journal of Planning Education and Research*, Vol. 26, pp. 271–283.

Holland, J.H. (1995) *Hidden Order: How Adaptation Builds Complexity*, Allison-Wesley, Reading, MA.

Holland, J.H. (1999) *Emergence: From Chaos to Order*, Perseus Books, Reading, MA.

Holmqvist, B. (2009) *Till relativismens försvar; Några kapitel ur relativismens historia – Boas, Becker, Mannheim och Fleck*, Symposion, Ågerup (Höör).

Holzinger, K., C. Knill, A. Lenschow (2006) *A Turn toward Soft Modes Governance in EU Environmental Policy – Viewed from a Policy Instruments Perspective*, paper prepared for the International Conference "Governance and Policy Making in the European Union", Osnabrück, 2–4 November.

Hooghe, L., G. Marks (2001) *Multilevel Governance and European Integration*, Rowman and Littlefield, Lanham, MD.

Hovik, S., M. Reitan (2004) National Environmental Goals in Search of Local Institutions, in: *Environment and Planning C: Government and Policy*, Vol. 22, pp. 687–699.

Huberts, L.W.J.C, S. Verberk e.a. (2005) *Overtredende overheden, Op zoek naar de omvang aan regelovertreding door overheden*, Boom Juridische Uitgevers, Den Haag.

Humblet, A.G.M., G. de Roo (eds.) (1995) *Afstemming door inzicht; Een analyse van gebiedsgerichte milieubeoordelingsmethoden ten behoeve van planologische keuzes*, Geo Pers, Groningen.

Hunt, S.D. (1991) Positivism and Paradigm Dominance in Consumer Research: Toward Critical Pluralism and Rapprochement, in: *Journal of Consumer Research*, Vol. 8, No. 8, pp. 32–44.

I&M Ministerie van Infrastructuur & Milieu (2014) *Modernisering Milieubeleid* (10 March 2014), Ministerie van Infrastructuur & Milieu, Den Haag.

ICLEI, International Council for Local Environmental Initiatives (2004) *ICLEI Position Paper on the Communication of the European Commission "Towards a Thematic Strategy on the Urban Environment" (Com (2004) 60)*, ICLEI, Freiburg.

IMPEL, EU Network of Implementation and Enforcement of Environmental Law (1998) *Interrelationship Between IPPC, EIA, SEVESO Directives and EMAS Regulation, Final Report December 1998*, IMPEL, Brussels.

302 Bibliography

Ingram, H.M., D.E. Mann (eds.) (1980) *Why Policies Succeed or Fail*, Sage, Beverly Hills.

Innes, J. (1996) Planning through Consensus Building – A New View of the Comprehensive Planning Ideal, in: *American Planning Association Journal*, Vol. 62, No. 4, pp. 460–472.

Innes, J., D. Booher (2010) *Planning with Complexity: An Introduction to Collaborative Rationality for Public Policy*, Routledge, Abingdon.

IPO, VROM, VNG, UvW en V&W (2003) *Nulmeting Milieuhandhaving, Nederland per 1 januari 2003*, Ministerie van VROM, Den Haag.

IPO, VROM, VNG, UvW en V&W (2005) *Eindmeting Professionalisering Milieuhandhaving, gebaseerd op cijfers per 12 mei 2005*, Ministerie van VROM, Den Haag.

Jänicke, M. (2008) Ecological Modernisation: New Perspectives, in: *Journal of Cleaner Production*, Vol. 16, No. 5, pp. 557–565.

Jänicke, M., H. Jörgens (2006) New Approaches to Environmental Governance, in: K. Jacob, M. Jänicke (eds.) *Environmental Governance in Global Perspective: New Approaches to Ecological and Political Modernisation*, Freie Universität Berlin, Berlin.

Jeppesen, T. (2000) EU Environmental Policy: Ever Closer to the Citizen? European Environment, in: *European Environment*, Vol. 10, No. 2. pp. 96–105.

Jessop, B. (1994) Post-Fordism and the State, in: A. Amin (ed.) *Post-Fordism: A Reader*, Blackwell, Oxford.

Jessop, B. (2002) *The Future of the Capitalist State*, Polity Press, Cambridge.

John, P. (2001) *Local Governance in Western Europe*, Sage, London.

Jones, C., W.S. Hesterly, S.P. Borgatti (1997) A General Theory of Network Governance: Exchange Conditions and Social Mechanisms, in: *Academic Management Review*, Vol. 22, No. 4, pp. 911–945.

Jorand, O., A. Perez-Uribe, H. Volken, A. Upegui, Y. Thoma, E. Sanchez, F. Mondada, P. Retornaz (2009) Noise and Bias for Free: PERPLEXUS as a Material Platform for Embodied thought-experiments, *Proceedings of the 2nd Symposium on Computing and Philosophy*, Edinburgh, April 2009, pp. 28–34.

Jordan, A. (2008). The Governance of Sustainable Development: Taking Stock and Looking Forwards, in: *Environment and Planning C*, Vol. 26, No. 1, 17–33.

Jordan, A., A. Lenschow (2008) *Innovation in Environmental Policy, Integrating the Environment for Sustainability*, Edward Elgar, Cheltenham.

Jordan, A., A. Lenschow (2010) Environmental Policy Integration: A State of the Art Review, in: *Environmental Policy and Governance*, Vol. 20, pp. 147–158.

Jordan, A.J. (1998) *The Politics of a Multi-Level Environmental Governance System: European Union Environmental Policy At 25*, CSERGE Working Paper, Norwich.

Jordan, A.J. (1999) *Subsidiarity and Environmental Policy: Which Level of Government Should Do What in the European Union?*, CSERGE Working Paper, Norwich.

Jordan, A.J. (2000) The Politics of Multilevel Environmental Governance: Subsidiarity and Environmental Policy in the European Union, in: *Environment and Planning A*, Vol. 32, No. 7, pp. 1307–1324.

Jordan, A.J. (2001) Environmental Policy (Protection and Regulation), in: N. Smelser, P. Baltes (eds.) *International Encyclopaedia of the Social and Behavioural Sciences*, Elsevier, Oxford.

Jordan, A.J. (ed.) (2002) *Environmental Policy in the European Union: Actors, Institutions and Processes*, Earthscan, London.

Jordan, A.J., J. Fairbrass (2002) *EU Environmental Governance: Uncomplicated and Predictable Policy-Making*, CSERGE Working Paper, Norwich.

Jordan, A.J., G. Pridham, M. Cini, D. Konstadakopulos, M. Porter, B. Flynn (2002) *Environmental Governance in Europe: An Ever Closer Ecological Union?* Oxford University Press, Oxford.

Bibliography 303

Jordan, A.J., R. Wurzel, A. Zito (2005) The Rise of 'New' Policy Instruments in Comparative Perspective: Has Governance Eclipsed Government?, in: *Political Studies*, Vol. 53, No. 3, pp. 477–496.

Jordan, G. (1990) Sub-governments, Policy Communities and Networks: Refilling Old Bottles?, in: *Journal of Theoretical Politics*, Vol. 2, No. 3, 319–338.

Kamphorst, D.A. (2006) *Veranderend milieubeleid: een onderzoek naar decentralisatie, doorwerking en integratie van milieubeleid in een stedelijke Context*, Geo Pers, Groningen.

Kastelein, J. (1990) *Modulair Organiseren, Tussen Autonomie en centrale beheersing*, Wolters-Noordhof, Groningen.

Kastelein, J. (1994) *Organisatiekunde tussen empirie, theorie en praktijk, Fac. Der PSCW, vakgroep Bestuurskunde*, afscheidscollege 22 oktober 1993, Universiteit van Amsterdam.

Kates, R. T. Parris, A. Leisorowitz (2005) What Is Sustainable Development?, in: *Environment*, Vol. 47, pp. 8–21.

Kauffman, S. (1995) *At Home in the Universe: The Search for Laws of Self-Organization and Complexity*, Oxford University Press, Oxford.

Kauffman, S.A. (1991) Antichaos and Adaptation, in: *Scientific American*, Vol. 265, No. 2, pp. 64–70.

Kearns, A., R. Paddison (2000) New Challenges for Urban Governance, in: *Urban Studies*, Vol. 37, No. 5–6, pp. 845–850.

Kemp, R., S. Parto, R.B. Gibson (2005) Governance for Sustainable Development: Moving from Theory to Practice, in: *International Journal of Sustainable Development*, Vol. 8, No. 1–2, pp. 12–30.

Kemp, R., J. Rotmans, D. Loorbach (2007). Assessing the Dutch Energy Transition Policy: How Does It Deal with Dilemmas of Managing Transitions?, in: *Journal of Environmental Policy & Planning*, Vol. 9, Nos. 3–4, pp. 315–331.

Kemp, R., S. Van den Bosch (2006) *Transitie-experimenten. Praktijkexperimenten met de potentie bij te dragen aan transities*, Publicatie 01. Kenniscentrum voor duurzame systeeminnovaties en transities.

Kenis, P., V. Schneider (1991) Policy Network and Policy Analysis: Scrutinizing a New Analytical Toolbox, in: B. Marin, R. Mayntz (eds.) *Policy Networks, Empirical Evidence and Theoretical Considerations*, Campus Verlag, Frankfurt am Main.

Keuning, D. (1978) *Management en organisatie*, Stenfert Kroese, Leiden.

Keynes, J.M. (1936) *The General Theory of Employment, Interest and Money*, Macmillan Cambridge University Press, Cambridge.

Kickert, W.J.M. (ed.) (1993) *Veranderingen in management en organisatie bij de rijksoverheid*, Samson H.D. Tjeenk Willink, Alphen aan den Rijn.

Kickert, W.J.M., E.H. Klijn, J.F.M. Koppenjan (eds.) (1997) *Managing Complex Networks: Strategies for the Public Sector*, Sage, London.

Klijn, E.H., C. Skelcher (2007) Democracy and Governance Networks: Compatible or Not?, in: *Public Administration*, Vol. 85, No. 3, pp. 587–609.

Klijn, E.J. (2008) Governance and Governance Networks in Europe, in: *Public Management Review*, Vol. 10, No. 4, pp. 505–525.

Kloppenberg, J.T. (1998) Pragmatism: An Old Name for Some New Ways of Thinking? in: M. Dickstein (ed.) *The Revival of Pragmatism*, Sage, London.

Knill, C., A. Lenschow (eds.) (2000) *Implementing EU Environmental Policy: New Directions and Old Problems*, Manchester University Press, Manchester.

Knill, C., A. Lenschow (2004) Modes of Governance in the European Union, Towards a Comprehensive Evaluation, in: J. Jordana, D. Levi-Faur, (eds.) *The Politics of Regulation, Institutions and Regulatory Reforms for the Age of Governance*, Elgar, Cheltenham.

304 *Bibliography*

Knorr-Cetina, K., M. Mulkay (eds.) (1983) *Science Observed*, Sage, London.

Kohler-Koch, B., V. Buth (2009) *Civil Society in EU Governance: Lobby Groups Like Any Other?*, TranState Working Papers, No. 108.

Kooiman, J. (ed.) (1993) *Modern Governance: New Government-Society Interactions*, Sage, London.

Kooiman, J. (2003) *Governing as Governance*, Sage, London.

Koppejan, J., M. Kars, H. van der Voort (2011) Politicians as Metagovernors – Can Metagovernance Reconcile Representative Democracy and Network Reality?, in: J. Torfing, P. Triantafillou (eds.) *Interactive Policymaking, Metagovernance and Democracy*, ECPR Press, Colchester.

Kreukels, A.M.J. (1980) *Planning en planningproces; Een verkenning van sociaalwetenschappelijke theorievorming op basis van ruimtelijke Planning*, VUGA bv, Den Haag.

Kronsell, A. (1997) The EU's Fifth Environmental Action Programme as a 'garbage-can', in: M.S. Andersen, D. Liefferink (eds.) *The Innovation of EU Environmental Policy*, Scandinavian University Press, Oslo.

Lachman, B.E., F. Camm, S.A. Resetar (2001) *Integrated Facility Environmental Management Approaches: Lessons from Industry for Department of Defense Facilities*, Rand, Santa Monica.

Lafferty, W.M., E. Hovden (2003) Environmental Policy Integration: Towards an Analytical Framework, in: *Environmental Politics*, Vol. 12, No. 3, pp. 1–22.

La Porte, T. (1975) *Organized Social Complexity. Challenge to Politics and Policy*, Princeton University Press, Princeton, NJ.

Lauber, V. (1997) Austria: A Latecomer which became a Pioneer, in: M.S. Andersen, D. Liefferink (eds.) *The Innovation of EU Environmental Policy*, Scandinavian University Press, Oslo.

Lawrence, P., J. Lörsch (1967) Differentiation and Integration in Complex Organizations, in: *Administrative Science Quarterly*, Vol. 12, pp. 1–30.

Leach, M., A.C. Stirling, I. Scoones (2010) *Dynamic Sustainabilities: Technology, Environment, Social Justice*, Earthscan, London.

Lemos, M.C., A. Agrawal (2006) Environmental Governance, in: *Annual Review of Environment and Natural Resources*, Vol. 31, No. 3, pp. 297–325.

Lempert, R.J., S.W. Popper, S.C. Bankes (2003) *Shaping the Next One Hundred Years: New Methods for Quantitative, Long-Term Policy Analysis*, RAND, MR-1626-RPC.

Levy, D, P. Newell (eds.) (2004) *The Business of Global Environmental Governance*, MIT Press, Cambridge, MA.

Lewin, R. (1992) *Complexity, Life at the Edge of Chaos*, Maxwell Macmillan International, New York, Oxford, Singapore, Sydney.

Liefferink, D. (1997) The Netherlands, a Net Exporter of Environmental Policy Concepts, in: M.S. Andersen, D. Liefferink (eds.) *The Innovation of EU Environmental Policy*, Scandinavian University Press, Oslo.

Liefferink, D., J. van Tatenhove Hajer (2002) The Dynamics of European Nature Policy: The Interplay of Front Stage and Back Stage by 2030, in: W. Kuindersma (ed.) *Bestuurlijke trends en het natuurbeleid, Planbureaustudies no.3*, Natuurplanbureau, Wageningen.

Liefferink, D., M. Van der Zouwen (2004) The Netherlands: The Advantage of Being 'Mr Average', in: A. Jordan, D. Liefferink (eds.) *Environmental Policy in Europe: The Europeanization of National Environmental Policy*, Routledge, London.

Lindblom, C.E. (1959) The Science of Muddling through, in: *Public Administration Review*, Vol. 2, pp. 79–88.

Lipsky, M. (1980) *Street-Level Bureaucracy*, Russell Sage Foundation, New York.

Bibliography 305

Liveable Cities (2005a) *Case Report Aalborg, Prepared for the Liveable Cities Workshop*, 8–10 September, 2005, Accessible on Request of the Author.

Liveable Cities (2005b) *Case Report Venice, Prepared for the Liveable Cities Workshop*, 1–3 December 2005, Accessible on Request of the Author.

Liveable Cities (2005c) *Case Report The Hague, Prepared for the Liveable Cities Workshop*, 31 March, 1 April 2005, Accessible on Request of the Author.

Liveable Cities (2006a) *Case Report Malmö, Prepared for the Liveable Cities Workshop*, 15–16 February 2006, Accessible on Request of the Author.

Liveable Cities (2006b) *Case Report Bourgas, Prepared for the Liveable Cities Workshop*, 10–12 May 2006, Accessible on Request of the Author.

Liveable Cities (2006c) *Case Report Bristol, Prepared for the Liveable Cities Workshop*, 6–8 September 2006, Accessible on Request of the Author.

Liveable Cities (2006d) *Copenhagen, Prepared for the Liveable Cities Workshop*, 15–18 February 2006, Accessible on Request of the Author.

Liveable Cities (2006e) *Case Report Lille, Prepared for the Liveable Cities Workshop*, 15–17 June 2006, Accessible on Request of the Author.

Liverman, D.M. (2004) Who Governs, at What Scale, and at What Price? Geography, Environmental Governance, and the Commodification of Nature, in: *Annals of the Association of American Geographers*, Vol. 94, No. 4, pp. 734–738.

Lõhkivi, E. (2001) Reconciling Realism and Relativism: A Study of Epistemological Assumptions in Relativist Sociology of Scientific Knowledge. Göteborg: Institutionen för idéhistoria och vetenskapsteori, Göteborgs Universitet, Rapport nr. 202.

Loorbach, D.A. (2007) *Transition Management: New Mode of Governance for Sustainable Development*, Erasmus University, Rotterdam.

Loorbach, D., (2010) Transition Management for Sustainable Development: A Prescriptive, Complexity-based Governance Framework, in: *Governance*, Vol. 23, No. 1, pp. 161–183.

Lowe, P., S. Ward (eds.) (1998) *British Environmental Policy and Europe, Politics and Policy in Transition*, Routledge, London.

Lucas, C. (1999) Quantifying Complexity Theory, CALResCo, Manchester, available online, www.calresco.org.

Lyall, C., J. Tait (2004) Foresight in a Multi-level Governance Structure: Policy Integration and Communication, in: *Science and Public Policy*, Vol. 31, No. 1, pp. 27–37.

Maas, H. (2005) *Gevolgen van decentralizatie op lokaal niveau*, Rijksuniversiteit Groningen, Groningen.

MacCallum, D., F. Moulaert, J. Hillier, S. Vicari Haddock (eds.) (2012) *Social Innovation and Territorial Development*, Ashgate, Farnham.

Mackay, E. (2005) *Procesarchitectuur, voorbeeldprojecten ontwikkelingsplanologie (Integrale Gebiedsontwikkeling)*, Cooper Feldman, Naarden.

Maddox, J. (1972) *The Doomsday Syndrome*, Macmillan, London.

Marsh, D., R.A.W. Rhodes (1992) New Directions in the Study of Policy Networks, in: *European Journal of Political Research*, Vol. 21, No. 1–2, pp. 181–205.

March, J.G., J.P. Olsen (1976) *Ambiguity and Choice in Organizations*, Universitetsforlaget, Bergen.

March, J.G., J.P. Olsen (1995) *Democratic Governance*, The Free Press, New York.

March, J.G., J.P. Olsen (2006) Elaborating the New Institutionalism, in R.A.W. Rhodes, S.A. Binder, B.A. Rockman (eds.) *The Oxford Handbook of Political Institutions*, Oxford University Press, Oxford.

Marks, G., L. Hooghe (2004) Contrasting Visions of Multi-Level Governance, in: I. Bache, M. Flinders (eds.) *Multi-level Governance*, Oxford University Press, Oxford.

306 Bibliography

Martens, K. (2007) Actors in a Fuzzy Governance Environment, in: G. de Roo, G. Porter (eds.) *Fuzzy Planning: The Role of Actors in a Fuzzy Governance Environment*, Ashgate, Aldershot.

Martin, R., J. Simmie (2008) Path Dependence and Local Innovation Systems in City-Regions, in: Innovation: Management, Policy & Practice, Vol. 10, Nos. 2–3, pp. 183–196.

Matland, R. (1995) Synthesizing the Implementation Literature: The Ambiguity-Conflict Model of Policy Implementation, in: *Journal of Public Administration Research and Theory*, Vol. 5, No. 2, pp. 145–174.

Matsuno, S. (1997) A Consideration on the Organizational Theory: From the Contingency Theory to the Post-Contingency Theory Research Reports of Ube National College of Technology, in: *Bulletin of Universities and Institutes*, Vol. 43, pp. 103–112.

Mayntz, R. (1980) *Gesetzgebung und Bürokratisierung, Wissenschafliche Auswertung der Anhörung zu Ursachen der Bürokratisierung in der öffentlichen Verwaltung*, Bundesministerium des Innern, Köln.

Mazamanian, D., P. Sabatier (1983) *Implementation and Public Policy*, Scott, Foresman, Glenview, IL.

Mazey, Richardson (2002) Environmental Groups and the EC: Challenges and opportunities, in: A. Jordan (ed.) *Environmental Policy in the European Union, Actors, Institutions & Processes*, Earthscan, London.

McCormick, J. (2001) *Environmental Policy in the European Union*, Basingstoke, Macmillan.

McElroy, M.W. (2000) Integrating Complexity Theory, Knowledge Management and Organizational Learning, in: *Journal of Knowledge Management*, Vol. 4, No. 3, pp. 195–203.

McGrath, R.G. (2008) Beyond Contingency: From Structure to Structuring in the Design of the Contemporary Organization, in: S.R. Clegg, C. Hardy, T.B. Lawrence, W.R. Nord (eds.) *The Sage Handbook of Organization Studies*, Sage, London.

McMillan, E. (2002) Considering Organisation Structure and Design from a Complexity Paradigm Perspective, in: G. Frizzelle, H. Richards, (eds.) *Tackling Industrial Complexity: The Ideas that Make a Difference, Institute of Manufacturing*, University of Cambridge, Cambridge.

Meadowcraft, J. (2009) What about the Politics? Sustainable Development, Transition Management and Long Term Energy Transitions, in: *Policy Science*, Vol. 42, pp. 323–340.

Meadows, D.H., D.L. Meadows, J. Panders, W.W. Behrens (1972) *The Limits to Growth: A Report for the Club of Rome's Project on the Predicament of Mankind*, Universal Books, New York.

Merton, R.K. (1968) *Social Theory and Social Structure*, Free Press, New York.

Meyer, A., A. Tsui, C.R. Hinings (1993) Configural Approaches to Organizational Analysis, in: *Academy of Management Journal*, Vol. 36, pp. 1175–1195.

Meyerson, M., E.C. Banfield (1955) *Politics, Planning, and the Public Interest*, The Free Press, Glencoe, IL.

Miers, D. (2001) Regulatory Reform Orders: A New Weapon in the Armoury of Law Reform, in: *Public Money & Management*, Vol. 21, pp. 29–34.

Milbrath, L.W. (1989) *Envisioning a Sustainable Society: Learning Our Way Out*, State University of New York Press, New York.

Miles, R.E., C.C. Snow (1978) *Organization Strategy, Structure, and Process*, McGraw-Hill, New York.

Miller, D. (1981) Toward a New Contingency Approach: The Search for Organizational Gestalts, in: *Journal of Management Studies*, Vol. 18, pp. 1–27.

Bibliography 307

Miller, D. (1986) Configurations of Strategy and Structure: Towards a Synthesis, in: *Strategic Management Journal*, Vol. 7, No. 3, pp. 233–249.

Miller, D. (1987) The Genesis of Configuration, in: *Academy of Management Review*, Vol. 12, pp. 686–701.

Miller, D. (1993) The Architecture of Simplicity, in: *Academy of Management Review*, Vol. 18, pp. 116–138.

Miller, D. (1994) What Happens After Success: The Perils of Excellence, in: *Journal of Management Studies*, Vol. 31, pp. 325–358.

Miller, D. (1996) Configurations Revisited, in: *Strategic Management Journal*, Vol. 17, No. 7, pp. 505–512.

Miller, D., G. de Roo (eds.) (2004) *Integrating City Planning and Environmental Improvement*, Second Edition, Ashgate Publishers Ltd. Aldershot, 3rd print.

Miller, D., G. de Roo (eds.) (2005) *Urban Environmental Planning; Policies, Instruments and Methods in an International Perspective*, Ashgate, Aldershot.

Miller, D., P. Friesen (1984) *Organizations: A Quantum View*, Prentice-Hall, Englewood Cliffs.

Mintzberg, H. (1973) Strategy Making in Three Modes, in: *California Management Review*, Vol. 16, No. 2, pp. 44–53.

Mintzberg, H. (1979) *The Structuring of Organizations*, Prentice-Hall, Englewood Cliffs, NJ.

Mintzberg, H. (1983) *Structure in Fives, Designing Effective Organisations*, Prentice Hall, London.

Mintzberg, H. (1990) Strategy Formation: Schools of Thought, in J.W. Fredrickson (ed.), *Perspectives on Strategic Management*, Harper Business, New York, pp. 105–235.

Mitlin, D., D. Satterthwaite (1996) Sustainable Development and Cities, in: C. Pugh (ed.) *Sustainability, the Environment and Urbanization*, Earthscan Publications, London, pp. 23–61.

Mobach, M.P., J.J.H. Rogier, A.C.J. de Leeuw (1998) *Fit and the Systems theory of Control, SOM Research Report*, University of Groningen, Groningen.

Mol, A., D. Sonnenfeld, G. Spaargaren (eds.) (2009) *The Ecological Modernisation Reader. Environmental Reform in Theory and Practice*, Routledge, London/New York.

Mol, A., G. Spaargaren (1992) Sociology, Environment, and Modernity: Ecological Modernization as a Theory of Social Change, in: *Society & Natural Resources: An International Journal*, Vol. 5, No. 4, pp. 323–344.

Mol, A.P.J. (1989) Hoofdlijnen van het Nationaal Milieubeleidsplan, in: E.C. van Ierland, A.P.J. Mol, W.A. Hafkamp (eds.) *Milieubeleid in Nederland; Reacties of het Nationaal Milieubeleidsplan*, Stenfert Kroese, Leiden.

Mol, A.P.J., V. Lauber, D. Liefferink (eds.) (2000) *The Voluntary Approach to Environmental Policy. Joint Environmental Policy-making in the EU and Selected Member States*, Oxford University Press, Oxford.

Moore, N., M. Scott (2005) *Renewing Urban Communities; Environment, Citizenship and Sustainability in Ireland*, Ashgate, Aldershot.

Morgeson F.P., D.A. Hofmann (1999) The Structure and Function of Collective Constructs: Implications for Multilevel Research and Theory Development, in: *The Academy of Management Review*, Vol. 24, No. 2, pp. 249–265.

MUE 25, Managing Urban Europe 25 (2007) *MUE-25 Guidance for Cities and Regions on Integrated Management*, Union of the Baltic Cities, Turku.

Mumby, D.K. (1997) Modernism, Postmodernism, and Communication Studies: A Rereading of an Ongoing Debate, in: *Communication Theory*, Vol. 7, pp. 1–28.

308   Bibliography

Nelissen, N.J.M. (1997) General Introduction, in: N. J. M. Nelissen, J. van der Straaten, L. Klinkers (eds.) *Classics in Environmental Studies: An Overview of Classic Texts in Environmental Studies*, International Books, Utrecht.

Nelissen, N.J.M. (2002) The Administrative Capacity of New Types of Governance, in: *Public Organization Review*, Vol. 2, No. 1, pp. 5–22.

Nelissen, N.J.M., A.J.A. Godfroij, P.J. Goede (eds.) (1996) *Vernieuwing van bestuur; Inspirerende visies*, Coutinho, Bussum.

Nelissen, N.J.M., J.B. Raadschelders (1999) *Renewing Government: Innovative and Inspiring Visions*, International Books, Utrecht.

Nelissen, N.J.M, M.L. Bemelmans-Videc, A. Godfroij, P. de Goede (1999) *Renewing Government, Innovative and Inspiring Visions*, International Books, Utrecht.

Nelissen, N., J. van der Straaten, J., L. Klinkers (1997) *Classics in Environmental Studies: An Overview of Classic Texts in Environmental Studies*, International Books, Utrecht.

Nevens, F., N. Frantzeskaki, L. Gorissen, D. Loorbach (2013) Urban Transition Labs: Co-creating Transformative Action for Sustainable Cities, in: *Journal of Cleaner Production*, Vol. 50, pp. 111–122.

Newig, J., D. Günther, C. Pahl-Wostl (2010) Synapses in the Network: Learning in Governance Networks in the Context of Environmental Management, in: *Ecology and Society*, Vol. 15, No. 4, p. 24.

Newman, J. (2001) *Modernising Governance. New Labour, Policy and Society*, Sage, London.

Nijkamp, P., W. Begeer, J. Berting (eds.) (1996) *Denken over Complexe Besluitvorming: een Panorama*, Sdu Uitgevers, Den Haag.

Norgaard, R. B. (1984). Coevolutionary Development Potential, in: *Land Economics*, Vol. 60, No. 2, 160–173.

Oates, W.E. (2001) *A Reconsideration of Environmental Federalism, Discussion Paper 01–54, Resources for the Future*, Washington, DC.

Offe, C. (1977) The Theory of the Capitalist State and the Problem of Policy Formation, in: L.N. Lindberg, A. Alford (eds.) *Stress and Contradiction in Modern Capitalism*, Heath, Lexington, KY, pp. 125–144.

OJEC (1973) *Programme of Action of the European Communities on the Environment, C. 112*, Office for Official Publications of the European Communities, Luxembourg.

Olmstead, S.M. (2013) Applying Market Principles to Environmental Policy, in: N.J. Vig, M.E. Kraft (eds.) *Environmental Policy: New Directions for the Twenty-first Century*, Eighth Edition, Sage, New York.

Olsen, J.P. (1970) Local budgeting – Decision Making or Ritual Act?, in: *Scandinavian Political Studies*, Vol. 5, pp. 85–118.

Olsson, P., L.H. Gunderson, S.R. Carpenter, P. Ryan, L. Lebel, C. Folke, C.S. Holling (2006) Shooting the Rapids: Navigating Transitions to Adaptive Governance of Social-ecological Systems, in: *Ecology and Society*, Vol. 11, No. 1, p. 18.

O'Riordan, T., H. Voisey (1998) *The Transition to Sustainability; The Politics of Agenda 21 in Europe*, Earthscan, London.

Osborne, D.E., T.A. Gaebler (1992) *Reinventing Government: How the Entrepreneurial Spirit is Transforming the Public Sector*, Addison-Wesley, Reading.

Osborne, S.P. (ed.) (2010) *The New Public Governance? Emerging Perspectives on the Theory and Practice of Public Governance*, Routledge, Abington.

Ostrom, E., B. Low, C. Simon, J. Wilson (2003) Redundancy and Diversity: Do they Influence Optimal Management?, in: F. Berkes, J. Colding, C. Folke (eds.) *Navigating*

Bibliography 309

Social-Ecological Systems: Building Resilience for Complexity and Change, Cambridge University Press, New York, pp. 83–114.

O'Toole, L.J. (2000) Research on Policy Implementation: Assessment and Prospects, in: *Journal of Public Administration Research and Theory*, Vol. 10, pp. 263–288.

Oxford English Dictionary (2009) available online, www.oed.com, accessed 6 May 2009.

Pahl-Wostl, C. (2009) A Conceptual Framework for Analysing Adaptive Capacity and Multi-level Learning Processes in Resource Governance Regimes, in: *Global Environmental Change*, Vol. 19, No. 3, pp. 354–365.

Papadopoulos, Y. (2000) Governance, Coordination and Legitimacy in Public Policies, in: *International Journal of Urban and Regional Research*, Vol. 24, No. 1, pp. 210–223.

Parsons, D.W. (1995) *Public Policy; An Introduction to the Theory and Practice of Policy Analysis*, Edward Elgar, Cheltenham.

Patomäki, H., C. Wight (2000) After Postpositivism? The Promises of Critical Realism, in: *International Studies Quarterly*, Vol. 44, No. 2, pp. 213–237.

Patton, M.Q. (2000) *Qualitative Evaluation and Research Methods*, Sage, London.

Payne, G.T. (2001) *Strategy and Structure Configurations, An Examination of Fit and Performance*, Texas Tech University, PhD Dissertation, Lubbock.

Pelkmans, J. (1997) *European Integration: Methods and Analysis*, Addison Wesley Longman Ltd., Harlow.

Persson, A. (2002) *Environmental Policy Integration, An Introduction*, Stockholm Environment Institute, Stockholm.

Pfeffer, J. (1982) *Organizations and Organization Theory*, Pitman, Marshfield, MA.

Phelan, S.E. (2001) What is Complexity Science, Really?, in: *Emergence*, Vol. 3, No.1, pp. 120–136.

Piattoni, S. (2009) Multi-level Governance: A Historical and Conceptual Analysis, in: *Journal of European Integration*, Vol. 31, No. 2, pp. 163–180.

Pierre, J. (1999) Models of Urban Governance, The Institutional Dimension of Urban Politics, in: *Urban Affairs Review*, Vol. 34, No. 3, pp. 372–396.

Pierre, J. (2009) Reinventing Governance, Reinventing Democracy?, in: *Policy & Politics*, Vol. 37, No. 4, pp. 591–609.

Pierre, J., B. Guy Peters (2000) *Governance, Politics and the State*, McMillan, London.

Pigliucci, M. (2000) Chaos and Complexity: Should We be Skeptical, in: *Skeptic*, Vol. 8, No. 3, pp. 62–70.

Pilot (2007) *Sustainable Urban Transport Planning SUTP Manual, Guidance for Stakeholders*, Pilot, Cologne.

Pinderhughes, R. (2004) *Alternative Urban Futures: Planning for Sustainable Development in Cities throughout the World*, Rowman and Littlefield, Lanham, MD.

Porter, G, J. Welsh Brown (1991) *Global Environmental Politics*, Westview Press, Boulder, CO.

Porter, L., S. Davoudi (2012) The Politics of Resilience for Planning: A Cautionary Note, in: *Planning Theory & Practice*, Vol. 13, No. 2, pp. 329–333.

Porter, M.E. (1980) *Competitive Strategy*, Free Press, New York.

Portugali, J. (2000) *Self-organization and the City*, Springer, Heidelberg.

Portugali, J. (2006) Complexity Theory as a Link Between Space and Place, in: *Environment and Planning A*, Vol. 38, pp. 647–664.

Portugali, J. (2010) Complexity Theories of Cities: First, Second or Third Culture of Planning?, in: G. de Roo, J. Hillier, J.E. van Wezenmael (eds.) (2012) *Complexity & Planning: Systems*, Assemblages and Simulations, Ashgate, Farnham.

310 *Bibliography*

Portugali, J. (2012) Complexity Theories of Cities: Achievements, Criticism and Potentials, in: J. Portugali, H. Meyer, E. Stolk, E. Tan (eds.) *Complexity Theories of Cities Have Come of Age: An Overview with Implications to Urban Planning and Design*, Springer-Verlag: Berlin Heidelberg.

Powell, W.W. (1990) Neither Market nor Hierarchy: Network Forms of Organizing, in: B. Staw, L.L. Cummings (eds.) *Research in Organizational Behavior*, JAI, Greenwich, CT, pp. 295–336.

Pozzebon, M. (2000) *Reviewing Contingency Arguments to Compose an IT Contingency Model, Cahier du GReSI no 0004*, HEC Montréal, Canada.

Pressman, J., A. Wildavsky (1973) *Implementation*, University of California Press, Berkeley.

Prigogine, I. (1997) *The End of Certainty*, The Free Press, New York.

Prigogine, I., I. Stengers (1984) *Orde uit chaos; De nieuwe dialoog tussen de mens en de natuur*, Uitgeverij Bert Bakker, Amsterdam.

Provan, K.G., P. Kenis (2007) Modes of Network Governance: Structure, Management, and Effectiveness, in: *Journal of Public Administration Research and Theory*, Vol. 18, pp. 229–252.

Prud'homme, R. (1994) *On the Dangers of Decentralization*, Policy Research Working Paper Series, TWUTD, No. 31005 (February), World Bank, Washington, DC.

Pugh, C. (ed.) (1996) *Sustainability, the Environment and Urbanization*, Earthscan, London.

Pugh, D.S. (1973) Measurement of Organization Structures, in: *Organizational Dynamics*, Vol. 1, No. 4, pp. 19–34.

Pugh, D.S., D. Hickson (eds.) (2000) *Writers on Organizations*, Fourth Edition, Penguin, Harmondsworth, UK.

Ramanujam, V., R. Varadarajan (1989) Research on Corporate Diversification: A Synthesis, in: *Strategic Management Journal*, Vol. 10, No. 6, pp. 523–551.

Reed, M., A.C. Evely, G. Cundill, I.R.A. Fazey, J. Glass, A. Laing, J. Newig, B. Parrish, C. Prell, C. Raymond, L. Stringer (2010) What Is Social Learning?, in: *Ecology and Society*, Vol. 15, No. 4, p. 1.

Reitan, M. (1997) Norway: A Case of 'Splendid Isolation', in: M.S. Andersen, D. Liefferink (eds.) *The Innovation of EU Environmental Policy*, Scandinavian University Press, Oslo.

Rhodes, R.A.W. (1981) *Control and Power in Central-local Government Relations*, Gower, Aldershot.

Rhodes, R.A.W. (1990) Policy Networks, A British Perspective, in: *Journal of Theoretical Politics*, Vol. 2, pp. 293–317.

Rhodes, R.A.W. (2000) *Understanding Governance: Policy Networks, Governance, Reflexivity and Accountability*, Open University Press, Buckingham/Philadelphia.

Rhodes, R.A.W. (2007) Understanding Governance: Ten Years on, in: *Organization Studies*, Vol. 28, No. 8, pp. 1243–1264.

Richardson, D. (1997) The Politics of Sustainable Development, in: S. Baker, M. Kousis, D. Richardson, S. Young (eds.) *The Politics of Sustainable Development: Theory, Policy and Practice within the European Union*, Routledge, London.

Richardson, J. (2006) *European Union, Power and Policy Making*, Third Edition, Routledge, Abington.

Rip, A., R. Kemp (1998) Technological Change, in: S. Rayner, E.L. Malone (eds) *Human Choice and Climate Change*, Battelle Press, Columbus, OH, pp. 327–399.

RMO, Raad voor Maatschappelijke Ontwikkeling (2002) *Bevrijdende kaders; Sturen op verantwoordelijkheid*, Sdu Uitgevers, Den Haag.

Bibliography 311

Rondinelli, D., J.R. Nellis (1986) Assessing Decentralization Policies in Developing Countries: A Case for Cautious Optimism, in: *Development Policy Review*, Vol. 4, No.1, pp. 3–23.

Rorty, R. (1980) *Philosophy and the Mirror of Nature*, Blackwell, Oxford.

Rorty, R. (1999) *Philosophy and Social Hope*, Penguin, New York.

Rosenthal, U. (1988) *Bureaupolitiek en Bureaupolitisme; om het behoud van een competitief overheidsbestel (inaugurele rede Leiden)*, Samsom H.D. Tjeenk Willink, Leiden.

Rosenthal, U., M.P.C.M. van Schendelen, A.B. Ringeling (1988) *Openbaar bestuur, organisatie, politieke omgeving en beleid, 4ᵉ druk*, Samsom H.D. Tjeenk Willink, Alphen a/d Rijn.

Rotmans, J., R. Kemp, M. van Asselt (2001) More Evolution than Revolution: Transition Management in Public Policy, in: *Foresight*, Vol. 3, No. 1, pp. 1–17.

Rowan, B. (2002) Teachers' Work and Instructional Management, Part II: Does Organic Management Promote Expert Teaching?, in: W.K. Hoy, C.G. Miskel (eds.) *Theory and Research in Educational Administration*, Information Age Publishing, Charlotte.

Rumelt, R.P. (1974) *Strategy, Structure, and Economic Performance*, Harvard University, Boston, MA.

Rydin, Y. (1997) Policy Networks, Local Discourses and the Implementation of Sustainable Development, in: S. Baker, M. Kousis, D. Richardson, S. Young (eds.) *The Politics of Sustainable Development: Theory, Policy and Practice Within the European Union*, London, Routledge.

Rydin, Y. (1998) The Enabling Local State and Urban Development: Resources, Rhetoric and Planning in East London, in: *Urban Studies*, Vol. 35, pp. 175–191.

Sabatier, P., H.C. Jenkins-Smith (eds.) (1993) *Policy Change and Learning: An Advocacy Coalition Approach*, Westview Press, Boulder, CO.

Sabatier, P.A. (1986) Top-Down and Bottom Up Approaches to Implementation Research: A Critical Analysis and Suggested Synthesis, in: *Journal of Public Policy*, Vol. 6, No.1, pp. 21–48.

Sager, T. (1994) *Communicative Planning Theory*, Avebury, Aldershot.

Sager, T. (2002) *Democratic Planning and Social Choice. Dilemmas. Prelude to Institutional Planning Theory*, Ashgate, Aldershot.

Saint-Andre, P. (2002) *The Ism Book: A Field Guide to Philosophy*, available online, www.ismbook.com, accessed Nov 2014.

Sassen, S. (2001) *The Global City: New York, London, Tokyo*, Princeton University Press, Princeton.

Sassen, S. (ed.) (2002) *Global Networks, Linked Cities*, Routledge, New York.

Sayer, A. (2000) *Realism and Social Science*, Sage, London.

Sbragia, A. (1996) The Push-Pull of Environmental Policy-Making, in: H. Wallace, W. Wallace (eds.) *Policy-Making in the European Union*, Third Edition, Oxford University Press, Oxford.

Scharpf, F.W. (1988) The Joint-Decision Trap. Lessons From German Federalism and European Integration, in: *Public Administration*, Vol. 66, No. 2. pp. 239–78.

Schön, D.A. (1983) *The Reflective Practitioner, How Professionals Think in Practice*, Basic Books, New York.

Schofield, J. (2001) Time for a Revival? Public Policy Implementation: A Review of the Literature and an Agenda for Future Research, in: *International Journal of Management Reviews*, Vol. 3, No. 3, pp. 245–263.

Scott, W.R. (1981) *Organizations: Rational, Natural and Open Systems*, Prentice-Hall, Englewood Cliffs, NJ.

312   *Bibliography*

Selman, P.H. (1996) *Local Sustainability: Managing and Planning Ecologically Sound Places*, St. Martin's Press, New York.

Senge, P.M. (1990) *The Fifth Discipline: The Art and Practice of the Learning Organisation*, Doubleday-Currency, New York.

Shaw, K. (2012) The Rise of the Resilient Local Authority?, in: *Local Government Studies*, Vol. 38, No. 3, pp. 281–300.

Shove, E., G. Walker (2007) CAUTION! Transitions Ahead: Politics, Practice and Sustainable Transition Management, in: *Environment and Planning A*, Vol. 39, pp. 763–770.

Siebert, H. (1987) *Economics of the Environment: Theory and Policy*, Springer-Verlag, Berlin.

Simmie, J. (2012a) Path Dependence and New Technological Path Creation in the Danish Wind Power Industry, in: *European Planning Studies*, Vol. 20, No. 5, pp. 753–772.

Simmie, J. (2012b) Special Issue: Path Dependence and New Path Creation in Renewable Energy Technologies, in: *European Planning Studies*, Vol. 20, No. 5, pp. 729–731.

Simmie, J. and R. L. Martin (2010) The Economic Resilience of Regions: Towards an Evolutionary Approach, *Cambridge Journal of Regions, Economy and Society*, Vol. 3, No. 1, pp. 27–43.

Simon, H. (1957) A Behavioral Model of Rational Choice, in: *Models of Man, Social and Rational: Mathematical Essays on Rational Human Behavior in a Social Setting*, Wiley, New York.

Simon, J. (1981) *The Ultimate Resource*, Princeton University Press, Princeton.

Simonis, U.E. (1988) International Environmental Problems and the Role of Legislators, in: A. Vlavianos-Arvanitis (ed.) *Biopolitics, the Bio-environment*, Vol. I, Editor. Primera Conferencia Internacional del B.I.O., May 1987, Athens.

Small, R. (1983) Nietzsche and a Platonist Idea of Cosmos: Centre Everywhere, Circumference Nowhere, in: *Journal of the History of Ideas*, Vol. 44, No. 1, pp. 89–104.

Smit, B., J. Wandel (2006) Adaptation, Adaptive Capacity and Vulnerability, in: *Global Environmental Change*, Vol. 16, pp. 282–292.

Smith, A., A. Stirling, F. Berkhout (2005) The Governance of Sustainable Socio-Technical Transitions, in: *Research Policy*, Vol. 34, No. 10, pp. 1491–1510. (Internal Report).

Smith, J. (1996) *Planning & Decision Making, An Active Learning Approach*, Blackwell, Oxford.

SNM (Stichting Natuur en Milieu) (2002) *Begin bij de fundering!, visie en discussie rond milieunormstelling voor de leefomgeving*, Stichting Natuur en Milieu (SNM), Utrecht.

SNM (Stichting Natuur en Milieu) (2005) *Bovenop handhaving, een actieve rol van gemeenteraden en provinciale staten bij de handhaving van milieuwetgeving*, Stichting Natuur en Milieu (SNM), Utrecht.

Sonnenfeld, D.A., A.P.J. Mol (2002) Globalization and the Transformation of Environmental Governance: An Introduction, in: *American Behavioral Scientist*, Vol. 45, No. 9, pp. 1417–1434.

Sorensen, A., P.J. Marcotullio, J. Grant (2004) *Towards Sustainable Cities, East Asian, North American and European Perspectives on Managing Urban Regions*, Ashgate, Aldershot.

Sørensen, E., J. Torfing (eds.) (2007) *Theories of Democratic Network Governance*, Palgrave Macmillan, Basingstoke.

Sørensen, E., J. Torfing (2007a) Introduction: Governance Network Research: Towards a Second Generation, in: E. Sørensen, J. Torfing (eds.) *Theories of Democratic Network Governance*, Palgrave Macmillan, Basingstoke.

Bibliography 313

Sørensen, E., J. Torfing (2009) Making Governance Networks Effective and Democratic through Metagovernance, in: *Public Administration*, Vol. 87, No. 2, pp. 234–258.

Spreeuwers, W.J., C. Zuidema, G. de Roo (2008) *De Basiskwaliteit voorbij . . .; een zoektocht naar milieukwaliteit op lokaal niveau*, Groningen, URSI.

Stacey, R.D. (1996) *Strategic Management and Organisational Dynamics*, Pitman Publishing, London.

Stavins, R.N. (2003) Experience with Market-based Environmental Policy Instruments, in: K. Mäler, J.R. Vincent (eds.) *Handbook of Environmental Economics, Economywide and International Environmental Issues*, Elsevier, Amsterdam, pp. 355–435.

Sterman, J.D. (2000) *Business Dynamics: Systems Thinking and Modeling for a Complex World*, Irwin/McGraw-Hill, Boston.

Steward, M. (2001*) Area-based Initiatives and Urban Policy*, paper presented at the Conference Area-based Initiatives in contemporary Urban Policy, Danish Building and Urban Research and European Urban Research Association, Copenhagen, May 2001.

Steward, P. (2001) Complexity Theories, Social Theory, and the Question of Social Complexity, in: *Philosophy of the Social Sciences*, Vol. 31, No. 3, September 2001, pp. 323–360.

Stoker, G. (1998) Governance as Theory: Five Propositions, in: *International Social Science*, No 155, pp. 119–131.

Stoker, G., K. Mossberger (2001) The Evolution of Urban Regime Theory, The Challenge of Conceptualization, in: *Urban Affairs Review*, Vol. 36, No. 6, pp. 810–835.

Susskind, L., J. Cruikshank (1987) *Breaking the Impasse: Consensual Approaches to Resolving Public Disputes*, New York, Basic Books.

Svedin, U., T. O'Riordan, A. Jordan (2001) Multilevel Governance for the Sustainable Transition, in: T. O'Riodan (ed.) *Globalism, Localism & Identity: Fresh Perspectives on the Transition to Sustainability*, Earthscan Publications Ltd, London.

Tarter, C.J., W.K. Hoy (1997) Toward a Contingency Theory of Decicion Making, in: *Journal of Education Administration*, Vol. 36, No. 3, pp. 212–228.

Taylor, F.W. (1911) *Principles of Scientific Management*, Harper & Brothers, New York and London.

Taylor, N. (1994) *Urban Planning Theory Since 1945*, Sage, London.

TCPA, Town and Country Planning Association (2004) *Response to the Communication 'Towards a Thematic Strategy on the Urban Environment'*, 15 April 2004.

Teisman, G.R. (1998) *Complexe besluitvorming; Een pluricentrisch perspectief op besluitvorming over ruimtelijke investeringen*, VUGA Uitgeverij bv, Den Haag.

Thompson, J.D., A. Tuden (1959) Strategies, Structures, and Processes of Organizational Decision, in J.D. Thompson, P.B. Hammond, R.W. Hawkes, B.H. Junker, A. Tuden (eds.) *Comparative Studies in Administration*, Pittsburgh University Press, Pittsburgh.

TK, Tweede Kamer (1989) *Nationaal Milieubeleidsplan 1989–1993, Kiezen of verliezen, Tweede Kamer, 1988–198, 21137, Nr. 1–2*, SdU Uitgevers, Den Haag.

TK, Tweede Kamer (1990) *Actieplan Gebiedsgericht Milieubeleid, Tweede Kamer, 1990–1991, 21896, Nr. 1–2*, SdU Uitgevers, Den Haag.

TK, Tweede Kamer (1993) *Nationaal Milieubeleidsplan 2; Milieu als maatstaf, Tweede Kamer, 1993–1994, 23560, Nr. 1–2*, SdU Uitgevers, Den Haag.

Toffler, A. (1970) *Future Shock*, Random House, New York.

Torfing, J. (2005) Governance Network Theory: Towards a Second Generation, in: *European Political Science*, Vol. 4, pp. 305–315.

Torfing, J., P. Triantafillou (eds.) (2011) *Interactive Policymaking, Metagovernance and Democracy*, ECPR Press, Colchester.

314 *Bibliography*

Trell, E.M., B. van Hoven (2010) Making Sense of Place: Exploring Creative and (Inter) Active Research Methods with Young People, in: *Fennia – International Journal of Geography*, Vol. 188, No. 1, pp. 91–104.

Turner, R.K. (1995) Environmental Economics and Managament, in: T. O'Riordan (ed.) *Environmental Science for Environmental Management*, Longman, Harlow, pp. 30–44.

Turok, I. (2004) The Rationale for Area-Based Policies: Lessons from International Experience, in: P. Robinson, J. McCarthy, C. Forster (eds.) *Urban Reconstruction in the Developing World: Learning Through an International Best Practice*, Sandown, Heinemann, pp. 405–412.

United Nations, Population Division (2003) *World Urbanization Prospects: The 2003 Revision, Data Tables and Highlights*, Population Division of the United Nations Secretariat.

Urwick, L.E. (1929) *The Meaning of Rationalisation*, Nisbet and Co., London.

Urwick, L.E. (1953) Administration in Theory and Practice, in: *British Management Review*, Vol. 8, pp. 37–59.

Urwick, L.E. (1956) The Manager's Span of Control, in: *Harvard Business Review*, May/June 1956.

Urwin, D. (1993) *The Community of Europe: A History of European Integration Since 1945*, Longman, London.

Van Ast, J.A., H. Geerlings (1995) *Milieukunde en milieubeleid, een introductie*, Samsom H.D. Tjeenk Willink, Alphen aan den Rijn.

Van der Graaf, H., R. Hoppe (1996) *Beleid en politiek; Een inleiding tot de beleidswetenschap en de beleidskunde*, Coutinho, Bussum.

Van de Ven, A.H., R. Drazin (1985) The Concept of Fit in Contingency Theory, in: B.M. Straw, L.L. Cummings (eds.) *Research in Organization Behavior*, Vol. 7. JAI Press, Greenwich.

Van der Valk, A. (1999) *Willens en wetens: Planning en wetenschap tussen wens en werkelijkheid*, Wageningen Universiteit, Wageningen.

Van Dijk, J. (1991) *The Network Society: Social Aspects of New Media*, Van Loghum, Houten.

Van Duinen, L. (2007) *Planning Imagery: The Emergence and Political Acceptance of Planning Concepts in Dutch National Planning*, University of Amsterdam, Amsterdam.

Van Gestel, R.A.J., M.L.M. Hertogh (2006) *Wat is regeldruk? Een verkennende internationale literatuurstudie*, Wolf Legal Publishers, Nijmegen.

Van Kersbergen, K., F. van Waarden (2004) 'Governance' as a Bridge between Disciplines: Cross-disciplinary Inspiration Regarding Shifts in Governance and Problems of Governability, Accountability and Legitimacy, in: *European Journal of Political Research*, Vol. 43, pp. 143–171.

Van Lange, P.A.M., W.B.G. Liebrand, D.M. Messick, H.A.M. Wilke (1992) Social Dilemmas: The State of the Art, in: W.B.G. Liebrand, D.M.M essick, H.A.M. Wilke (eds.) *Social Dilemmas: Theoretical Issues and Research Findings*, Pergamon, London, pp. 3–28.

Van Meter, D.S., C.E. Van Horn (1975) The Policy Implementation Process: A Conceptual Framework, in: *Administration and Society*, Vol. 6, No. 4, pp. 445–488.

Van Tatenhove, J. (1993) *Milieubeleid onder dak, beleidsvoeringsprocessen in het Nederlandse milieubeleid in de periode 1970–1990*, nader uitgewerkt voor de Gelderse Vallei, Pudoc, Wageningen.

Van Tatenhove, J., B. Arts, P. Leroy (eds.) (2000) *Political Modernisation and the Environment; The Renewal of Environmental Policy Arrangements*, Kluwer Academic Publishers. Dordrecht.

Vattimo, G. (2002) *Nietzsche: An Introduction*, The Athlone Press, London.

Bibliography 315

Venkatraman, N., J. Prescott, J. (1990) Environment-strategy Coalignment: An Empirical Test of its Performance Implications, in: *Strategic Management Journal*, Vol. 11, pp. 1–23.

Verbong, G., F. Geels (2007) The Ongoing Energy Transition: Lessons from a Sociotechnical, Multi-level Analysis of the Dutch Electricity System (1960–2004), in: *Energy Policy*, Vol. 35, pp. 1025–1037.

Verma, N. (1998) *Similarities, Connections, Systems: The Search for a New Rationality for Planning and Management*, Lexington Books, Lanham, MD.

VI (VROM-inspectie) (2003a) *Daadkracht in Handhaving – waar draait het om bij VROM*, Ministerie van VROM, Den Haag.

VI (VROM-inspectie) (2005a) *Naar een selectiever preventief toezicht op ruimtelijke plannen*, Ministerie van VROM, Den Haag.

VI (VROM-inspectie) (2005b) *Landelijke Rapportage VROM-brede gemeenteonderzoeken 2003–2004*, Ministerie van VROM, Den Haag.

VI (VROM-inspectie) (2006a) *Landelijke Rapportage VROM-brede gemeenteonderzoeken 2005*, Ministerie van VROM, Den Haag.

VI (VROM-inspectie) (2007a) *Landelijke Rapportage VROM-brede gemeenteonderzoeken 2006*, Ministerie van VROM, Den Haag.

VI (VROM-inspectie) (2007b) *Handreiking Ruimtelijke Ordening en Milieu, voor ruimtelijke plannen*, Ministerie van VROM, Den Haag.

Vig, N.J., M.E. Kraft (eds.) (2013) *Environmental Policy: New Directions for the Twenty-first Century*, Eighth Edition, Sage, New York.

Visser, J., C. Zuidema (2007) *De Milieuatlas; de leefomgeving in kaart*, SdU Uitgevers, Den Haag.

Vlek, C., L. Steg (2007) Human Behavior and Environmental Sustainability: Problems, Driving Forces and Research Topics, in: *Journal of Social Issues*, Vol. 63, No. 1, pp. 1–19.

VNG, Vereniging Nederlandse Gemeenten (1986) *Bedrijven en Milieuzonering, Groene Reeks, No. 80*, VNG Uitgeverij, Den Haag.

VoMIl, Ministerie van Volksgezondheid en Milieu (1972) *Urgentienota Milieuhygiëne*, Den Haag.

VoMIl, Ministerie van Volksgezondheid en Milieu (1976) *Nota Milieuhygiënische Normen*, Den Haag.

Voogd, H. (1995) *Methodologie van Ruimtelijke Planning*, Coutinho, Bussum.

Vranken, J. (2001) Unravelling the Social Strands of Poverty: Differentiation, Fragmentation, Inequality, and Exclusion, in: H.T. Andersen, R. Van Kempen (eds.) *Governing European Cities: Social Fragmentation, Social Exclusion and Urban Governance*, Ashgate, Aldershot, pp. 71–88.

VROM, Ministerie van Volkshuisvesting, Ruimtelijke Ordening en Milieubeheer (1983) *Plan Integratie Milieubeleid (PIM), Tweede Kamer, 1982–1983, 17931, No. 6*, Den Haag.

VROM, Ministerie van Volkshuisvesting, Ruimtelijke Ordening en Milieubeheer (1984a) *Meer dan de Som der Delen, Eerste Nota Over de Planning van het Milieubeleid, Tweede Kamer, 1983–1984, 18292, No. 2*, Den Haag.

VROM, Ministerie van Volkshuisvesting, Ruimtelijke Ordening en Milieubeheer (1984b) *Indicatief Meerjarenprogramma Milieubeheer 1985–1989, Tweede Kamer, 1984–1985, 18602, No. 2*, Den Haag.

VROM, Ministerie van Volkshuisvesting, Ruimtelijke Ordening en Milieubeheer (1989a) *Indicatief Meerjarenprogramma Milieubeheer 1985–1989, Tweede Kamer, 1984–1985, 18602, No. 2*, Den Haag.

316   *Bibliography*

VROM, Ministerie van Volkshuisvesting, Ruimtelijke Ordening en Milieubeheer (1989b) *Projectprogramma Cumulatie van bronnen en integrale milieuzonering, Integrale milieuzonering deel 2*, Directie Geluid, DGM, Den Haag.

VROM, Ministerie van Volkshuisvesting, Ruimtelijke Ordening en Milieubeheer (2001) *Een wereld en een wil; werken aan duurzaamheid. Nationaal Milieubeleidsplan 4*, VROM, Den Haag.

VROM, Ministerie van Volkshuisvesting, Ruimtelijke Ordening en Milieubeheer (2002) *Vaste waarden, nieuwe vormen, Milieubeleid 2002–2006*, Den Haag.

VROM, Ministerie van Volkshuisvesting, Ruimtelijke Ordening en Milieubeheer (2003) *Naar een optimale leefkwaliteit. Methoden voor milieukwaliteit in lokale plannen'* Den Haag.

VROM, Ministerie van Volkshuisvesting, Ruimtelijke Ordening en Milieubeheer (2004a) *Liveable Cities: A Dutch Recipe for Environmental Policy and Spatial Planning in the City & Environment Project*, Dutch Ministry of Spatial Planning and the Environment, The Hague.

VROM, Ministerie van Volkshuisvesting, Ruimtelijke Ordening en Milieubeheer (2004b) *Milieu in de Leefomgeving*, MILO Handreiking, Den Haag.

VROM, Ministerie van Volkshuisvesting, Ruimtelijke Ordening en Milieubeheer (2006) *Informatie Over de leefomgeving*, Actieprogramma Gezondheid en Milieu, Den Haag.

VROM, Ministerie van Volkshuisvesting, Ruimtelijke Ordening en Milieubeheer (2007a) *Commissie Herziening Handhavingsstelsel VROM-regelgeving*, VROM, Den Haag.

Vroom, C.W. (1981) Organisatie, in: L. Rademaker (ed.) *Sociologische grondbegrippen 1: Theorie en analyse*, Het Spectrum, Utrecht.

Vroom, V.H., P.W. Yetton (1973) *Leadership and Decision Making*, University of Pittsburgh Press, Pittsburgh, PA.

Walberg, H.J., S.J. Paik, A. Komukai, K. Freeman (2000) Decentralization: An International Perspective, in: *Educational Horizons*, Vol. 78, No. 3, pp. 153–164.

Waldrop, M.M. (1994) *De Rand van Chaos, Over complexe systemen*, Uitgeverij Contact, Amsterdam.

Walker, B., J.A. Meyers., (2004) Thresholds in Ecological and Social–ecological Systems: A Developing Database, in: *Ecology and Society*, Vol. 9, No. 2, p. 3.

Walker, B., S.R. Carpenter, J. Anderies, N. Abel, G.S. Cumming, M. Janssen, L. Lebel, J. Norberg, G.D. Peterson, R. Pritchard (2002) Resilience Management in Social–Ecological Systems: A Working Hypothesis for a Participatory Approach, in: *Conservation Ecology* (6), available online, www.ecologyandsociety.org/vol6/iss1/art14/print.pdf.

Wallace, W., H. Wallace (1997) *Policy-Making in the European Union*, Oxford University Press, Oxford.

Ward, S., R. Williams (1997) From Hierarchy to Networks? Sub-central Government and EU Urban Environmental Policy, in: *Journal of Common Market Studies*, Vol. 35, pp. 439–464.

Warner, W. (2010) Ecological Modernisation Theory: Towards a Critical Ecopolitics of Change?, in: *Environmental Politics*, Vol. 19, No. 4, pp. 538–556.

Wätli, S. (2004) How Multilevel Structures Affect Environmental Policy in Industrialized Countries, in: *European Journal of Political Research*, Vol. 43, pp. 599–634.

Watt, James H., Alicia J. Welch (1983) Effects of Static and Dynamic Complexity on Children's Attention and Recall of Televised Instruction, in: J. Bryant, D.R. Anderson (eds.) *Children's Understanding of Television: Research on Attention and Comprehension*, Academic Press, New York, pp. 69–102.

Bibliography 317

WCED, World Commission on Environment and Development (1987) *Our Common Future*, Oxford University Press, Oxford.

Weale, A. (1992) The New Politics of Pollution Control, Manchester University Press, Manchester.

Weber, M. (1922) The Nature of Social Action, in W.G. Runciman (1991) *Weber: Selections in Translation*, Cambridge University Press, Cambridge.

Weick, K.E. (1993) *Sensemaking in Organizations*, Sage, London.

Weidner, H., M. Jänicke (eds.) (2002) *Capcity Building in National Environmental Policy, A Comparative Study of 17 Countries*, Springer-Verlag, Berlin.

Weill, P., M.H. Olson (1989) An Assessment of Die Contingency Theory of Management Information Systems, in: *Journal of Management Information Systems*, Vol. 6, No. 1, pp. 59–85.

Westin, S. (2014) *The Paradoxes of Planning: A Psycho-Analytical Perspective*, Ashgate, Farnham.

Wijen, F. (2002) *Stakeholder Influence and Organizational Learning in Environmental Management*, Center Thesis Publication, Tilburg University, Tilburg.

Wimberley, R., L. Morris, G. Fulkerson (2007) World Population Takes a Turn, in: *The News and Observer*, 23 May.

Winsemius, P. (1986) *Gast in eigen huis. Beschouwingen Over milieumanagement*, Samson H.D. Tjeenk Willink, Alphen aan den Rijn.

Winsemius, P. (2001) *De Maatschap Nederland*, Ministerie van Economische Zaken, Den Haag.

Wolfe, C. (1994) Making Contingency Safe for Liberalism: The Pragmatics of Epistemology in Rorty and Luhmann, in: *New German Critique*, No. 61, Special Issue on Niklas Luhmann (Winter, 1994), pp. 101–127.

Wolfram, S. (2002) *A New Kind of Science*, Wolfram Media Ltd, Champaign, IL.

Woltjer, J. (2000) *Consensus Planning*, Ashgate, Burlington.

Woodward, J. (1958) *Management and Technology, Problems and Progress in Industry 3*, H.M.S.O., London.

Woodward, J. (1965) *Industrial Organization: Theory and Practice*, Oxford University Press, Oxford.

World Health Organization (2007) *Country Profiles of the Environmental Burden of Disease*, WHO.

World Health Organization (2010) *Urban Planning, Environment and Health; From Evidence to Policy Action, Meeting Report*, WHO Regional Office for Europe, Copenhagen.

WRR (Wetenschappelijke Raad voor het Regeringsbeleid) (1998) *Ruimtelijke ontwikkelingspolitiek, rapporten aan de regering, nr. 53*, Sdu Uitgevers, Den Haag.

Yanow, D. (2000) *Conducting Interpretive Policy Analysis*, Sage, Newbury Park.

Zandvoort, B, C. Zuidema (2007) Van lucht naar kwaliteit, in: *Tijdschrift Lucht*, Vol. 3, No. 3, pp. 6–9.

Zito, A.R. (2000) *Creating Environmental Policy in the European Union*, Macmillan, Basingstoke.

Zolo, D. (1992) *Democracy and Complexity: A Realist Approach*, Polity Press, Cambridge.

Zuidema, C. (2004) *Urban Sustainability in the European Union, Dealing with a Paradoxical Concept*, paper presented at the IUPEA Conference, Louisville, KY, USA, 4–8 September 2004.

Zuidema, C. (2005) *Changes in European Union Environmental Policy-making; Multilevel Governance from a Local Urban Perspective*, paper presented at the Aesop Conference, Vienna, 13–17 July 2005.

318   *Bibliography*

Zuidema, C. (2011) *Stimulating Local Environmental Policy, Making Sense of Decentralization in Environmental Governance*, Wöhrmann, Zutphen.

Zuidema, C., G. de Roo (2004) Complexiteit als planologisch begrip, in: *Rooilijn*, Vol. 37, No. 10, pp. 485–490.

Zuidema C., G. de Roo (2007) *Naar leefbare steden; voortgang in de Europese stedelijke agenda, 'Europa, je beste kennis – Europa op het grensvlak van recht en beleid'*, Hessel, B, J. Verwoert (red.), SdU Uitgevers, Den Haag.

Zuidema, C., G. de Roo (2009) Towards Liveable Cities, Progress in the EU Urban Agenda, in: *European Planning Studies*, Vol. 17, No. 9, September 2009, pp. 1405–1419(15).

Zuidema C., G. de Roo, J. Visser (2005) *Complexiteit en planologische besluitvorming; Over de betekenis van complexiteit in planologische vraagstukken*, paper presented at the conference 'Lof der verwarring', Rotterdam, 19 May 2005.

Index

ability: constraints on 144, 157; local 11, 97, 134
actor-consulting 110
adaptive governance 283
administrative experimentation 4, 27–8
area-based: approach 59–62, 71, 138, 152; Dutch approach 226, 256, 261, 265
argumentative model *see* governance through argumentation

base quality 226–7
bureaucracy 30, 41, 153–5, 183

central government control 1, 6, 158, 220, 268, 272
centralization 20, 53, 56–8, 66, 124
City & Environment 225
coercive tools 69–70, 148
co-evolution 47, 51–2, 64, 104
command and control 36, 54, 165, 168
common denominator bargaining 167
common market 34, 163–4
common policy formats 71
communicative turn 23, 53
competitive model *see* governance through competition
complex adaptive systems 44, 48, 103, 284
complexity: definition 128; degree of 75, 94, 101–5, 112, 122–3; dynamic 103–4, 282, 285; sciences 37, 44, 47–8, 282–5; societal 136; static 102–3, 282, 285
complicated systems 102
contingency between structure and function 75, 95, 113–20
contingency theory: criticism on 98–9; definition of 13, 74, 93; innovations in 100–1; rise of 96–8; SARFIT (structural adaptation to regain fit) 283

coordinative model *see* governance through coordination
critical realism 88–90
cross-sectoral 6, 10, 136, 206, 247

decentralization: dangers of 59, 66–72, 133, 143, 151–2; horizontal 59; motives for 10, 20, 53, 56–9, 60–5, 135–9; vertical 59
Dewey, J. 85–7
duality (realism–relativism) 92, 262

eco-labels 55
ecological modernization 56
EEC (Environmental Action Plan), first 161
EEC (Environmental Action Plan), sixth 161
enabling local governance 152
environmental limit values *see* environmental standards
environmental management plan (EU) 162, 173, 176–8, 180–7, 197, 205
environmental management system (EU) 162, 173, 174–7, 183–4, 197, 205, 210
environmental policy: Dutch 17, 215–18; Dutch (changes) 224–8; Dutch (current situation) 228–33, 262–4; EU 161, 164, 168–9; new instruments 20, 24, 53, 56; proactive 7–11, 45–7, 56; reactive 7, 43–5, 136, 279; reactive (Dutch) 165, 171; reactive (EU) 216, 218; renewal 2, 20, 29, 53–9, 131
environmental policy integration 46
environmental revolution 33, 34, 165, 216
environmental standards: deviate from 225; EU 165, 195; indicative 227; the Netherlands 216–20; use of 43, 141
environmental zone 221

320 *Index*

EU Expert group on the Urban
Environment 170
evolving rationale 283
external effects 69–71, 147–9, 158
external integration 220–2
externalities 32

fragmentation: Dutch environmental
policies 217, 229, 234, 244, 266; EU
environmental policies 172, 177, 181–2,
188, 190–1, 211; institutionalization
217; jurisdictions 176–7, 190, 208, 249,
260; policies 3, 6, 35, 42, 156
function 75, 95, 113–20

generic standards: EU 168, 204–5, 213;
the Netherlands 217, 221–3, 227, 259,
280; use of 39, 65–72, 141–4
governance: hybrid governance 20, 24,
56, 71, 144; modes of 4, 22–7, 52, 282;
networks 24–6, 52, 56, 60; renewal
2, 4–6, 27–9, 53–7, 268–70; shifts
in governance 2, 22–7, 53; though
coordination 20, 22–7, 29–32, 72, 124;
through argumentation 23–4, 55, 91,
263; through competition 24, 53, 91,
105; through coordination (improving)
41–7; through coordination (problems)
34, 39, 48–50, 62–5
government (style of steering) 21–2
Green Paper on the Urban Environment
(EU) 9, 161, 169–70
guarantee & stimulate 150

Habermas, J. 82–8
harmonization 163
hierarchical control 30–1, 35, 124, 140

implementation (policy) 150–2
Integrated Environmental Zoning 222–3
internal integration 218–19
intersubjective oriented focus 13, 75, 92,
112–14

legitimacy 31, 40, 106, 193
Limits to Growth 219
Liveable Cities project 162, 184–97
logical positivism 78

management overload 155, 197
market-based instruments 24, 53–6, 151
matching configurations 119, 126
matter of choice 111–15

matter of degree (complexity) *see*
complexity
Milieu in de Leefomgeving (MILO) 227
Monnet Method 165
multilevel governance: and transitions
65; concept of 22, 25, 71–2; in the EU
165–6
multiple composite goals 121–4
multiscalar issues 148
Municipal Health Departments 237–8,
253

National Collaboration Programme on Air
Pollution 255
National Environmental Policy Plan, First
[NMP1] 219–20
National Environmental Policy Plan,
Fourth [NMP4] 226–8
National Environmental Policy Plan,
Second [NMP2] 224
neoliberalism 22
neoliberal turn 23, 53
networks *see* governance networks
niches (transitions) 64
non-linearity 37–9, 102–7, 282–5

object oriented focus 13, 75, 92, 112–14
optimal living quality 251
Our Common Future 141

participative approaches 21, 27, 49, 61,
124, 224
path dependency 51
performance: EU 208; local governments
66, 134, 140–4, 147–52, 154; the
Netherlands 235
perspectivism 79, 81–4
plurality: plural governance landscape
6, 12–13, 27, 73, 77, 127–31, 273;
theoretical plurality 5, 28, 73, 74, 270
post-contingency: explanation of 92–6,
111–15; introduction 13–14, 173–5;
relevance 273; summary 128–31
postmodernism 5, 35, 77
post-positivism 5, 28, 75–9, 81–5, 90,
116, 270
power dispersal 3, 40, 77, 116, 136, 196
pragmatism 85–8
proactive (environmental policies) *see*
environmental policy
public-private divide 60–1, 179, 196, 250

quality of the living environment 226

Index 321

rationality: communicative 82–4, 107–9; rational planning model 31; spectrum of 105, 109, 121; technical/instrumental 105–6; uniform 37, 78, 106, 273
reactive (environmental policies) *see* environmental policy
realism 77–80
regimes (transitions) 65
regional environmental agencies 249, 255
regulatory based 1–2, 6, 21, 33, 220
relativism 80, 81–4
relativist epistemology 88–9
representative democracy 40–1, 61, 136
resilience 16, 47–53, 61–2, 104
risk society 284
Rorty, R. 85–8
routine implementation 144, 156

scenario planning 109
self-organization 48, 65, 285
single fixed goals 121–4
social dilemma 10, 68–70, 148, 154, 158
social fragmentation 2–3, 41, 76–7, 116, 136
socially constructed 5, 85, 96, 99
span of control 31, 35, 60
specialization (functional) 6, 30, 35, 106, 116, 130, 137
stimulate & guarantee 150
structure 75, 95, 113–20

subsidiarity 58, 71–2, 169–70, 175
sustainable development: challenge of 7–8, 35, 52–6, 269; EU 173, 176, 184, 189, 201; implementation 62–4; meaning of 45–8; the Netherlands 219
sustainable urban management plan 178
sustainable urban transport plan (EU) 162, 173, 176–8, 180–7, 197, 205

tailor-made (policies) 10, 59–60, 95, 144, 226, 286
tapering effect 221
Thematic Strategy on the Urban Environment: background 172; finalized 197–202; technical advice 197–8, 200; towards a 172–3, 174–80
transitions 50–2, 64
Treaty of Rome 163

uncertainty 37, 61, 94, 101–7, 109–11, 128
uniform: issues 81, 257; policies 17, 31–3, 68–9, 95, 164–9, 217

voluntary agreements 24, 53–5

weak profile 10, 68, 149–51, 158
willingness: constraints on 144, 157; local 11, 97, 134